Natural Extracts Using SUPERCRITICAL CARBON DIOXIDE

Natural Extracts Using SUPERCRITICAL CARBON DIOXIDE

Mamata Mukhopadhyay
Professor of Chemical Engineering
Department of Chemical Engineering
Indian Institute of Technology, Bombay
Powai, Mumbai, India

CRC Press
Boca Raton London New York Washington, D.C.

Library of Congress Cataloging-in-Publication Data

Mukhopadhyay, Mamata.
Natural extracts using supercritical carbon dioxide / Mamata Mukhopadhyay.
p. cm.
Includes bibliographical references and index.
ISBN 0-8493-0819-4 (alk. paper)
1. Supercritical fluid extraction. Carbon dioxide. 3. Natural products. I. Title.
TP156.E8 M84 2000
660′.284248—dc21 00-039733
CIP

Direct all inquiries to CRC Press LLC, 2000 N.W. Corporate Blvd., Boca Raton, Florida 33431.

International Standard Book Number 0-8493-0819-4
Library of Congress Card Number 00-039733
Printed in the United States of America 1 2 3 4 5 6 7 8 9 0
Printed on acid-free paper

Preface

The great success story of the 20th Century has been the evolution of a system that is increasingly more efficient at directly translating knowledge into technology and commercial products. Utilization of supercritical carbon dioxide for production of natural extracts is such a system that has evolved to keep the wheel of development rolling. There has been considerable interest in the last decade in switching from synthetic to natural substances in the food and pharmaceutical industries and accordingly significant research and development efforts have been focussed on newer processes and products. Supercritical carbon dioxide extraction is such a novel process that can produce tailor-made natural extracts in concentrated form, free from any residual solvents, contaminants, or artifacts.

The development of any new food or pharmaceutical product involves usage of natural ingredients for appropriate flavor, color, and consistency. With world-wide concern for environmental issues, there is a phenomenal growth in the utilization of supercritical extraction technology for these new products. Several large-scale, supercritical carbon dioxide extraction plants have already come into commercial operation in the last two decades for applications that include decaffeination of coffee and tea, denicotinization of tobacco, and recovery of flavor, fragrance, and pharmaceuticals from botanicals. What I have attempted in this book is to present the recent developments, not only in these areas of applications, but also in newer areas that can utilize supercritical carbon dioxide for production of ever-widening ranges of concentrated natural extracts, including high-purity, life-saving pharmaceuticals from a host of natural products.

For thousands of years, the diverse agro-climatic zones have made India a land of biodiversity with a huge reserve of flora and fauna. Consequently occurrence and cultivation of a large variety of medicinal plants in India have resulted in development of the oldest medical sciences, known as Ayurveda and Unani. Supercritical CO_2 extraction facilitates recovery of bioactive thermolabile natural molecules from these medicinal plants without any degradation. That has encouraged me to mention in this book a few recently developed and increasingly consumed herbal products that have been used in India and the Far East for ages.

This book is divided into eleven chapters, encompassing the fundamentals of supercritical fluids, basic concepts of phase equilibria and transport processes for design, operation, and optimization of supercritical fluid extraction process plants in the first three chapters, and the recent developments in eight broad classes of natural extracts using supercritical CO_2 extraction technology in the subsequent chapters. Some natural overlaps in later chapters are due to the multitude of the common attributes of these natural extracts.

As the scope of this book is deliberately broad and detailed on both processes and products, it was felt necessary to include a huge data base collated on both fundamental and application aspects of potential supercritical extraction processes and products. Elaborate comparison of the performance of supercritical carbon dioxide extraction technique with contemporary processes practiced by various industries, has been presented in this book with a view to alleviating the apprehension about the feasibility of using this technology commercially.

The information in this book has been collected from various international conferences, meetings, and symposia, and books and journals published in the past decade on supercritical fluids. This is in addition to my personal experience over the last two decades in research, design, and development in this area.

The book is intended to be used by students, researchers, and industrialists in the disciplines of chemical and biochemical engineering and food technology, in addition to house managers, herbalists, nutritionists, and those interested in natural products, including Ayurvedic and Unani medicines. It is hoped that the book will widen the scope of commercial success for the potential applications of supercritical CO_2 extraction technology in the new millennium.

Mamata Mukhopadhyay

The Author

Mamata Mukhopadhyay is a Professor of Chemical Engineering at the Indian Institute of Technology, Bombay, India (phone: 091.22.576.7248/8248; fax: 091.22.5726895; e-mail: mm@che.iitb.ernet.in). She received her B.Ch.E. degree from Jadavpur University, Calcutta; M. Tech. degree from Indian Institute of Technology, Kharagpur, India; and Ph.D. degree from Ohio State University, Columbus, OH, all in the chemical engineering discipline. She was awarded a Senior Research Fellowship by the U.S. National Science Foundation (NSF) while working for her Ph.D. in the area of thermodynamics, under the supervision of Prof. Webster B. Kay.

For the last 30 years Dr. Mukhopadhyay has been a member of the faculty of Chemical Engineering in three reputed Institutes of Technology in India, namely, IIT–Kapur, IIT–Delhi, and now in IIT–Bombay. She has taught thermodynamics, separation processes, cryogenics, and supercritical fluid technology at the undergraduate and graduate levels of these institutes. Her research interests encompass even wider areas including supercritical extraction of natural products, reactions in supercritical fluids, environmental protection, and food process engineering. She has supervised numerous M.Tech. and Ph.D. projects as well as several externally funded R&D projects. Recently, a technology development mission project on Supercritical Fluid Extraction Systems Design was successfully completed by her, in which a commercial prototype of an SCFE plant with 10-liter extractors was designed and commissioned for demonstration; the SCFE technology so developed has been transferred for commercialization.

Dr. Mukhopadhyay was honored with three awards recognizing her outstanding contributions to the field of chemical engineering. She is the author of more than 75 technical papers in renowned international journals and proceedings of conferences. Two of her innovations in supercritical fluid extraction (SCFE) processes have been patented. She is a member of the American Institute of Chemical Engineers, Indian Institute of Chemical Engineers, and International Society for Advancement of Supercritical Fluids.

Acknowledgment

My inspiration to write this book originated from the fascinating lustrous green natural surroundings amidst lakes and hills where I worked and lived for years. Many people have contributed to sustaining my motivation to compile this book. I am indebted to all of my doctoral and other graduate students and research engineers who contributed to this book directly or indirectly by way of their experimental measurements, theoretical analysis, and mathematical modeling. References have been made to their works in this book. I am indebted to my colleagues Professor Sandip Roy and Dr. Swapneshu Baser for useful discussions and constructive comments during proofreading of the manuscript. I am thankful to Dr. Niyati Bhattacharya for her continuous support and creative suggestions when I needed them most. I am thankful to Sangita Prasad, Sanjay Singh, and Krishna Tej for their assistance in compiling information. I sincerely acknowledge the tolerance and perseverance of Mr. N.K. Bhatia, who typed the manuscript a number of times and the technical support provided by Mr. Sunil Ladekar and Mr. T.K.M. Nair for graphics.

My appreciation goes to my husband for putting up with me, and a special mention is due to our daughter, Anasuya, for her constant encouragement and moral support throughout the preparation of the manuscript, without which this work would have remained incomplete.

I would like to acknowledge the financial assistance provided by the Curriculum Development Program Cell of Indian Institute of Technology, Bombay towards the preparation of the manuscript. Efforts made by Ms. Lourdes Franco, Editor of CRC Press LLC in providing all necessary guidance and answering endless questions during the preparation of the manuscript are sincerely acknowledged.

The book is dedicated to my Mother
Who inspired me from Heaven that
I pursue and accomplish her desire
And explore the wonders of Nature.

Table of Contents

Appendices

1 Introduction

1.1 IMPORTANCE OF NATURE'S CURE

Good health is everybody's concern and "health — the natural way" is everybody's favorite. Correction of physical disorders by restoring and maintaining health through substances freely available in nature is an age-old practice. Nature has immense curative power, but it needs to be tapped selectively. Nature's cure is like mother's gentle care and works wonders by restoring the equilibrium and harmony in the entire body system. Nature's cure acts on two fundamental principles, namely, it regulates nutrition and stimulates the vitality of the body (Bakhru, 1991).

Nutrition plays a vital role in strengthening the body's immune system. A strong immune system is essential to keep all diseases at bay. It is believed that there are at least 43 chemical components, called essential nutrients, which must be present in our food. Deficiency of any one of them in our food creates imbalance in our systems and leads to disease or death. Nutritional deficiencies may occur during preservation, processing, transportation, contamination, or degradation. "Mother Nature" provides us with all the essential nutrients for human growth and health. It is known that seeds, nuts, grains, spices, fruits, and vegetables constitute excellent natural sources of nutrients, including vitamins A, B, C, D, E, and K, that are needed for stimulating health and prevention of premature aging. However, as some of the nutrients are lost or depleted by thermal, natural, or bacteriological degradation between the time of harvesting and the time of ingestion, it becomes absolutely essential to supplement our foods with these nutrients, particularly vitamins and antioxidants. These nutrients ought to be close to their original form so that they create the least disturbance in body systems. Consequently it is preferable that they be derived or concentrated from natural sources, in the form of natural extracts. Besides, some of the toxins or harmful components present need to be removed or depleted from our food before consumption, e.g., caffeine, cholesterol, gossipol, and aflatoxins.

Some medicinal herbs with specific bioactivities are recommended in Ayurveda to cure disorders, such as nervous and physical exhaustion, lack of concentration, muscular tremors, loss of memory, tension headaches, and restlessness. Some herbs contain bioactive molecules which even promote awareness and strength of the mind (Peterson, 1995). Some herbal medicines are used for curing and preventing chronic diseases like cancer, AIDS, diabetes, hypertension, and even liver and cardiac disorders (Debelic, 1990).

1.2 NATUROPATHY

As prescribed by the old adage, "An ounce of prevention is better than a pound of cure." Nature's care or naturopathy is an excellent preventive therapy. It can also eliminate certain diseases, such as high blood pressure and diabetes, which other therapies merely control. In fact, naturopathy is an ancient system of healing which is known to be effective in almost all illnesses from colds to cancer. The basic principle of naturopathy is to eliminate the accumulated toxicity in the body and purify the system internally. Thus, naturopaths recommend intake of health food or food freed from toxic substances and all medicines from nature.

Health food is food grown organically. It is consumed in its natural form or after it is transformed by using natural substances into a high-nutrition value product, such as natural concentrates. For example, naturopaths prescribe herbal tea or natural stimulants instead of normal tea or coffee as such, or decaffeinated tea or coffee. In recent years, aroma therapy and herbal skin care have gained immense popularity, which are also a form of naturopathy. For example, the floral fragrance of jasmine is used for relieving stress. The pure essence of jojoba, a unique product of nature, is almost identical to that of natural skin oil and is recommended as an herbal beauty oil. Its efficacy is due to the occurrence of a potent active ingredient. Similarly, natural oils from almonds, wheat grass, olives, aloe vera, etc. also have characteristic properties which beautify and moisturize skin for making it naturally soft and for maintaining it youthful over the years. Scientific research into biological activity of these natural products and their healing potential has confirmed their therapeutic uses for cosmetic benefits. This has facilitated enormous increase in the demand for these products and further search for more applications. To cite an example, the demand for tea tree oil has enormously increased in the past few years after successful test trials and is presently hailed in Australia as the "antiseptic" of the future. The huge demand for tea tree oil has led to new plantations in California in recent years.

1.3 NATURAL EXTRACTS

Over the past decade, several noteworthy consumer trends have emerged, such as enhanced concern for the quality and safety of foods and medicines, regulations for nutritive and toxicity levels, and increased preference for "natural" as opposed to synthetic substances. Furthermore, the present popular belief that everything "natural" is good, provides a positive incentive towards growth of the natural products industry, particularly in the food, flavoring, perfumery, and pharmaceutical sectors. "Mother Nature" is considered a highly efficient synthesizer of desirable blends of constituents ideally suitable for human consumption. For example, the subtle nuances and characteristic notes possessed by natural extracts or concentrates have not yet been matched by mixtures of their major ingredients produced synthetically, although considerable efforts are made in mimicking the natural molecules.

No doubt, safety of both producers and consumers is now a major requirement of any new product or process. Accordingly, compelling regulations on the usage of hazardous, carcinogenic, or toxic solvents, as well as high energy costs for solvent regeneration have curtailed the growth of the natural extract industries. To suppress the competitive edge of synthetic materials, alternative extraction methodologies that comply with both consumer preference and regulatory control and that are cost effective, are becoming more popular. One of such major technologies that has

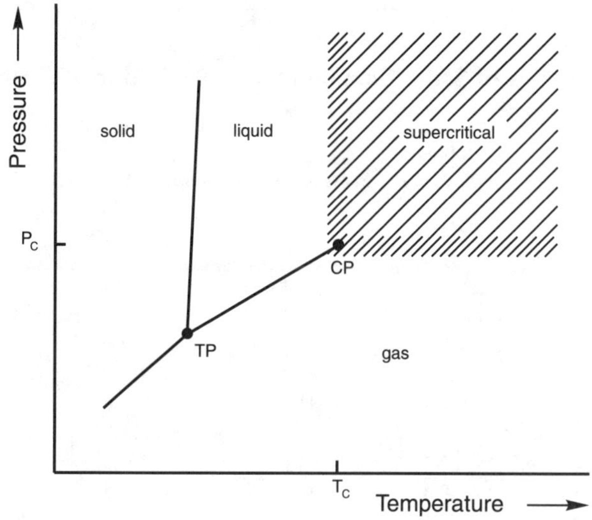

FIGURE 1.1 Pressure-temperature diagram for a pure component.

emerged over the last two decades as the alternative to the traditional solvent extraction of natural products is the supercritical fluids extraction technique. It uses a clean, safe, inexpensive, nonflammable, nontoxic, environment-friendly, nonpolluting solvent, such as carbon dioxide. Besides, the energy costs associated with this novel extraction technique are lower than the costs for traditional solvent extraction methods. Supercritical fluid extraction technology is thus increasingly gaining importance over the conventional techniques for extraction of natural products. The principle behind this technique is outlined in the next section.

1.4 CO_2 AS AN SCF SOLVENT

When a gas is compressed to a sufficiently high pressure, it becomes liquid. If, on the other hand, the gas is heated beyond a specific temperature, no amount of compression of the hot gas will cause it to become a liquid. This temperature is called the critical temperature and the corresponding vapor pressure is called the critical pressure. These values of temperature and pressure define a critical point which is unique to a given substance. The state of the substance is called supercritical fluid (SCF) when both the temperature and pressure exceed the critical point values as schematically described in a pressure-temperature phase diagram (Figure 1.1). This "fluid" now takes on many of the properties of both gas and liquid. It is the region where the maximum solvent capacity and the largest variations in solvent properties can be achieved with small changes in temperature and pressure. It offers very attractive extraction characteristics, owing to its favorable diffusivity, viscosity, surface tension and other physical properties. Its diffusivity is one to two orders of magnitude higher than those of other liquids, which facilitates rapid mass transfer and faster completion of extraction than conventional

TABLE 1.1
Physical Properties of Some Common Solvents Used in SCF State

Fluid	Normal Boiling Point (°C)	Critical Constants Pressure (bar)	Temperature (°C)	Density (g/cm³)
Carbon dioxide	–78.5	73.8	31.1	0.468
Ethane	–88.0	48.8	32.2	0.203
Ethylene	–103.7	50.4	9.3	0.20
Propane	–44.5	42.5	96.7	0.220
Propylene	–47.7	46.2	91.9	0.23
Benzene	80.1	48.9	289.0	0.302
Toluene	110.0	41.1	318.6	0.29
Chlorotrifluoromethane	–81.4	39.2	28.9	0.58
Trichlorofluoromethane	23.7	44.1	196.6	0.554
Nitrous oxide	–89.0	71.0	36.5	0.457
Ammonia	–33.4	112.8	132.5	0.240
Water	100.0	220.5	374.2	0.272

Adapted from Klesper, 1980.

liquid solvents. Its low viscosity and surface tension enable it to easily penetrate the botanical material from which the active component is to be extracted. The gas-like characteristics of SCF provide ideal conditions for extraction of solutes giving a high degree of recovery in a short period of time. However, it also has the superior dissolving properties of a liquid solvent. It can also selectively extract target compounds from a complex mixture. Sometimes the target compound is the active ingredient of interest. At other times, it may be an undesirable component which needs to be removed from the final product. The strong pressure and temperature (or density) dependence of solubility of certain solutes in an SCF solvent is the most crucial phenomenon that is exploited in supercritical fluid extraction (SCFE). Many of the same qualities which make SCF ideal for extraction, also make them good candidates for use as a superior medium for chemical reactions offering enhanced reaction rates and preferred selectivity of conversion. Once such a reaction is over, the fluid solvent is vented to precipitate the reaction product. A comparison of physical properties of some SCF solvents are given in Table 1.1.

The most desirable SCF solvent for extraction of natural products for foods and medicines today is carbon dioxide (CO_2). It is an inert, inexpensive, easily available, odorless, tasteless, environment-friendly, and GRAS (generally regarded as safe) solvent. Further, in SCF processing with CO_2, there is no solvent residue in the extract, since it is a gas in the ambient condition. Also, its near-ambient critical temperature (31.1°C) makes it ideally suitable for thermolabile natural products. Due to its low latent heat of vaporization, low energy input is required for the extract separation system which renders the most natural smelling and natural-tasting extracts. Further, the energy required for attaining supercritical (SC) state of CO_2 is often less than the energy associated with distillation of conventional organic solvent. In general, the

TABLE 1.2
Solubility of Botanical Ingredients in Liquid and SC CO_2

Very Soluble	Sparingly Soluble	Almost Insoluble
Nonpolar and slightly polar low M.W. (<250) Organics, e.g., mono and sesquiterpenes, e.g., thiols, pyrazines, and thiazoles, acetic acid, benzaldehyde hexanol, glycerol, acetates	Higher M.W. organics,(<400), e.g., substituted terpenes and sesquiterpenes, water, oleic acid, glycerol, decanol, saturated lipids up to C_{12}	Organics with M.W. above 400, e.g., sugars, proteins, tannins, waxes, inorganic salts, chlorophyll, carotenoids, citric, malic acids, amino acids, nitrates, pesticides, insecticides, glycine, etc.

Moyler, 1993.

extractability of the compounds with supercritical CO_2 depends on the occurrence of the individual functional groups in these compounds, their molecular weights, and polarity. Table 1.2 presents a classification of natural ingredients as very soluble, sparingly soluble, and almost insoluble in supercritical CO_2.

For example, hydrocarbons and other organic compounds of relatively low polarity, e.g., esters, ethers, aldehydes, ketones, lactones, and epoxides, are extractable in SC CO_2 at a lower pressure in the range of 75 to 100 bar, whereas moderately polar substances, such as benzene derivatives with one carboxylic and two hydroxyl groups, are moderately soluble. The highly polar compounds, such as the ones with one carboxylic and three or more hydroxyl groups, are scarcely soluble. For the extraction of a certain class of products, a cosolvent or an entrainer is often injected into SCF CO_2 to increase its polarity and hence its solvent power. Ethanol, ethyl acetate, and possibly water are the best natural entrainers for food-grade products.

Commercial CO_2 required for supercritical fluid extraction process is already present in the environmental system, obtained as a by-product from the fermentation process or the fertilizer industry. So its use as an extractant does not cause any further increase in the amount of CO_2 present in the earth's atmosphere. Therefore, there is no additional "green house effect" from using CO_2 as the SCF solvent.

1.5 SCFE PROCESS

Figure 1.2 shows a schematic diagram of the supercritical fluid extraction (SCFE) process. The four primary steps involved are extraction, expansion, separation, and solvent conditioning, and the four corresponding critical components needed are a high pressure extractor, a pressure reduction valve, a low pressure separator, and a pump for intensifying the pressure of the recycled solvent. Other ancillary equipment include heat exchangers, condenser, storage vessels, fluid make-up source, etc. The feed, generally ground solid, is charged into the extractor. CO_2 is fed to the extractor through a high-pressure pump (100 to 350 bar). The extract-laden CO_2 is sent to a separator (120 to 50 bar) via a pressure reduction valve. At reduced temperature and pressure conditions, the extract precipitates in the separator, while CO_2, free of any

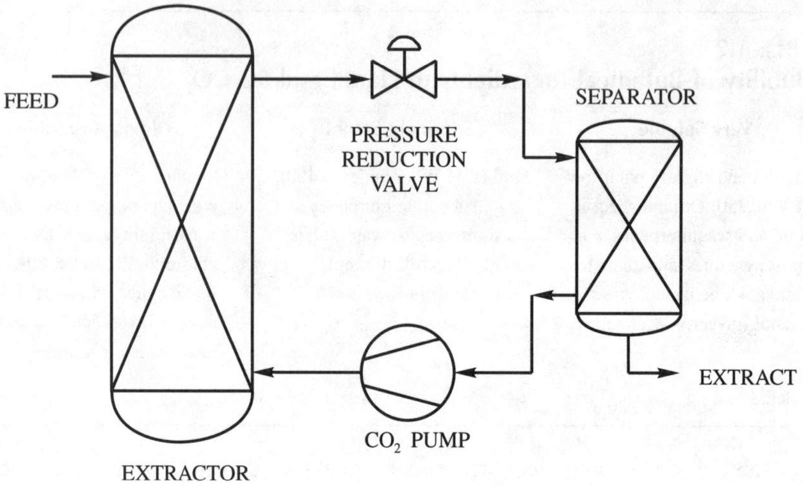

FIGURE 1.2 A schematic diagram of the SCFE process.

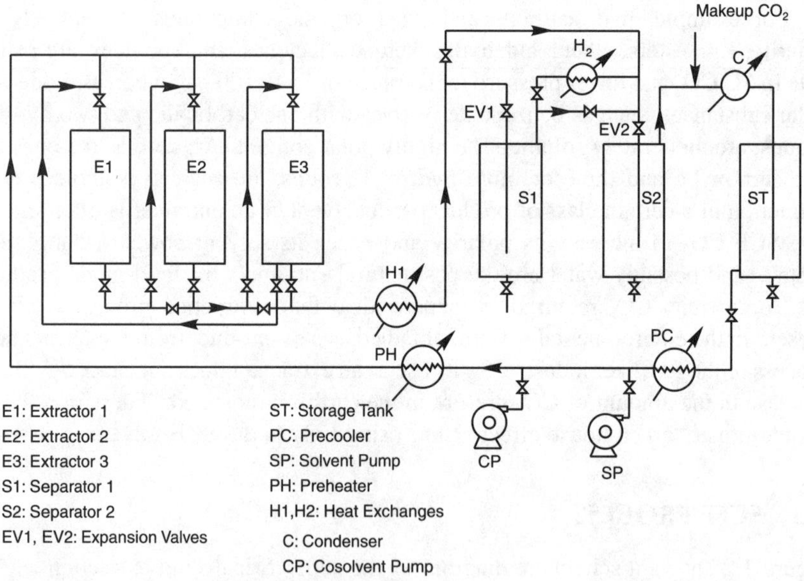

FIGURE 1.3 A schematic diagram of an SCFE plant on a commercial scale.

extract, is recycled to the extractor. SCFE for solid feed is a semibatch process in which carbon dioxide flows in a continuous mode, whereas the solid feed is charged in the extractor basket in batches. For making the process semicontinuous at the commercial scale, multiple extraction vessels are sequentially used, such that when one vessel is on loading or unloading, the other vessels are kept in an uninterrupted extraction mode, as shown in Figure 1.3. An entrainer or cosolvent is often pumped and mixed with the high-pressure CO_2 for enhancing the solvent power or selectivity

TABLE 1.3
SCFE Applications in Natural Products and Food Industries

Decaffeination of coffee and tea
Spice extraction (oil and oleoresin)
Deodorization of oils and fats
Extraction of vegetable oils from flaked seeds and grains
Flavors, fragrances, aromas, and perfumes
Hops extraction for bitter
Extraction of herbal medicines
Stabilization of fruit juices
Lanolin from wool
Deoiling of fast foods
Decholesterolization of egg yolk and animal tissues
Antioxidants from plant materials.
Food colors from botanicals
Natural pesticides
Denicotinization of tobacco

of separation of the specified components. Separation is often carried out in stages by maintaining different conditions in two or three separators for fractionation of the extract, depending on the solubilities of the components and the desired specifications of the products. Similarly, by varying the pressure, it is possible to alter the solvent power of the extractant, the effect of which is equivalent to changing the polarity of an extraction solvent. Thus a production plant can have flexible operating conditions for multiple natural products and it is also possible to obtain different product profiles from a single botanical material by merely using a single solvent, namely, supercritical CO_2.

1.6 SCFE APPLICATIONS

SCFE using CO_2 is today a popular technology for rapid, contamination-free extraction in the food and pharmaceutical industries. Large-scale supercritical CO_2 extraction has been in commercial operation since the late 1970s for decaffeination of coffee and tea, refining of cooking oils, and recovering flavors and pungencies from spices, hops, and other plant materials. Table 1.3 summarizes some of the known applications of the SCFE technique.

Some of the newer applications which are presently in the developmental stage or are being explored include deoiling of crude lecithin, recovery of flavors from fermented sea weeds, or production of oil and protein from corn germ, wheat germ, etc. Some applications of natural extracts using supercritical or liquid CO_2 are listed in Table 1.4.

A commercial plant was built in the 1980s for the separation of ethanol from water in combination with other preliminary processes, such as pervaporation, distillation, etc. There are some proven applications of SCFE using solvents other than

TABLE 1.4
Some Applications of CO_2 Extracts

Extract	Source	Commercial Applications
Ginger oil	*Zingiber officinalis*	Oriental cuisines, beverages
Pimento berry oil	*Pimenta officinalis*	Savory sauces, oral hygiene
Clove bud oil and oleoresin	*Eugenia caryophyllata*	Meats, pickles, oral hygiene
Nutmeg oil	*Myristica fragrans*	Soups, sauces, vegetable juices
Juniper berry oil	*Juniperus officinalis*	Alcoholic beverages, gin
Celery seed oil	*Apium graveolens*	Soups, vegetable juice (tomato)
Vanilla absolute	*Vanilla fragrans*	Cream liqueurs, pure dairy foods
Cardamom oil	*Elletaria cardamomum*	Meats, pickles, spice blends
Aniseed oil	*Illicium verum*	Liqueurs, oral hygiene
Coriander oil	*Coriander sativum*	Curry, chocolate, fruit flavors
Pepper oleoresin and oil	*Piper nigrum*	Spices, meat, salad dressing
Cinnamon bark oil	*Cinnamonum zeylendium*	Baked goods, sweet products
Cumin oil	*Cuminium cyminum*	Mexican and Indian cuisines and pharmaceuticals
Marjoram oil	*Majorlana hortensis*	Soups, savoury sauces
Savory oil	*Satureja hortensis*	Soups, savoury sauces
Rosemary oil	*Rosemary officinalis*	Antioxidant, soups
Sage oil	*Salva officinalis*	Meat, sauces, soups
Thyme oil	*Thymus vulgaris*	Meat, pharmaceutical products
Paprika color (oleoresin)	*Capsicum annum*	Soup, sauces, sweets

Moyler, 1994.

CO_2, for example, the processing of heavy hydrocarbons, such as deasphalting petroleum fractions, recovery and purification of lube oils, coal liquefaction, etc., which have already been commercialized. Other applications of SCFE include chemical separations and purification, polymer processing, regeneration of activated carbon and other adsorbents, supercritical fluid chromatography, comminution via precipitation from supercritical fluids, deposition of materials in microporous substrates, critical point drying, cleaning of micro-electronics, etc. which are described in detail elsewhere (McHugh and Krukonis, 1994). SCF have been investigated as reactants as well as solvent medium for several important chemical reactions and their advantages are detailed elsewhere (Mukhopadhyay and Srinivas, 1994).

However, in spite of extensive research and development work done all over the world on a wide sprectum of applications of the SCF technology in the past 20 years, it is the supercritical CO_2 extraction process which has received the most significant acceptance for commercial exploitation, especially for high-value, low-volume natural product extracts. This book will deal with some of the proven and potential products that can be obtained using liquid and supercritical CO_2 for extraction from natural materials. These naturally extracted concentrates, having high value addition, offer remarkable benefits from the commercial, environmental, and health points of view.

REFERENCES

Bakhru, H. K., *A Complete Handbook of Nature Cure*, Jaico Publishing House, Bombay, India, 1991.

Debelic, I. N., Ed., *Magic and Medicine of Plants*, Reader's Digest Association, Pleasantville, New York, 1990.

Klesper, E., Chromatography with supercritical fluids, in *Extraction with Supercritical Gases*, Schneider, G. M., Stahl, E., and Wilke, G., Eds., Verlag Chemie, Weinheim, Germany, 1980.

McHugh, M. A. and Krukonis, V. J., *Supercritical Fluid Extraction: Principles and Practice*, 2nd ed., Butterworth-Heinmann, Stoneham, MA, 1994.

Moyler, D. A., Extraction of flavours and fragrances, in *Extraction of Natural Products Using Near-Critical Solvents*, King, M. B. and Bott, T. R., Eds., Blackie Academic and Professional, an imprint of Chapman & Hall, Glasgow, 1993.

Moyler, D. A., Oleoresins, tinctures and extracts, in *Food Flavourings*, Ashrurst, P. R., Ed., Blackie Academic and Professional, an imprint of Chapman & Hall, Glasgow, 1994.

Mukhopadhyay, M. and Srinivas, P., Oxidation of cyclohexane in supercritical CO_2, *Ind. Eng. Chem. Res.*, 33, 3118–3124, 1994.

Peterson, N., *Herbal Remedies*, Blitz Editions, Amazon Publishing Ltd., Middlesex, U.K., 1995.

2 Fundamentals of Supercritical Fluids and Phase Equilibria

The principal objective of this chapter is to present thermodynamic principles essential to understanding the supercritical fluid extraction (SCFE) technique and the interdependence of relevant process parameters. In addition, systematic calculation procedures and predictive thermodynamic models are reviewed for quantitative understanding of the process, selection of the process parameters, and design, operation and optimization of the SCFE systems. Knowledge of phase equilibrium behavior and generation of phase equilibrium data, such as solubility, distribution coefficient, and selectivity of separation of the extractables in the supercritical fluid (SCF) solvent, are required at the extraction and separation conditions. In order to fully appreciate the phase diagrams and theoretical developments presented in this chapter, the reader may require familiarity with the basic concepts of molecular thermodynamics. Nevertheless, for brevity, a quick glance at the trends of thermodynamic property variations will surely give some insight into the background principles. However, subsequent chapters have been written with the general reader in mind and the materials discussed therein may be understood without recourse to the advanced concepts presented in this chapter.

2.1 PROCESS SCHEMES AND PARAMETERS IN SCFE

The unique solvent properties of supercritical fluids (SCF), i.e., substances above their critical temperature and critical pressure, were first observed more than a century ago (Hannay and Hogarth, 1879, 1880). However, it has been only in the past two decades that supercritical fluid extraction (SCFE) has evolved as a novel separation technique. In the SCF state, a solvent displays properties which are intermediate to those of liquid and gaseous states. Table 2.1 compares the orders of magnitudes of some common physical properties of SCF solvents with those corresponding to the liquid and gaseous states. The liquid-like density of an SCF solvent provides its high solvent power, whereas the gas-like viscosity and diffusivity, together with zero surface tension, impart excellent transport properties to the SCF solvent, which in turn enhance the rates of transfer from the original botanical substrate to the SCF solvent as compared to that to liquid organic solvents.

Of the various possible SCF solvents listed earlier (see Chapter 1, Table 1.1), carbon dioxide is by far the most widely used SCF solvent because it is inexpensive, nontoxic, nonflammable, and available in abundance and at high purity. Supercritical CO_2 is becoming increasingly popular due to its inherent potential for high recoverability and selectively fractionating superior-grade natural extracts from a variety of biomaterials.

TABLE 2.1
Orders of Magnitude of Physical Properties of Solvents in Different States

Property	State: Gas	SCF	Liquid
Density, ρ (g/cm^3)	10^{-3}	0.3	1
Diffusivity, D(cm^2/s)	10^{-1}	10^{-3}	5×10^{-6}
Viscosity, η(g/cm·s)	10^{-4}	10^{-4}	10^{-2}

Klesper, 1980.

TABLE 2.2
Combination of Extraction and Separation Conditions

	Pressure	Temperature
Case 1	$P_1 > P_c > P_2$	$T_1 > T_c > T_2$
Case 2	$P_1 > P_c > P_2 > P_3$	$T_1 \geq T_2 > T_c > T_3$
Case 3	$P_1 > P_2 > P_c > P_3$	$T_1 \geq T_2 > T_c > T_3$
Case 4	$P_1 = P_2 > P_c > P_3$	$T_1 < T_2 > T_c > T_3$
Case 5	$P_1 > P_2 > P_c > P_3$	$T_1 \geq T_2 > T_3 > T_c$

Adapted from Hubert and Vitzthum, 1980.

Technoeconomic viability of the SCFE process largely depends on the identification of the appropriate process scheme and the selection of process parameters from many possible alternative extraction protocols. For example, there are several alternative combinations of extraction and separation conditions that can be selected to design an SCFE process, as outlined in Table 2.2, where P_1 and T_1 are the conditions of pressure and temperature of the extractor, and P_2, T_2 and P_3, T_3 are the pressure and temperature conditions of the separators S_1 and S_2, respectively.

As described earlier in Section 1.5 (see Chapter 1, Figure 1.3), for commercial scale of operation, two or more extractors may be used sequentially or parallelly with simultaneous fractionation of extracts in one or two separators. Since it is desirable to recover the extractables to the maximum extent possible followed by precipitation of the extracts by depressurization into separate fractions, case 2 or 3 or 5 is often chosen rather than case 1 for practical advantages. In these cases, separator temperature, T_2 (of S_1) would be between the extractor temperature T_1 and the critical temperature, T_c, whereas separator temperature T_3 (of S_2) could be less or more than T_c. Yet another possible scheme, namely case 4, would be beneficial from the consideration of energy savings because of the isobaric method of operation, where separation is facilitated by increasing temperature. However, one needs to ascertain whether separation is complete or whether thermolabile components can withstand the rise in temperature beyond the extractor temperature, T_1. The basis of selection of these conditions will be discussed in the next section.

Separation can also be carried out by adding a third substance to decrease the solubility in order that the extract is precipitated out. But this requires separation of the same substance from the SCF solvent for its recirculation or reuse. Alternatively, an adsorbent or a selective semipermeable membrane may be used for achieving the desired separation. The adsorbent or membrane may be regenerated and reused.

It is thus necessary to execute careful integration and intelligent synthesis of the multistage SCFE process for multiproduct systems toward a technoeconomically viable commercial venture. Accordingly extensive experimental investigations and theoretical analysis are required with three clearly defined objectives, namely, understanding of:

1. The fundamentals of thermodynamic principles of phase equilibrium behavior and molecular interactions involved in the supercritical fluid extraction (SCFE) process and separation systems
2. Kinetics of mass and heat transfer for design and operation of SCFE plant
3. Design and development of supercritically extracted products with desired specifications for selected applications

While the first objective is considered in this chapter, the second objective is dealt with in Chapter 3, and the third objective is covered in subsequent chapters.

2.2 THERMODYNAMICS OF SCF STATE

Thermodynamically, SCF is a state where the pressure and temperature are beyond the critical point values. In practice, an SCF solvent is mostly used as an extractant in the approximate range of temperature up to 1.2 times the critical temperature, T_c, and pressure up to 3.5 times the critical pressure, P_c. This range of operating conditions provides liquid-like densities as can be seen from Figure 2.1. Further, as is indicated in Figure 2.2, the variation of density, ρ, and static dielectric constant, ε, with pressure show similar trends. For example, for CO_2, both ρ and ε rise sharply between 70 and 200 bar. It is important to note that at around 200 bar and beyond, both parameters attain values similar to those for liquids. This provides an explanation as to why SCF CO_2 exhibits high solvent power above a certain pressure, depending on what needs to be dissolved, and thus can be used as a good solvent in place of conventional organic solvents. The CO_2 molecule has no net dipole moment, i.e., it is nonpolar and hence in the SCF state, it serves as a good solvent for natural molecules that are nonpolar. However, it has a quadruple moment for which it can also dissolve slightly polar and some polar substances at relatively high pressures (>250 bar).

2.2.1 Variability of Density with P and T

The solvent capacity at the supercritical fluid state is density dependent and it is the sharp variability of density with pressure and temperature in this state that provides uniqueness to an SCF solvent. Figure 2.3 illustrates such variation of density with pressure at different temperatures in terms of the reduced parameters of T_r (= T/T_c) and P_r (= P/P_c). At $P > P_c$, the variation of density with an increase in temperature

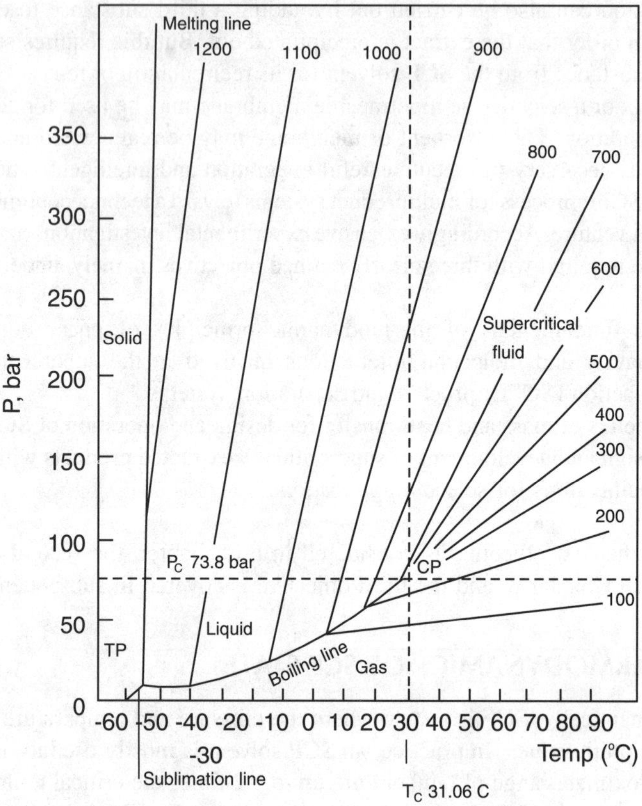

FIGURE 2.1 P–T diagram of CO_2 at densities from 100 to 1200 g/L.

from subcritical to supercritical condition is monotonic. For example, at a reduced temperature, T_r in the range of 0.9 to 1.2, the reduced solvent density, ρ_r (or ρ/ρ_c), can increase from 0.1 to 2.0, as the reduced pressure, P_r increases in the range from 1 to 2. However, the density of the SCF solvent reduces as T_r is increased to 1.55 and reduced pressures greater than 10 are needed to attain liquid-like densities. By regulating the pressure and temperature, it is thus possible to alter density, which in turn regulates the solvent power of the SCF solvent. The variability of density in the vicinity of the critical point is better explained in terms of isothermal compressibility, K_T, which is defined as:

$$K_T = -\frac{1}{\rho}\left(\frac{\partial \rho}{\partial P}\right)_T \tag{2.1}$$

where ρ is the molar density.

As can be seen from Figure 2.3, the compressibility of any fluid (or the product of reciprocals of P_c and reduced density, and the slope of the isotherm) is very high in the vicinity of its critical point and diverges to infinity at its critical point. In other

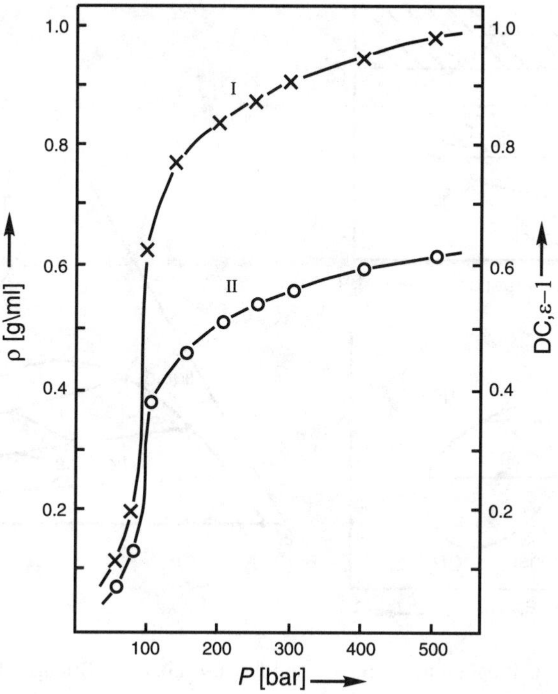

FIGURE 2.2 Variations with pressure of density, ρ(I), and the dielectric constant (II) of CO_2 at 40°C (Hubert and Vitzhum, 1980).

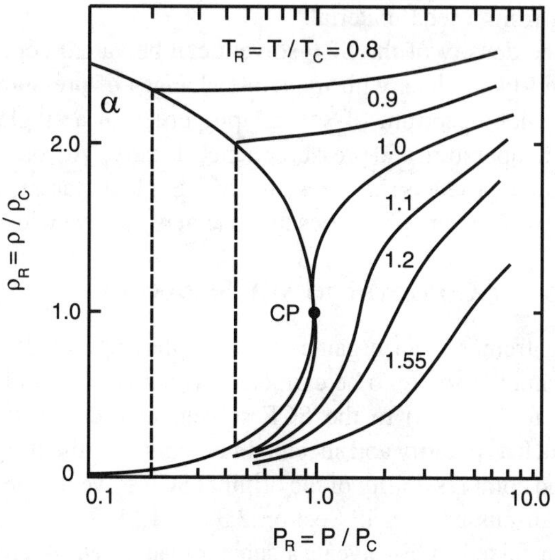

FIGURE 2.3 Variation of the reduced density (ρ_r) of a pure component near its critical point (McHugh and Krukonis, 1994).

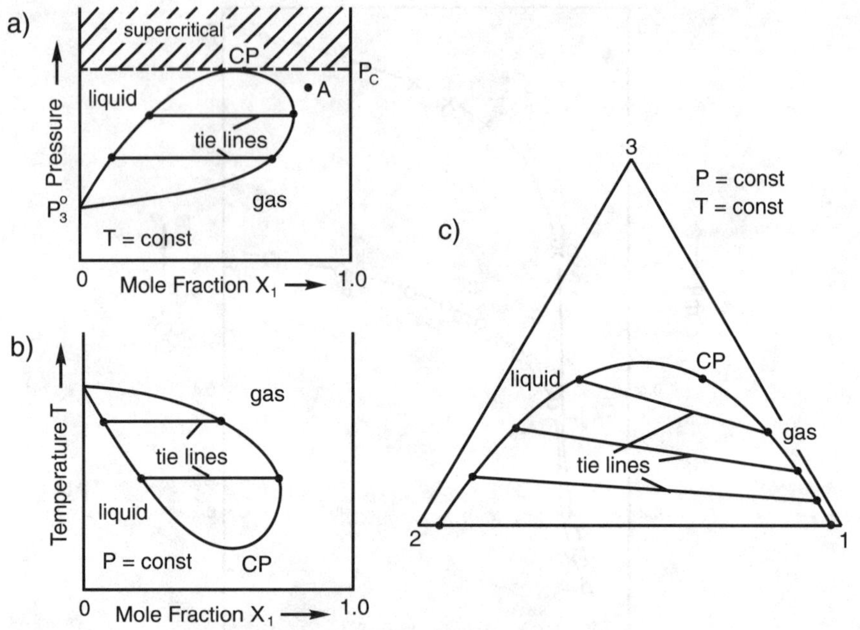

FIGURE 2.4a–c Critical points in binary and ternary mixtures (Brunner, 1994).

words, an SCF solvent is highly compressible and essentially gas-like, in contrast to liquid solvents. This high compressibility of the SCF solvent facilitates the alteration of density and hence the solvent power. This also allows fine tuning of the solvent power for selective separation of one or more active constituents out of the total extractables from a feed material.

In short, as the density of the SCF solvent can be varied continuously between gas-like and liquid-like values with moderate changes of pressure, it is possible to make avail of a wide spectrum of solvent properties in a single SCF solvent by simply changing temperature and pressure (hence density). Accordingly, it is possible to produce multiple products in the same SCFE plant, taking advantage of the possible variability of solvent properties that one may achieve with supercritical CO_2.

2.2.2 Addition of Cosolvent to SCF Solvent

A cosolvent or entrainer is an organic substance having volatility intermediate to the SCF solvent and the solute to be extracted, which is often added in a very small concentration (1 to 5 mol%) to the SCF solvent in order to change the solvent characteristics, such as polarity and specific interactions, without significantly changing the density and compressibility of the original SCF solvent. The role of cosolvents in SCFE will be discussed later in Section 2.3.5.

The cosolvent mixed SCF solvent is supercritical when its pressure is above its mixture critical pressure and its temperature is above its mixture critical temperature for a particular composition, which are usually not very different from the critical values of the pure SCF solvent. As shown in Figure 2.4a, the mixture critical pressure is always the highest pressure on an isothermal P–x diagram of the binary mixture, beyond which there is no two-phase region for a particular temperature. The mixed SCF solvent is supercritical at all pressures above its mixture critical pressure, as shown by the hatched area. At a pressure less than the critical pressure, such as the one at point A, it corresponds to a gaseous state as it is outside the two-phase region. On the other hand, as can be seen from Figure 2.4b, the two-phase vapor–liquid region may extend beyond the mixture critical temperature for a particular pressure. Hence a mixture is not necessarily supercritical at a temperature above its mixture critical temperature. The critical point (CP) differentiates the gaseous phase from the liquid phase.

When a binary mixture of SCF solvent (1) and a cosolvent (3) is employed beyond its binary mixture critical pressure to solubilize a liquid solute (2), then the system is represented by a ternary diagram (Figure 2.4c). In such cases all three components are usually distributed both in the liquid and in the SCF phases. The extent of solubilization of the component in the two phases is characterized by the distribution coefficient, which is given by the ratio of the concentration of the component in the fluid phase to that in the liquid phase as represented by the two end points of a tie line.

2.3 SOLUBILITY BEHAVIOR IN SCF SOLVENT

Among the unique features characteristic of the solubility behavior of a solute in an SCF solvent are the exponential solubility enhancement and retrograde solubility behavior, which are attributed to the divergence of partial molar volumes and partial molar enthalpies in the vicinity of the critical point of the solvent. The solubility behavior of a solute in an SCF solvent is analyzed in terms of pressure, temperature, and cosolvent effects on solubility isotherms and isobars.

2.3.1 Solubility Isotherms and Crossover Phenomena

As shown in Figure 2.5 for a typical binary solid–fluid system, the solubility isotherms exhibit a remarkable pressure-dependent behavior. It can be seen that the solubility of a solid solute initially decreases, reaches a minimum, and then exponentially increases with pressure in the vicinity of the CP in the SCF state. The increase in solubility with further increase in pressure is less pronounced. The region between pressures marked as P_L^* and P_U^* (called lower and upper cross over pressures, respectively) where the various isotherms seem to converge, is of phenomenological importance. In this region, which is found to be characteristic of a solid–fluid system, an isobaric increase in temperature leads to a decrease in solubility. This phenomenon, commonly referred to as the "retrograde solubility behavior" is traditionally observed in the high compressibility region of the SCF solvent.

Figure 2.6a depicts a typical solubility behavior, namely that of solid naphthalene (melting point of 80.2°C) in supercritical ethylene (T_c = 9.3°C, P_c = 50.5 bar) near the critical point of pure ethylene (McHugh and Krukonis, 1994). Along the 12°C isotherm (T_r = 1.01), the solubility of solid naphthalene sharply increases as the pressure is increased to 50 bar, i.e., near the CP of ethylene. At pressures greater than 90 bar, the solubility of naphthalene in ethylene asymptotically reaches a value

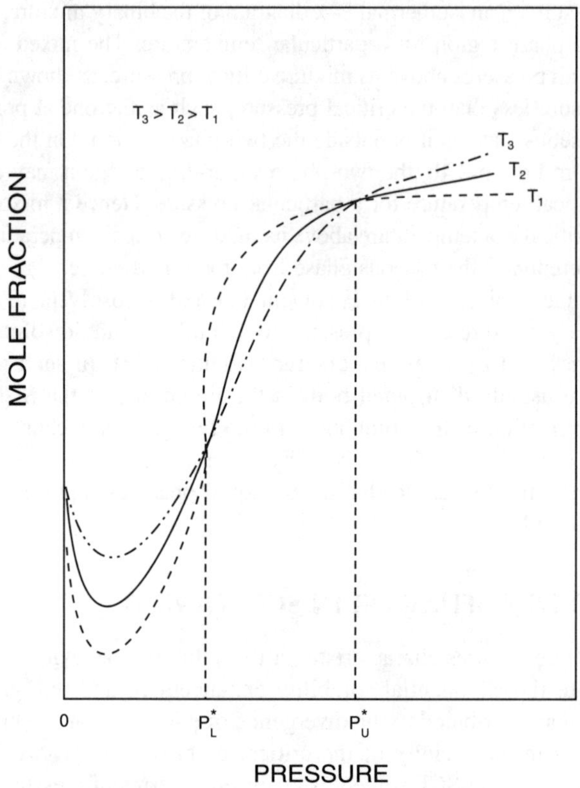

FIGURE 2.5 Solubility behavior of a solid solute in an SCF solvent.

of 1.5 mol%, whereas at pressures below 50 bar, the solubility of naphthalene in ethylene is extremely low, as it would be in a low pressure gas. It is interesting to note that the variation of the isothermal solid solubility with pressure in an SCF solvent, as shown in Figure 2.6a, has the same characteristic shape as the reduced density isotherm, as in Figure 2.3, corresponding to $T_r = 1.0$. Similar conclusion can be arrived at from Figure 2.6b for the isothermal solubility of phenanthrene in ethylene. The similarity between the density isotherm and the solubility isotherm suggests *apriori* that the solubility in an SCF solvent is density dependent. As can be seen from Figure 2.6b, the density effect on solubility sharply increases at a higher density. It implies that at higher densities, the molecular interactions between the solvent and the solute are enhanced and as a result, more solute is dissolved. However, density alone does not give the complete explanation of solubility enhancement. Another factor, the volatility of the solid solute is also responsible for contributing to the solubility behavior. As a matter of fact, the retrograde solubility behavior is explained by the relative influence of the density effect and the volatility effect. An isobaric increase in temperature decreases density of the SCF solvent and hence decreases the solubility by the density effect. On the other hand, the same increase in temperature increases the volatility of the solute and hence increases the

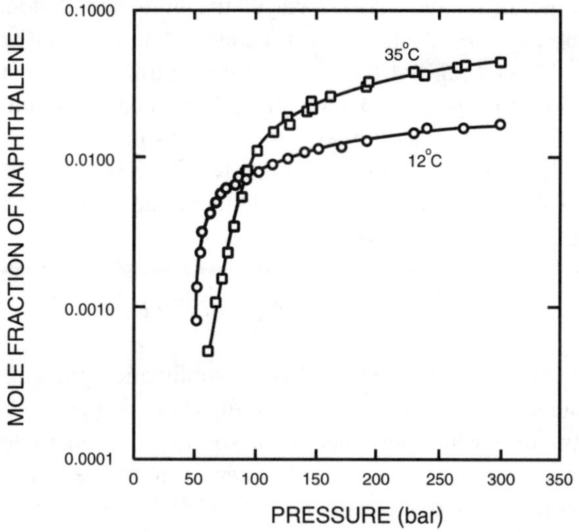

FIGURE 2.6a Isothermal solubility behavior of solid naphthalene in supercritical ethylene with pressure (McHugh and Krukonis, 1994).

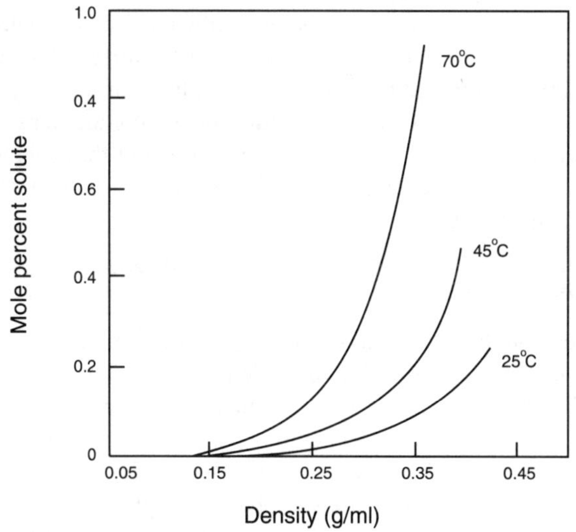

FIGURE 2.6b Typical exponential behavior of solubility isotherms of phenanthrene in SCF ethylene with density (Brennecke and Eckert, 1989).

solubility by the volatility effect. At a pressure less than the crossover pressure P_u, the density effect is more pronounced than the volatility effect, facilitating an increase in solubility by an isobaric decrease in temperature or by a corresponding increase in density. On the other hand, beyond the crossover pressure, P_u, the volatility effect is more pronounced than the density effect, resulting in an increase in solubility with an increase in temperature.

In order to get an indirect assessment of the influence of nonideality of SCF mixture on solubility, let the solubility of a solid solute be simply calculated in an ideal gas, which is nothing but the ratio of the sublimation pressure of the solid solute to the pressure of the gas. Incidentally, this explains why solubility decreases with pressure at lower pressures. An estimate of solid solubility, y_2 in an SCF solvent can be obtained from the following relation, which is derived based on the thermodynamic criteria for equilibrium of a nonvolatile solute (2) between the pure solid phase and the SCF phase as:

$$y_2 = (P_2^s / P\bar{\phi}_2)\exp((v_2^s P)/RT) \tag{2.2}$$

where subscript 2 refers to the solute, P_2^s is the sublimation pressure of the solid, v_2^s is the molar volume of the solid, and $\bar{\phi}_2$ is the fugacity coefficient of component 2 in the dilute SCF mixture, which accounts for the specific molecular interaction between the solute and the solvent molecules and which is highly sensitive to the pressure variation. For an ideal gas, $\bar{\phi}_2$ equals one. At 12°C and 100 bar pressure, the ratio of the experimentally observed solubility, y_2^{exp} in supercritical ethylene to the calculated value of ideal gas-solubility of naphthalene (assuming $\bar{\phi}_2 = 1$), was found to be 16,156 (McHugh and Krukonis, 1994). This large value which is equal to the reciprocal of $\bar{\phi}_2$, is due to the strong attractive interaction between the solvent and the solute molecules causing high nonideality of the SCF ethylene mixture with naphthalene.

The crossover pressures are distinctly different for different solutes and accordingly the retrograde solubility behavior of two solutes in a binary mixture can be utilized for purification or selective separation of the components. Chimowitz and Pennisi (1986) and Kelly and Chimowitz (1989) utilized this unique feature of the SCF systems to develop a generic process for separating multicomponent mixtures into their pure constituents.

2.3.2 Solubility Isobars

In order to understand the variation of isobaric solubility with temperature, solubility of naphthalene in supercritical carbon dioxide has been chosen as the model system as shown in Figure 2.7a. As can be seen, there is no crossover in the solubility isobars, unlike solubility isotherms (Figure 2.5). Depending upon the level of pressure, the temperature effects are seen to be different. For example, at 300 bar, an increase in temperature increases the solubility of naphthalene in CO_2, whereas at a lower pressure of 90 bar, an increase in temperature decreases the solubility. This is due to the fact that the density decreases whereas the vapor pressure increases with temperature and at a higher pressure (>110 bar), the increase in vapor pressure more than compensates the decrease in density.

A similar trend is observed in the solubility isobars of naphthalene in ethylene (Figure 2.7b) and those of soybean oil in CO_2 (Figure 2.7c). The solubility of soybean oil in CO_2 decreases with temperature at 150 bar and becomes negligible at 80°C. On the other hand, the solubility of soybean oil increases with temperature at pressures of 300 bar and above. The solubility of soybean oil increases by five times at 80°C by doubling the pressure from 300 to 600 bar.

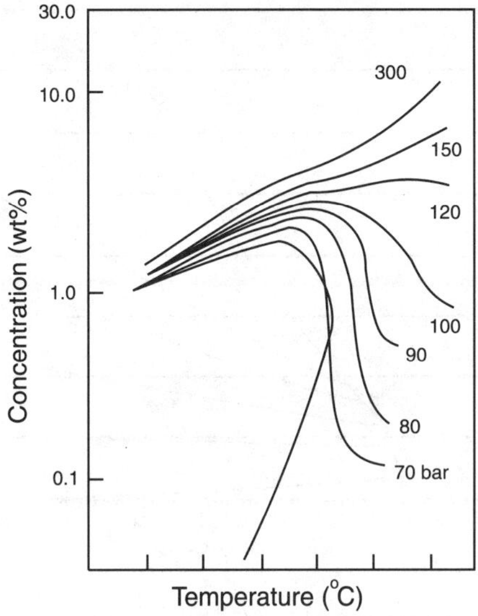

FIGURE 2.7a Behavior of solubility isobars of naphthalene in SC CO_2 with temperature (McHugh and Krukonis, 1994).

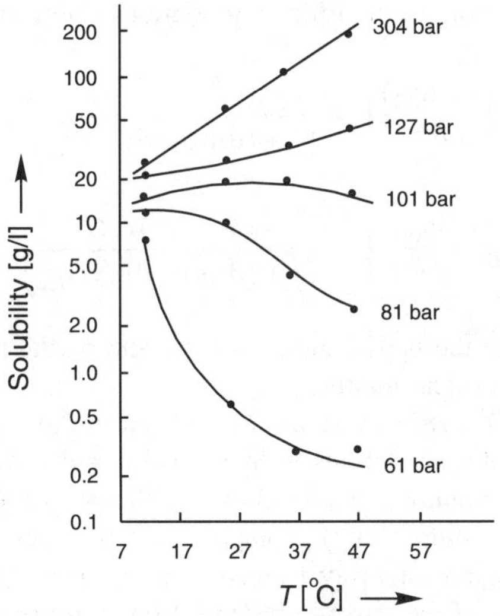

FIGURE 2.7b Solubility isobars of naphthalene in compressed ethylene as a function of temperature (Hubert and Vitzthum, 1980).

2.3.3 Pressure and Temperature Effects

The pressure and temperature effects on solubility behavior in an SCF solvent can be analyzed in terms of the fundamental macroscopic thermodynamic properties, such as partial molar volumes and partial molar enthalpies (Kim et al., 1985). The

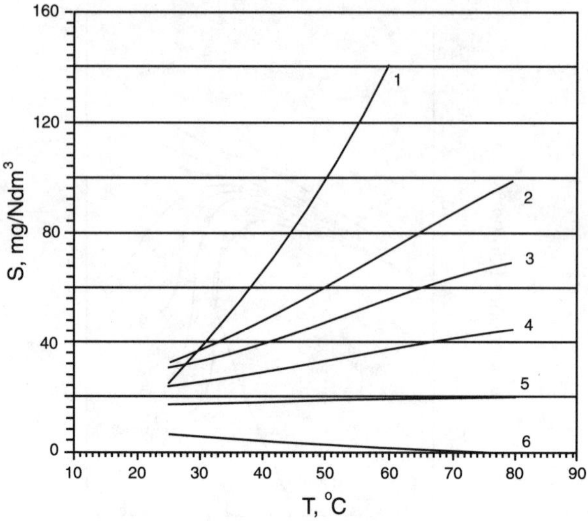

FIGURE 2.7c Solubility of soybean oil in SC CO_2 as a function of temperature: (1) 1000 bar; (2) 600 bar; (3) 500 bar; (4) 400 bar; (5) 300 bar; (6) 150 bar (Reverchon and Osseo, 1994).

pressure and temperature derivatives of the solubility are directly related to the partial molar properties according to the following relations (Prausnitz et al., 1986):

$$\left[\frac{\partial(\ln y_2)}{\partial P}\right] = \frac{(v_2^s - \bar{v}_2)/RT}{[1 + (\partial \ln \bar{\phi}_2/\partial \ln y_2)_{T,P}]} \tag{2.3}$$

$$\left[\frac{\partial(\ln y_2)}{\partial T}\right] = \frac{-(h_2^s - \bar{h}_2)/RT^2}{[1 + (\partial \ln \bar{\phi}_2/\partial \ln y_2)_{T,P}]} \tag{2.4}$$

where $\bar{v}_2$ and $\bar{h}_2$ denote the partial molar volume and partial molar enthalpy of the solute (2) in the supercritical mixture.

Eckert et al. (1983, 1986) were the first to measure the partial molar volumes of the solute in very dilute solutions in SCF. Their results revealed very large and negative partial molar volumes. For example, as shown in Figure 2.8a, the infinite dilution partial molar volumes ($\bar{v}_2^\infty$) of naphthalene in supercritical CO_2 are small and positive in the region far away from the critical pressure, while it is largely negative in the vicinity of the critical pressure. However, the large negativity of $\bar{v}_2^\infty$ decreases with increasing temperature away from the critical region, as shown in Figure 2.8b. This is attributed to strong attractive dispersion forces between the solvent and solute molecules resulting in a decrease in the total pressure. Since the compressibility is very high in the near critical region, the restoration of the system pressure requires a large decrease in the volume making $\bar{v}_2^\infty$ largely negative.

Further, the partial molar volume data can be used to interpret the pressure effects on solubility and to predict the solubility extrema (Kim et al., 1985). At low pressures $\bar{v}_2^\infty >> v_2^s$ and solubility decreases with pressure. At pressures below the solvent's

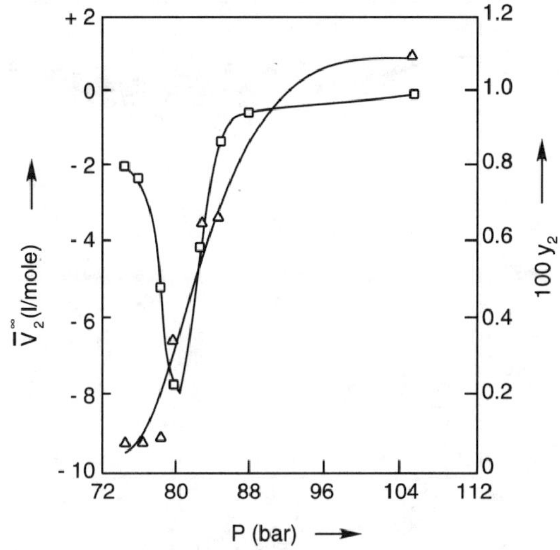

FIGURE 2.8a Solubility (△) and $\bar{v}_2^{\infty}$ (□) vs. pressure for naphthalene dissolved in CO_2 at 35°C (Kim et al., 1985).

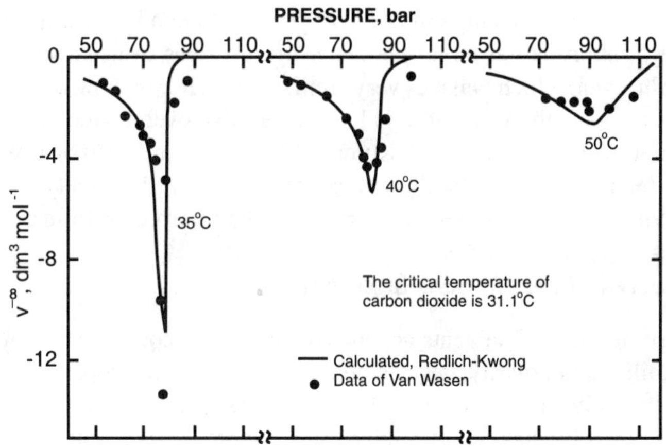

FIGURE 2.8b Partial molar volumes of naphthalene infinitely dilute in SC CO_2 (Prausnitz et al., 1986).

critical pressure, Pc_1 $\bar{v}_2$ approaches $\bar{v}_2^s$ and solubility goes through a minimum at a pressure where $\bar{v}_2 = \bar{v}_2^s$. At very high pressures $\bar{v}_2$ again exceeds $\bar{v}_2^s$ due to increased repulsive forces and hence solubility, y_2 goes through a maximum and then decreases with pressure. Another interesting feature is that the greatest rate of isothermal solubility increase with pressure occurs at the pressure corresponding to the minimum value of $\bar{v}_2^{\infty}$ (Shim and Johnston, 1991).

Similarly, the solute partial molar enthalpies can be utilized to understand the temperature effects of the isobaric solubilities, as expressed by Equation 2.4. The

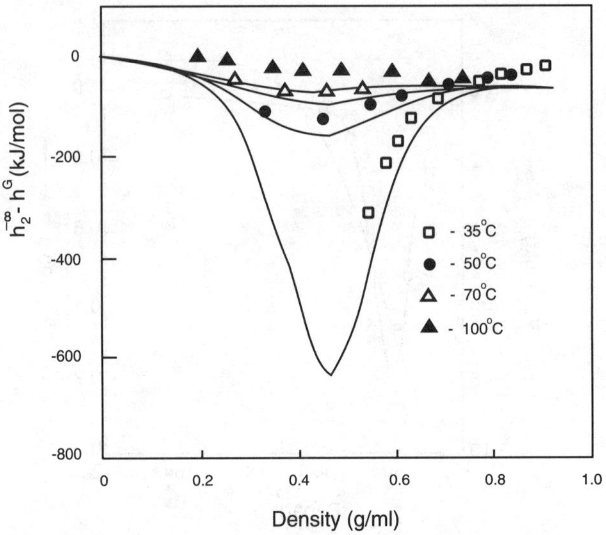

FIGURE 2.9 Partial molar enthalpies of naphthalene in CO_2 (Shim and Johnston, 1991).

numerator of the right hand side term of this equation is the heat effect associated with the dissolution of a solute in an SCF solvent. As can be seen from Figure 2.9, the infinite dilution partial molar enthalpy, $\bar{h}_2^\infty$, diverges to negative infinity in the near critical region, which behaves very similar to $\bar{v}_2^\infty$. For the low solubility systems (for which $\bar{h}_2^\infty \cong \bar{h}_2$), the value of $\bar{h}_2$ is largely negative over a wide range of pressure and temperature (Debendetti and Kumar, 1988). At the crossover region, where $\bar{h}_2 << h_2^s$, the retrograde solubility behavior is observed, namely, the solubility increases with a decrease in temperature as can be interpreted from Equation 2.4.

2.3.4 Solvent Capacity and Selectivity

Solvent capacity of SCF solvents depends on their physicochemical properties, such as polarizability and polarity, besides other thermodynamic properties. It is interesting to find that solvents having critical temperatures in a close range exhibit different capacities to dissolve low volatile substances. Figure 2.10 compares solvent capacities of palm oil in various supercritical solvents having different as well as similar critical temperatures. For example, ethane, CO_2, N_2O, and chlorotrifluoromethane (Freon 13), which all have critical temperatures close to 30°C, dissolve very different amounts of palm oil. The difference in the solvent capacities cannot be attributed merely to the difference in the densities of the solvents. The density of ethane is much lower, about 0.37 g/cm^3 at 70°C and 200 bar, than that of CO_2 which is 0.645 g/cm^3 under the same conditions of pressure and temperature. Even the molar density of ethane is 25% lower than that of CO_2. The difference in solvent capacity cannot be explained by polarity either, since ethane is nonpolar, N_2O has a dipole moment of 0.2 Debye (D), and that of Freon 13 is 0.5 D. It can be seen that solubility of palm oil is less in a solvent having lower critical temperature, e.g., N_2.

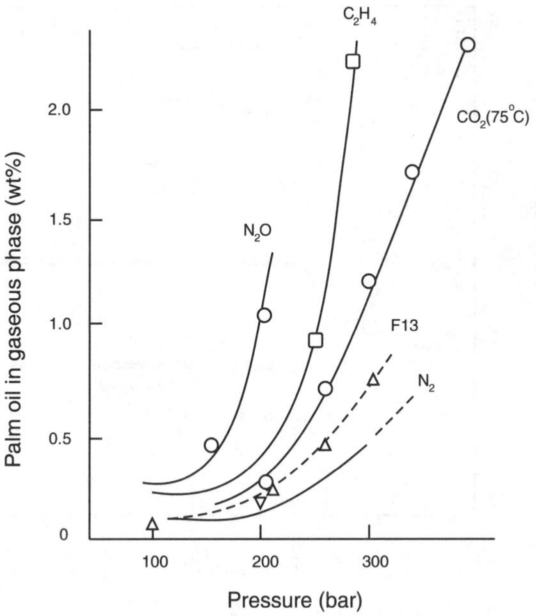

FIGURE 2.10 Solubility of palm oil in various gases (Brunner and Peter, 1982).

The solvent capacity of the SCF solvent is also related to its solubility in the liquid phase. The higher the solubility of the SCF solvent in the liquid phase, the more liquid component is dissolved in the SCF phase. The solvent capacity for low volatile components is higher if the solvent is in the SCF state rather than in the liquid state. However, chemically similar solutes dissolve in an SCF solvent according to their respective volatility (vapor pressure). On the contrary, solutes having similar volatility but having different polarity have widely different solubilities, e.g., water and ethanol dissolve differently in CO_2. Further, solubility of components in a mixture of solutes is different from those of the pure components. As shown in Figure 2.11, the solubilities of both solute components in benzoic acid + naphthalene + CO_2 system may be enhanced by more than 100% (Kurnik and Reid, 1982). With low solubilities in the range of 10^{-2} to 10^{-3} in mole fraction units, the attractive dispersion forces appear to pull both the solutes in the SCF phase, resulting in a synergestic behavior. However, the solubilities of the components in phenanthrene + anthracene mixture behave in a different way. Kwiatkowski et al. (1984) observed that the solubility of anthracene (about 10^{-5} in mole fraction) increased while that of phenanthrene (about 10^{-3} in mole fraction) decreased slightly relative to the respective pure component solubilities. This suggests that anthracene finds attractive forces from the more concentrated phenanthrene, but anthracene is sufficiently dilute not to affect phenanthrene. In other words, the components having lower solubilities are enhanced to a greater extent than the components having higher solubilities in the synergystic solubility enhancements of the components in a mixture. These observations indicate that the unusual solute–solute and solute–solvent interactions are of prime considerations in deciding the solvent

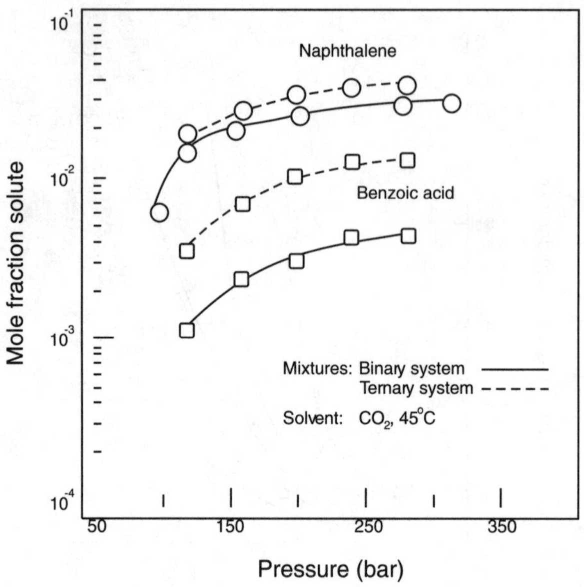

FIGURE 2.11 Synergistic solubility behavior of a mixed solute system (Brennecke and Eckert, 1989).

capacities for different components in a mixture of solutes and should be considered for the design of the SCF separations.

A striking feature of an SCF solvent is that it can selectively dissolve more amounts of certain specific compounds from a mixture of compounds having even similar volatility but different chemical structures. In general, the selectivity of separation depends on the mixture composition, temperature, pressure, and solvent characteristics such as molecular structure and molecular properties. Figure 2.12a illustrates the variation of selectivity of separation of octadecane (2) from hexadecanol (1) with pressure by various SCF solvents. It is mostly observed that selectivity is related to solubility in that the higher the solubility, the lower the selectivity. But in SCF solutions, the selectivity–solubility behavior is little more complex. For many high molecular weight compounds, as can be seen from Figure 2.12b that selectivity increases with pressure at lower pressures, attains a maximum, and then decreases with pressure. This behavior is observed at any temperature, though at a higher level of selectivity at higher temperature.

2.3.5 Cosolvent Effects

A new dimension was added to SCFE when it was established that the solubilities in SCF could be greatly enhanced by the addition of small amounts (1 to 5 mol%) of a cosolvent or entrainer (Dobbs, 1986). The role of cosolvent in SCFE has been to increase the polarity and solvent strength while retaining the sensitivity of solubility with respect to pressure and temperature. Additionally, a cosolvent can improve the selectivity of separation by preferentially interacting with one or more components and facilitating

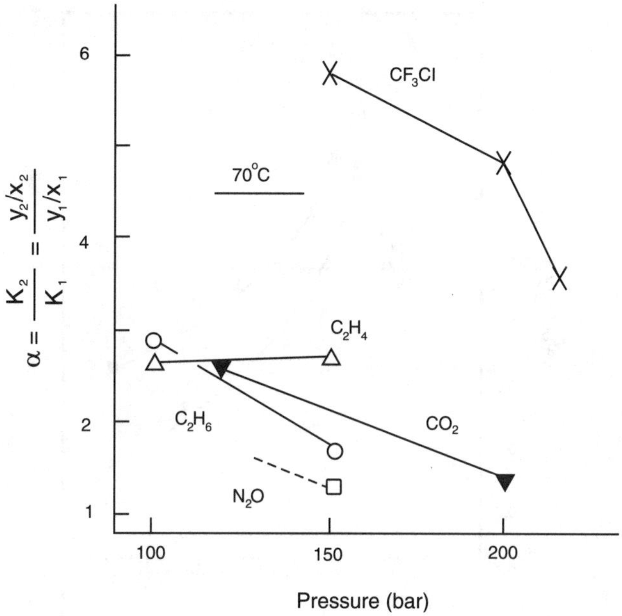

FIGURE 2.12a Selectivity of the components in various gases in an equimolar mixture of hexadecanol and octadecane (Brunner and Peter, 1982).

selective fractional separation. Dobbs (1986) studied the effects of nonpolar as well as polar cosolvents on the solubility behavior of organic solutes. He observed that the solubilities of nonfunctional aromatics could be considerably increased with nonpolar cosolvents, whereas for systems consisting of polar solutes, solubility enhancements were significantly greater with polar cosolvents due to dipole–dipole interactions and hydrogen bonding. In addition to the enhanced solubility, Dobbs observed that the cross-over pressure of a solid solute in an SCF solvent in the presence of a cosolvent was lower than its value in a pure SCF solvent. This suggests the possibility of separating the components at a lower pressure by a suitable process, synthesis-based on the cross-over phenomenon as proposed by Chimowitz and Pennisi (1986).

In general, nonpolar cosolvents increase the solubility of nonfunctional aromatic hydrocarbons, up to several hundred percent. For n-alkane cosolvents, the solubility enhancement is directly related to the chain length, i.e., the longer the chain, the greater the enhancement. Also the greater the concentration of the cosolvent, the greater is the enhancement of the solubility. For a mixture of nonpolar solutes, the solubility enhancements with a nonpolar cosolvent are more or less similar and are relatively independent of density and accordingly no real improvement in selectivity can be achieved by temperature and pressure variations.

However, this is not the case with a polar cosolvent in enhancing the solubilities of polar solutes in a mixture. Moreover, the solubility and selectivity enhancements are even greater than those found in the case of nonpolar solutes. For example, for 2-aminobenzoic acid, the addition of 3.5 mol% methanol increases the solubility by 620% (Dobbs et al., 1987). This suggests that the cosolvent can facilitate selective separation

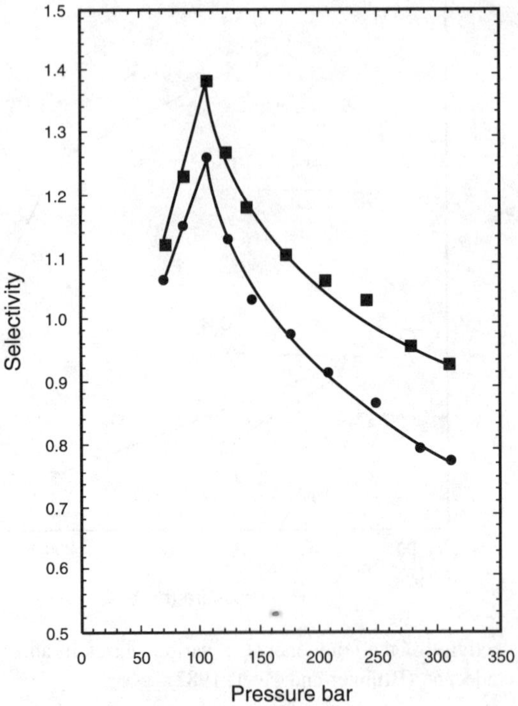

FIGURE 2.12b Selectivity of cholesterol with respect to triglyceride in SC CO_2-AMF system at (●) 40°C; (■) 60°C (Rizvi et al., 1994).

of the solutes having different polarities, hydrogen bonding, and abilities for association or complexation. Wong and Johnston (1986) experimentally verified the solubilities of three sterols of similar polarity and found that though the selectivities were not improved, the solubilities were enhanced by an order of magnitude. This is an important consideration for process design in view of very low solubilities of the sterols in SCF carbon dioxide. A striking example of the cosolvent effect was cited by Johnston et al. (1989), who reported an increase of solubility of hydroquinone in CO_2 by 2 to 3 orders of magnitude with the addition of small amounts of the tributylphosphate (TBP), as shown in Figure 2.13a. Further, as can be seen from Figure 2.13b, the solubility of sunflower oil in CO_2 is enhanced at the rate 5 g/kg CO_2 for each percent (by weight) ethanol addition to SC CO_2 at 300 bar and 42°C. Even the crossover pressure is lowered to 350 bar by addition of 10% ethanol to SC CO_2 as can be seen from Figure 2.13c. Thus a careful choice of cosolvent could be used to separate the components, not just on the basis of polarity, but also on the basis of functional groups and the ability to have specific interactions (Raghuram Rao et al., 1992).

2.4 SCF PHASE EQUILIBRIUM BEHAVIOR

The foundation of all emerging and existing SCFE applications that include food, flavor, fragrance, and pharmaceuticals lies in the basic understanding of high pressure,

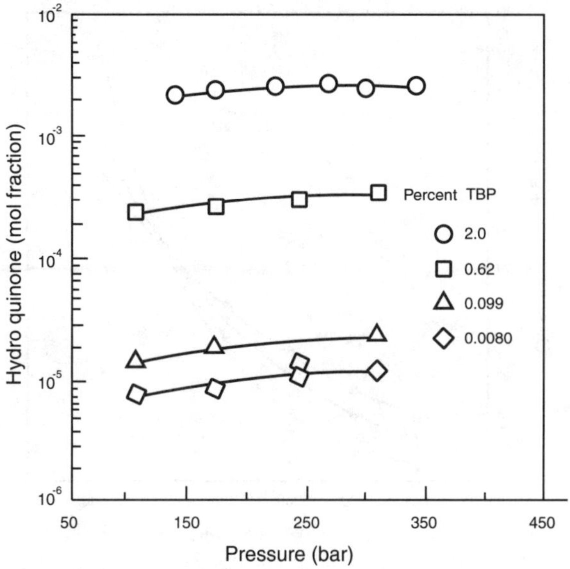

FIGURE 2.13a Effect of TBP as entrainer on the solubility enhancement of hydorquinone in SC CO_2 (Brennecke and Eckert, 1989).

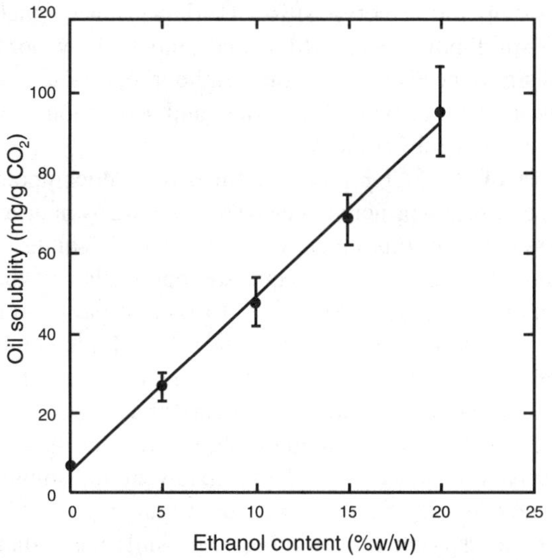

FIGURE 2.13b Solubility of sunflower oil in CO_2, modified with ethanol at 42°C and 300 bar (Cocero and Calvo, 1996).

fluid-phase, equilibrium behavior. In an SCFE process, when an SCF solvent is contacted with solute at conditions near the critical point of the solvent plus solute mixture, there may be occurrence of multiple phases involving vapor, liquid, or solid phases, depending on the mixture composition and temperature and pressure

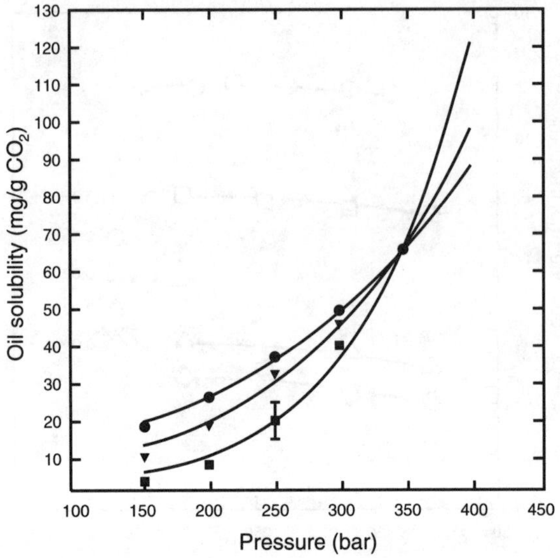

FIGURE 2.13c Effect of pressure on solubility of sunflower oil in SC CO_2 mixed with 10% ethanol at 42°C (●); 60°C (▼); and 80°C (■) (Cocero and Calvo, 1996).

conditions. For practical reasons, it is desirable to avoid regions of multiple phases in the pressure-temperature-composition (P–T–x) space, such as liquid–liquid–vapor (LLV), solid–liquid–vapor (SLV) and solid–solid–vapor (SSV) equilibria. Accordingly a designer needs to select the neighborhood of temperature and pressures at which processing of natural materials and separation of their extracts are technically and economically attractive.

Although most of the SCFE processes involve multicomponent systems, the important phase equilibrium principles can be understood by interpreting the limiting case of a binary system constituting an SCF solvent and a solute, either in the form of a solid or a liquid. There are two sequential approaches to the analysis of the SCF phase equilibrium behavior, namely, (1) to study the pressure effects in the proximity of the critical point on the occurrence of multiple equilibrium phases in P–T space, based on the experimental observations of different types of systems, and (2) to predict the phase equilibrium behavior for a specified condition in terms of specific interactions between the solute–solute and solute–solvent molecules.

In case a cosolvent is added, the solvent–cosolvent and solute–cosolvent interactions are also included. In this section an understanding of SCF phase behavior will be initiated by classifying various types of possible phase diagrams applicable for mixtures of interest.

2.4.1 Liquid–Fluid Phase Equilibria

Based on the shape of the critical locus in the P–T space of three-dimensional P–T–x diagrams, van Konynenberg and Scott (Ekart et al., 1991) classified all of the experimentally observed liquid–fluid phase behavior into six schematic P–T diagrams, corresponding to six fundamental classes (I–VI), as shown in Figure 2.14. The dark

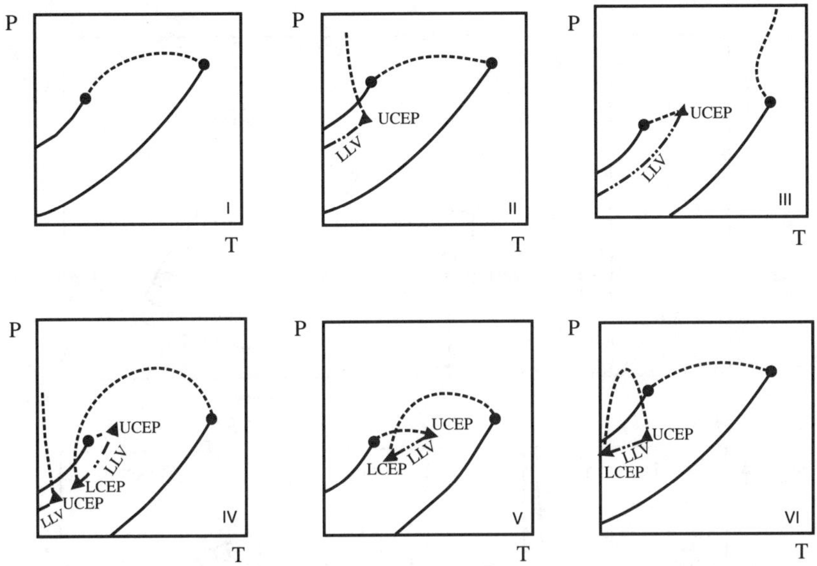

FIGURE 2.14 Van Konynenburg and Scott classification of P–T phase diagrams (Ekart et al., 1991).

circles represent the critical points of components 1 and 2, respectively; the solid curves represent the pure component phase boundaries; the dashed curve in each figure represents the locus of the mixture critical points for the binary mixture; and the triangles (Δ) are the critical end points, lower critical points (LCEP), and upper critical points (UCEPs), which occur at the intersection of the LLV lines and the mixture critical curves. At critical end points, two phases critically merge into a single phase in the presence of another phase.

The locus of the upper critical solution temperature (UCST) is represented by the line at which two liquid phases critically merge to form a single phase as the temperature is increased. The UCST is at a lower temperature than the lower critical solution temperature (LCST) which is the locus of temperature at which two liquid phases critically merge to form a single liquid phase as the system temperature is decreased at constant pressure.

As shown in Figure 2.14, I for class I systems, the components are completely miscible with a continuous mixture (gas–liquid) critical line. The mixture critical line can have a variety of shapes including a maximum or minimum in temperature and/or with a maximum in pressure. Class I systems are not too far from ideal.

For class II systems, the (gas–liquid) critical line is continuous, but at low temperatures liquid–liquid phase separation occurs. As shown in Figure 2.14, II, the intersection of the UCST line and a 3-phase (LLV) line is called the upper critical end point (UCEP), where two liquid phases critically merge to form a single liquid phase in equilibrium with a vapor phase as the temperature is increased.

As shown for class III and IV systems (Figures 2.14, III and IV), UCEP is also the point of intersection of the LLV line and the lower temperature part of the locus

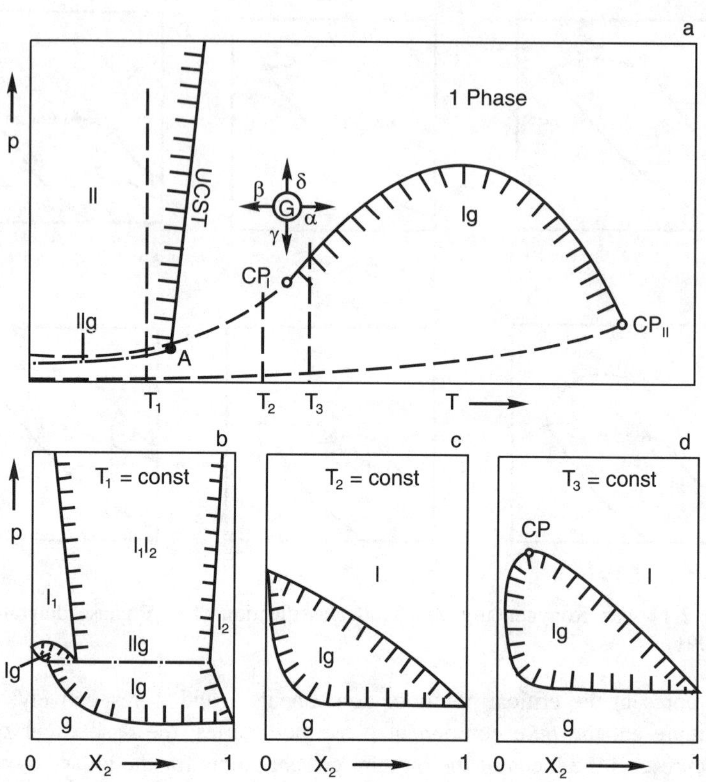

FIGURE 2.15a P–T diagram with selected P–x isotherms for class I and class II systems (Schneider, 1980).

of mixture critical curve. For class III systems, the higher temperature part of the mixture critical curve rises rapidly to a very high pressure or may have a maximum or minimum pressure and/or a minimum in temperature. In the case of a class IV system, there are however three critical end points, namely, UCEP and LCEP at the intersections of two parts of the mixture critical curve and the LLV line at the two ends, respectively, and one more UCEP at the intersection of LLV and UCST lines.

For class V systems, LCEP and UCEP occur in a close temperature range, as in the case of class IV systems. Only difference is that there is no liquid–liquid immiscibility at a temperature less than LCEP in class V systems. The class VI system has a continuous GL critical curve and a LL critical curve that intersects an LLV line at the two ends, namely, LCEP and UCEP, respectively. The class VI type of equilibrium occurs in the case of hydrogen bonding systems.

Figure 2.15a–c show schematic P–T diagrams of a binary system along with P–x isotherms of class II, class III, and class IV systems, respectively. It can be noticed that in the homogeneous region beyond the mixture critical curve, the components 1 and 2 are completely miscible in all proportions. For example, at point "G," the low-volatile component 2 can be completely dissolved in the SCF solvent

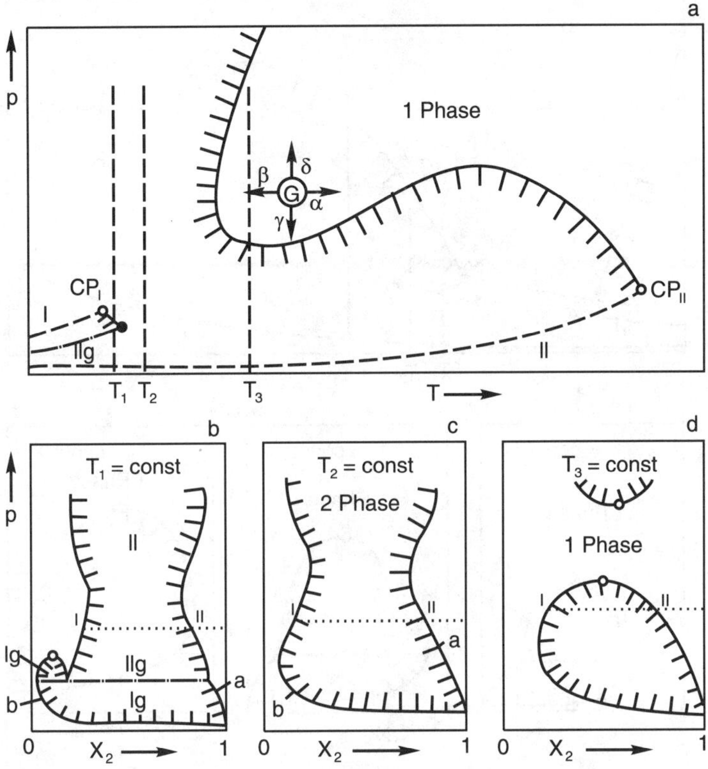

FIGURE 2.15b P–T diagram for class III system with selected P–x isotherms (Schneider, 1980).

(1). For precipitation of the dissolved substance in the form of extract, it is necessary to enter the 2-phase region in one of the following ways (Schneider, 1980):

1. By increasing temperature, i.e., by following the direction α
2. By decreasing temperature, i.e., by following direction β
3. By decreasing pressure, i.e., by following the direction γ
4. By increasing pressure, i.e., by following the direction δ

2.4.2 Solid–Fluid Phase Equilibria

In many solid–fluid systems, the critical temperature of the SCF solvent is less than the triple point of the solid solute and there is no common range of temperature in which two pure components can exist in the liquid state. Figure 2.16a represents the schematic P–T diagram for the solid-fluid phase behavior of systems which correspond to class I liquid-fluid systems. With increasing difference between the two critical temperatures of the pure components, this solid–fluid phase behavior gets transformed to the one represented by Figure 2.16b which also corresponds to class IV liquid-fluid systems (Streett, 1983).

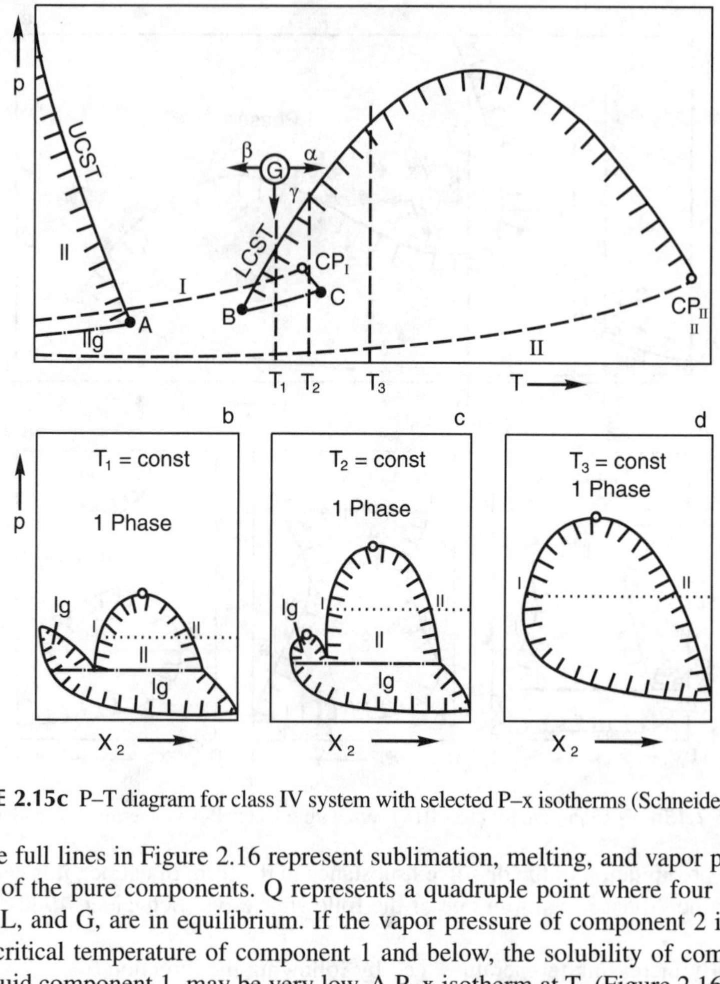

FIGURE 2.15c P–T diagram for class IV system with selected P–x isotherms (Schneider, 1980).

The full lines in Figure 2.16 represent sublimation, melting, and vapor pressure curves of the pure components. Q represents a quadruple point where four phases, S_I, S_{II}, L, and G, are in equilibrium. If the vapor pressure of component 2 is small at the critical temperature of component 1 and below, the solubility of component 2 in liquid component 1, may be very low. A P–x isotherm at T_2 (Figure 2.16b), just above the nose on the right side of the diagram (Figure 2.16c), indicates the dramatic increase in the solubility of component 2 in the SCF component 1 with an increase in pressure, but at much higher pressures the trend of enhancement of solubility may be reversed (Streett, 1983).

2.4.3 Polymer–SCF Phase Equilibria

In recent years there has been significant interest in understanding the high pressure phase equilibrium behavior in polymer solvent mixtures. In the analysis of polymer–SCF phase equilibria, it is customary to treat polymer as a pseudo-single component and amorphous polymer that does not crystallize. The polymer–fluid phase equilibria can be schematically represented by Figure 2.17 A–D for two classes (III and IV) of systems (McHugh and Krukonis, 1994).

Figure 2.17A for small molecule systems shows that V–L curves for two pure components end in their respective critical points, c_1 and c_2, as was shown earlier

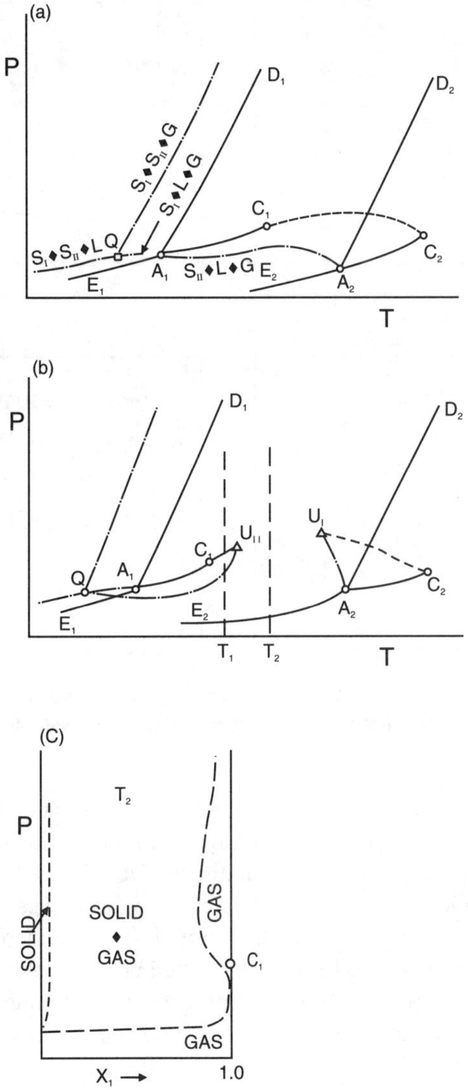

FIGURE 2.16 P–T diagram for solid–fluid phase equilibria for class I and class IV systems and P–x isotherm at a selected temperature (Streett, 1983).

for the class IV system in Figure 2.14 IV. The point shown here is that the critical mixture curve starting from c_2 exhibits a maximum pressure and then ends up intersecting with the LLV line at the LCEP at a condition close to the critical point of the more volatile component (1). This portion of the mixture critical curve is termed as LCST curve, as the two phases coalesce into a single phase on lowering the temperature isobarically.

Figure 2.17B represents the schematic P–T diagram for a polymer-solvent mixture belonging to the class IV system, which is very similar to Figure 2.17A, except

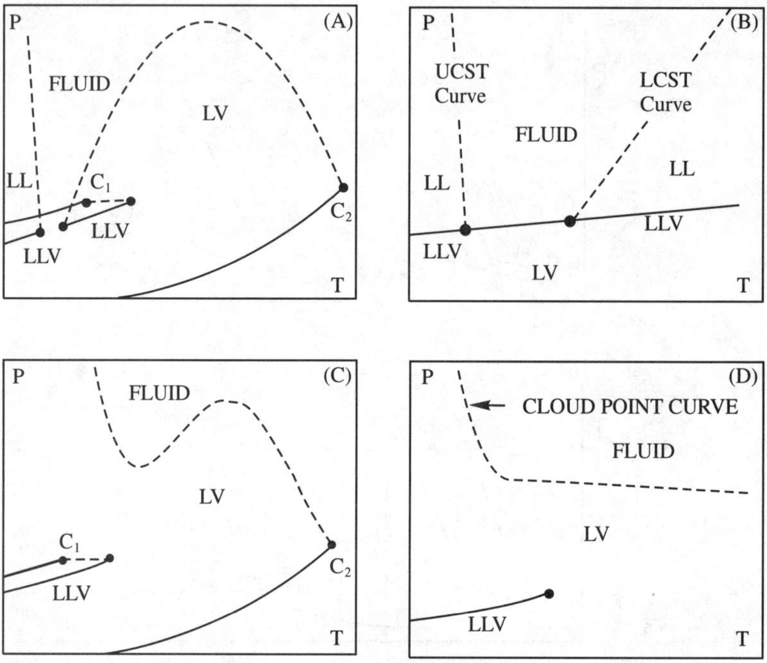

FIGURE 2.17 (A–D) Comparison of P–T diagrams (A) and (C) for small molecules with P–T diagrams; (B) and (D) for polymer–solvent in type IV systems and type III systems, respectively (McHugh and Krukonis, 1994).

for the fact that a pure polymer does not have a critical point or vapor-pressure curve and accordingly high temperature portion of the mixture critical curve is missing in Figure 2.17B. For polymer–solvent systems, the UCST and LCST branches of the mixture critical curve are often termed as cloud point curves. The striking feature of the polymer–solvent equilibria is that with an isobaric increase in temperature, the two liquid phases merge to form a single fluid phase across the UCST curve and on a further rise of temperature, phase splitting occurs resulting in the reappearance of the two liquid phases across the LCST curve. This phenomenon of phase splitting which is known as the LCST phenomenon is usually attributed to the large difference in the thermal expansion or free volume of the polymer and solvent molecules. When a polymer solution is heated, the solvent expands at a much faster rate than the polymer, causing phase separation. It is therefore expected that the location of the LCST curve should depend on the molecular weight and chemical nature of the polymer and the critical temperature of the solvent.

Figure 2.17C shows a schematic P–T diagram for a class III type of small molecule system, whereas Figure 2.17D shows its transformation for a polymer–solvent system belonging to the class III type. The cloud point curve in Figure 2.17D can be approximated as the combination of LCST and UCST types of transitions at higher and lower temperatures, respectively. This kind of behavior is depicted in a number of poly(ethylene-comethyl acrylate)-hydrocarbon solvent mixtures (McHugh and Krukonis, 1994).

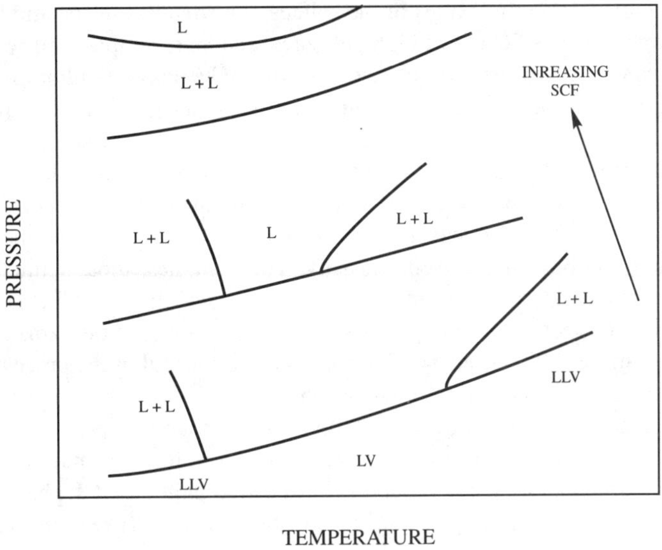

FIGURE 2.18 Schematic representation of the effect of an SCF additive on the P–T behavior of a polymer solution (McHugh and Krukonis, 1994).

As shown in Figure 2.17B, a single phase solution exists at temperature and pressure conditions between the UCST and LCST border curves. It is therefore possible to split the polymer solution by heating it beyond the LCST and separate the polymer-rich fraction from the polymer-lean solution. But it is necessary to heat the solution to a temperature near the solvent critical temperature. For cyclohexane (T_c = 280.3°C) as the polymer solvent, the temperature required for phase splitting could be as high as 260°C at which thermal degradation may take place. In order to obviate the problem, it is customary to add an SCF solvent to the solution to shift the LCST curve to lower temperatures. The SCF solvent dilates the polymer solvent without raising its temperature and lowers the temperature required for phase splitting, as illustrated in Figure 2.18 (McHugh and Krukonis, 1994). Without any SCF solvent (e.g., ethane), the LCEP for polystyrene–toluene system occurs at 284°C. With addition of ethane (17.8 wt%), the LCEP is lowered to 122°C. It can be further reduced to 53°C by increasing the SCF ethane content to 22.5 wt%, and finally the LCST and UCST lines coalesce, with SCF ethane content of 24.9 wt% (McHugh and Krukonis, 1994). Thus it is possible to separate the polymers with purity close to 100% from their solution without thermal degradation while using less energy as compared to conventional solvent evaporation technique (Irani and Cozewith, 1986).

2.5 THERMODYNAMIC MODELING

Some of the phase behavior variations outlined in the previous section reveal interesting features of the SCF phase equilibria that are needed for the process design of SCF-based separations. However, the task to model and predict such phase behavior for developing both qualitative and quantitative understanding poses a serious challenge due to the molecular complexities of the solutes, uncertainties in specific interactions in dilute supercritical solutions at high pressures, and high compressibility of the SCF solvent. There are two aspects of modeling the SCF phase equilibria. One is the prediction of the various critical boundaries such as gas–liquid, liquid–liquid, liquid–liquid–vapor, solid–gas, solid–liquid–gas, etc., in the phase diagrams. These predictions are useful in deciding the operating regions where complexities of phase splitting can be circumvented. The other aspect is the prediction of equilibrium solubilities of solutes in the SCF solvent, their selectivity of separation, and the conditions for optimum recovery and solvent capacity. These studies, though fundamental in nature, facilitate making a decision regarding the applicability of SCFE to the system in question and selection of multiple products for processing in the same SCFE plant to make it commercially attractive. They are also later useful in the process design of the extractors, separators, and heat exchangers.

In the recent past, several investigators have ventured experimental measurements and thermodynamic modeling in search of more fundamental understanding. The most common approach to modeling has been to treat the SCF phase as a dense gas which can be represented using an equation of state (EOS) to calculate the fugacity coefficients. In this approach the results are commonly very sensitive to the mixing rules for the interaction energy and size parameters of the solute and the solvent molecules. Another approach has been to treat the SCF phase as an expanded liquid. There is also a large number of semiempirical correlations as well as molecular models based on computer simulations.

2.5.1 The Equation of State (EOS) Approach

The accuracy of the solubility predictions is essentially dependent upon the accuracy of the fluid density and solute fugacity calculations. The supercritical mixtures are highly compressible and strongly asymmetric with respect to the size and attractive energy of the constituents. Numerous investigators applied different thermodynamic models to solid–fluid, solid–fluid–cosolvent, and liquid–fluid systems as well as multiphase systems, as was summarized by Johnston et al., (1989). Most widely used models for fugacity calculations treating the SCF solvent as a dense gas are based on cubic equations of state (CEOS), such as the Peng-Robinson (PR; Peng and Robinson, 1976) and the Soave-Redlich-Kwong (SRK; Soave, 1972) equations. However, it is observed that compared to several other equations, the PR EOS performs as well as the more complicated purturbed hard sphere van der Waals equations (Brennecke and Eckert, 1989). With certain limitations, the PR EOS gives a good qualitative picture of all types of phase behavior and reasonable quantitative representation for a variety of systems (Hong et al., 1983). It is for this reason that only the PR EOS will be used in the following discussions. The PR equation is given in the form:

$$P = \frac{RT}{v-b} - \frac{a(T)}{v(v+b)+b(v-b)} \tag{2.5}$$

or in terms of the compressibility factor Z as:

$$Z^3 - (1 - B)Z^2 + (A - 3B^2 - 2B)Z - (AB - B^2 - B^3) = 0 \tag{2.5a}$$

where $A = aP/(R^2T^2)$, $B = bP/(RT)$, and $z = Pv/(RT)$, v is the molar volume, a accounts for the intermolecular attractive energy, and b accounts for the size or covolume of the molecule. For pure components, a and b are calculated from the critical constants, T_c, P_c, and acentric factor, ω, as:

$$b = 0.07780\ R\ T_c/P_c \tag{2.6}$$

$$a(T) = a\ (Tc)\ \alpha(T_r, \omega) \tag{2.7}$$

$$a(T_c) = 0.45724\ R^2\ T_c^2/P_c \tag{2.8}$$

$$\alpha = [1 + k(1 - T_r^{1/2})]^2 \tag{2.9}$$

$$k = 0.37464 + 1.54226\ \omega - 0.26992\ \omega^2 \tag{2.10}$$

$$\omega = -1.000 - \log_{10}\ (P^s/P_c)_{Tr = 0.7} \tag{2.11}$$

When the PR EOS is extended to mixtures, the mixture parameters, a and b, are calculated from the quadratic van der Waals (VDW) mixing rule as:

$$a = \sum_i \sum_j y_i y_j a_{ij} \tag{2.12}$$

$$b = \sum_i \sum_j y_i y_j b_{ij} \tag{2.13}$$

where a_{ij} and b_{ij} are calculated from the combining rules:

$$a_{ij} = (a_{ii}a_{jj})^{0.5}\ (1.0 - k_{ij}) \tag{2.14}$$

$$b_{ij} = 0.5\ (b_i + b_j)\ (1.0 - n_{ij}) \tag{2.15}$$

where k_{ij} and n_{ij} are the adjustable binary interaction parameters which are usually determined by regressing experimental phase equilibrium data and are normally assumed to be independent of temperature, pressure, and composition. In most cases, n_{ij} is taken to be zero, unless otherwise stated.

2.5.2 Solid–Fluid Equilibrium Calculations

For prediction of solubility isotherms using the dense gas approach, the solubility of a solid solute in a supercritical fluid solvent is obtained by using an equation of state. For the solid–fluid equilibria for a binary system, the solid phase is considered essentially pure and the region in P–T space is considered to lie between LCEP and UCEP in a class IV system (Figure 2.16). From the thermodynamic criteria for solid–fluid phase equilibrium, the solubility, y_2 of a solid solute (2) in an SCF solvent (1) is expressed as:

$$y_2 = \frac{P_2^s \exp[v_2^s(P - P_2^s)/RT]}{\bar{\phi}_2 P} \tag{2.16}$$

where P_2^s is the saturation vapor pressure, v_2^s is the molar volume of the solid solute, and the fugacity coefficient, $\bar{\phi}_2$, is obtained from a CEOS from the relation:

$$\ln\bar{\phi}_2 = \frac{1}{RT}\int_v^{\infty}\left[\left(\frac{\partial P}{\partial n_2}\right)_{T,V,n_1} - \frac{RT}{v}\right]dv - \ln\left(\frac{Pv}{RT}\right) \tag{2.17}$$

The PR EOS with the VDW mixing rules give good agreement with the experimental data for nonpolar components having energy and size parameters in a close range. However, for highly asymmetric systems such as in solid–fluid systems, k_{ij} of the VDW mixing rule is found to be temperature and pressure dependent (Gangadhara Rao and Mukhopadhyay, 1989a) and the solubility calculations are found to be highly sensitive to small variations in its value. Thus any correlations developed for the regressed values of k_{ij} in terms of physical properties of the components are found ineffective for solubility predictions of nonvolatile solid solutes in SCF solvent (Wong et al., 1985; Gangadhara Rao, 1990).

2.5.2.1 Mixing Rules

To obviate the above limitations and to increase the range of applicability of the PR EOS, several modifications were suggested to the VDW mixing rule (Johnston et al., 1989). Even with all these modifications, which invariably included additional adjustable parameters, the predictions were still extremely sensitive to the binary interaction parameters.

Vidal (1978) developed an empirical mixing rule for the ratio of the energy parameter to the size of the molecule, namely, a/b, which was expressed as:

$$a/b = \sum_i y_i(a_{ii}/b_{ii}) - g_\infty^E/\ln 2 \tag{2.18}$$

The excess Gibb's energy, g_∞^E in the above relation is calculated from a modified NRTL model. This model requires three adjustable parameters and hence it is less useful for predictive purposes. More general mixing rules were introduced by Smith (1972) by extending the isotropic conformal solution method as:

$$a^{\alpha}b^{\beta} = \sum\sum \; y_i y_j \; a_{ij}^{\alpha} \; b_{ij}^{\beta} \tag{2.19}$$

$$a^{\gamma}b^{\delta} = \sum\sum \; y_i y_j \; a_{ij}^{\gamma} \; b_{ij}^{\delta} \tag{2.20}$$

The four exponents in the above equations are determined empirically. Radosz et al. (1982) used the values for $\alpha = 0$, $\beta = 1$, $\gamma = 1$, and $\delta = -0.25$ for VLE of light hydrocarbon mixtures. But they failed for higher molecular weight solutes dissolved in small solvent molecules.

In order to reduce the sensitivity of the binary interaction parameters and to make them predictive while restricting the number of adjustable parameters, a covolume dependent (CVD) mixing rule was developed (Gangadhara Rao and Mukhopadhyay, 1988), which is given as:

$$a = \sum_i\sum_j y_i y_j a_{ij}(b/b_{ij})^m \tag{2.21}$$

where m was considered as an adjustable parameter varying between 0.0 and 4.0. However an optimum value of 1.5 was found to give best results for the systems considered in the analysis (Gangadhara Rao and Mukhopadhyay, 1988 and 1989b). The sensitivity reductions were manifested in the successful development of correlation of the binary interaction parameters, as will be seen later. This CVD mixing rule with m as the adjustable parameter was applied to the cosolvent mixed SCF solvent as well, for prediction of solubilities of mixed solutes with success.

Subsequently, the CVD mixing rule was modified to widen its scope of applicability and predictability of solid solubilities (Mukhopadhyay and Raghuram Rao, 1993) as:

$$a = \sum_i\sum_j y_i y_j a_{ij}(b/b_{ij})^{m_{ij}} \tag{2.22}$$

where $m_{ii} = m_{jj} = 1.0$, whereas m_{ij} is evaluated by regressing the experimental solubility data.

The other quantities in Equation 2.22 are

$$b = \sum \; y_i \; b_{ii} \tag{2.23}$$

$$a_{ij} = (a_{ii} \; a_{jj})^{1/2} \tag{2.24}$$

$$b_{ij} = (b_{ii} \; b_{jj})^{1/2} \tag{2.25}$$

Using this new CVD mixing rule, the expression for the fugacity coefficient is given as:

$$\ln\bar{\phi}_2 = (Z-1)B_1 - \ln(Z-B) - \frac{a}{2\sqrt{2}RTb}(A_1 + A_2 - B_1)\ln\left[\frac{Z+2.414B}{Z-0.414B}\right] \quad (2.26)$$

where $A = aP/(RT)^2$, $B = bP/RT$

$B_1 = b_2/b$

$$A_1 = 2/a \sum_i y_i a_{i2}(b/b_{i2})^{m_{i2}}$$

$$A_2 = 1/a(B_1 - 1)\sum_i \sum_j y_i y_j a_{ij} m_{ij}(b/b_{ij})^{m_{ij}}$$

2.5.2.2 CS and GC Methods

The pure component, like parameters a_{ii} and b_{ii} for the PR EOS, are calculated by either (1) the corresponding states (CS) method using Equations 2.6 to 2.11, or (2) a group contribution (GC) method using heats of sublimation and molecular volume, etc. (Mukhopadhyay and Raghuram Rao, 1993). The latter method is preferable as critical constants are not required which are not readily available for large molecular weight nonvolatile solutes. The GC method for calculation of a_{22} and b_2 are

$$b_2 = 1.71\ b_{vdw} + 1.28 \quad (2.27)$$

$$a_{22} = 6666.6\ [\Delta H_2^s/RT]^{0.123}\ [M_2/b_{vdw}^{0.25}]^{2.98} \quad (2.28)$$

where b_{vdw} is the van der Waal's volume and ΔH_2^s is the heat of sublimation of the solid solute calculated by Bondi's group contribution method (Bondi, 1968). The empirical correlations, i.e., Equations 2.27 and 2.28, were developed (Raghuram Rao, 1992) based on the behavioral trends with size and cohesive energy parameters for 21 solids. The values of a_{22} and b_{22} calculated by the two methods, namely, CS and GC, differ by an average deviation of 9.6% and 8.7%, respectively, between the two methods. These deviations are meager considering the uncertainties in the reported values of the critical constants.

For testing the capability of the new CVD mixing rule (Equation 2.22), solubilities in supercritical carbon dioxide for 21 solids with varying polarities were calculated using m_{12} as the adjustable parameter. The calculated solubilities were found to be in very good agreement with the corresponding experimental data within 11.9% and 11.7% of overall AARD by the CS and GC methods (for b_2 and a_{22}), respectively. For certain complex solids, such as hexamethyl benzene, benzoic acid, and pyrene, the GC method gave better results than the CS method and thus it is preferable to use the GC method, even if the critical constants are available. The improvement by the new CVD mixing rule (Mukhopadhyay and Raghuram Rao, 1993) over the conventional VDW mixing rule with the adjustable parameter, k_{ij} for a_{ij}, is demonstrated in Figure 2.19a–c for three solid solutes with different functional groups.

For modeling solubilities of solid solute (2) in an SCF solvent (1) mixed with a cosolvent (3), an additional adjustable parameter, m_{23} needs to regressed from the experimental solubility data, while m_{13} is taken as 1.0 and m_{12} is regressed from

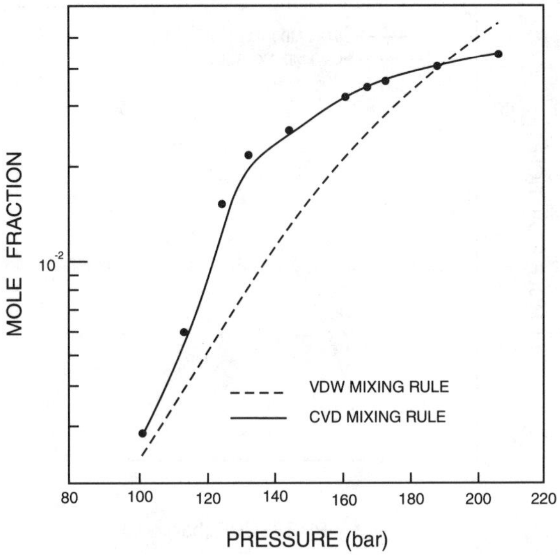

FIGURE 2.19a Comparison of predicted (Mukhopadhyay and Raghuram Rao, 1993) solubility of naphthalene in SC CO_2 at 45°C with experimental data of McHugh and Paulaitis (1980).

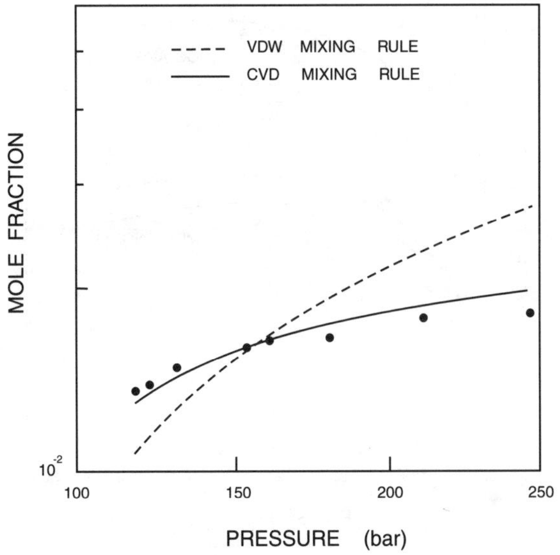

FIGURE 2.19b Comparison of predicted (Mukhopadhyay and Raghuram Rao, 1993) solubility of phenol in SC CO_2 at 36°C with experimental data of Van Leer and Paulaitis (1980).

the binary solubility data. The CVD mixing rule using both CS and GC methods was found to give very good agreement with the corresponding experimental solubility data for eight solids in cosolvent mixed SC CO_2 with percentage AARD of 10.3 and 9.1, respectively. Figure 2.19d illustrates the agreement of the calculated

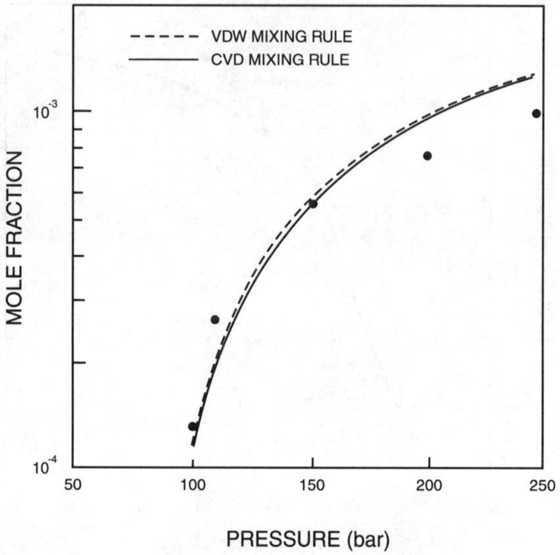

FIGURE 2.19c Comparison of predicted (Mukhopadhyay and Raghuram Rao,1993) solubility of palmitic acid in SC CO_2 at 40°C with experimental data of Bamberger et al. (1988).

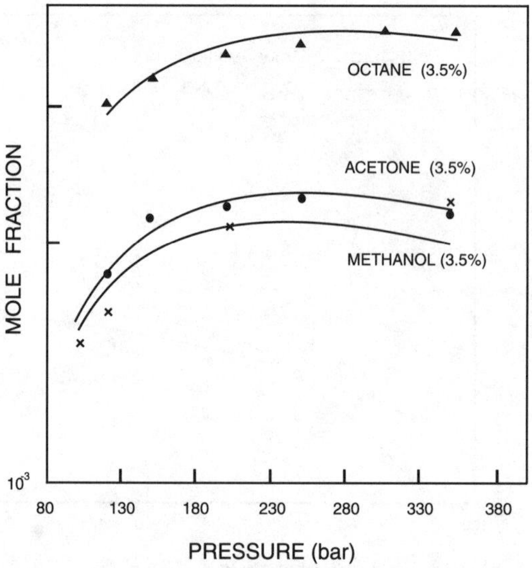

FIGURE 2.19d Comparison of predicted solubilities (Mukhopadhyay and Raghuram Rao, 1993) of hexamethyl benzene in SC CO_2 mixed with cosolvents at 35°C with experimental data of Dobbs (1986).

solubilities of hexanethyl benzene in SC CO_2 mixed with cosolvents with the corresponding experimental data.

The new CVD mixing rule was found to predict the solubilities of mixed solid solutes in pure SC CO_2 and also in SC CO_2 mixed with cosolvent as well. The

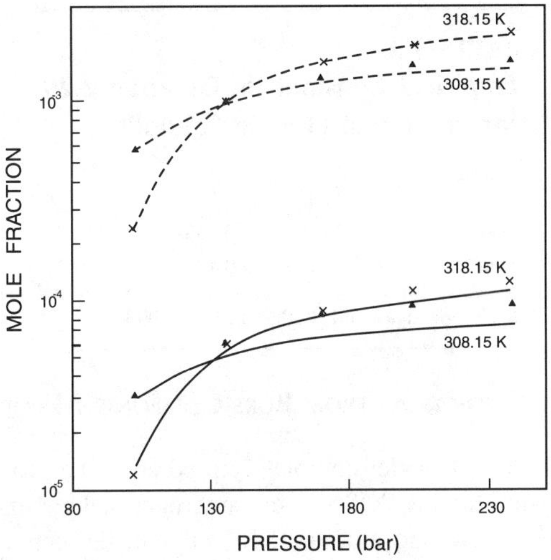

FIGURE 2.20a Comparison of solubility of anthracene and phenanthrene mixture in SC CO_2 predicted by the new CVD mixing rule (Mukhopadhyay and Raghuram Rao, 1993), with experimental data of Kosal and Holder (1987). Upper curves are for Phenanthrene; lower curves are for Anthracene.

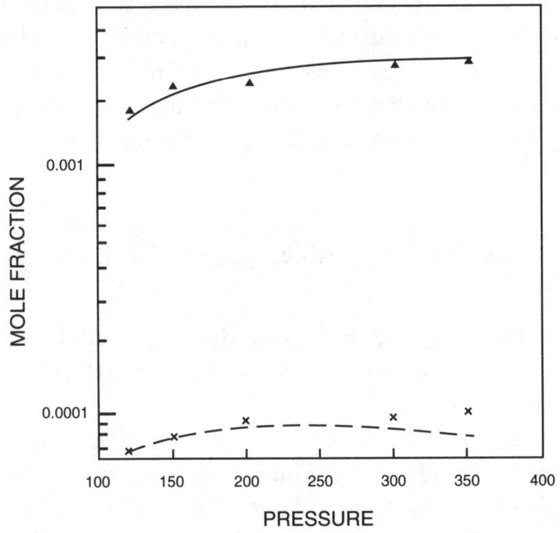

FIGURE 2.20b Comparison of solubility of 2-naphthol + anthracene in SC CO_2 predicted by the new CVD mixing rule (Mukhopadhyay and Raghuram Rao, 1993), with experimental data of Dobbs and Johnston (1987). The upper curve is for 2-Naphthol; the lower curve is for Anthracene.

agreement was within an overall percentage AARD of 19% for 14 isotherms in pure SC CO_2 and within 16% in mixed solvent without having to use the adjustable parameter for the solute–solute interactions, i.e., m_{22} is taken as 1.0. The success of the present model is clearly evident from Figure 2.20(a-b) as even the cross-over pressure could be predicted for binary solid mixtures with less than 4% AARD.

TABLE 2.3
Empirical Constants in Equation 2.29 for Prediction of Solid Solubilities

Solids	α	β	γ
Type I	–0.282	0.743	–0.045
Type II	0.886	–0.409	0.680

Mukhopadhyay and Raghuram Rao, 1993.

2.5.3 Solubility Predictions from Pure Component Properties

In order to make the CVD model completely predictive, i.e., to be able to predict the solubility data using the PR EOS without any regressed parameters, like m_{12} and m_{23} for a pure SCF solvent and a mixed SCF solvent, respectively, it is necessary to have a correlation function for the binary interaction energy parameters in terms of the easily available pure component properties. Attempts toward this objective were faced with limited success (Johnston et al., 1982; Wong et al., 1985; Dobbs, 1986) due to the inadequacy of the mixing rule to reduce the high sensitivity of the adjustable parameters and their temperature dependency. However, this could be achieved by using the new CVD mixing rule, as it was possible to reduce the sensitivity of even the single adjustable parameter employed and to make it insensitive to temperature variation. For this, instead of m_{12}, the whole term, $a_{12}/b_{12}^{m_{12}}$, called the CVD binary interaction parameter, was the adjustable parameter considered in the mixing rule, which was correlated (Mukhopadhyay and Raghuram Rao, 1993) as:

$$a_{12}/b_{12}^{m_{12}} = [\alpha + \beta\,(M_2/M_1)\,[(b_1 b_{2,VDW})^{1/2}/v_2^s] + \gamma(\mu_1\mu_2)^{1/2}] \times 10^6 \qquad (2.29)$$

where μ_1 and μ_2 are the dipole moments, and M_1 and M_2 are the molecular weights. The empirical constants, α, β, and γ for the oxygen-free (I) and oxygen-containing (II) solids, are given in Table 2.3.

The overall percentage AARD of the predicted solubilities utilizing the PR EOS with the temperature independent CVD binary interaction parameters from the above correlation for 52 isotherms was found to be 23.6%, which is even lower than the overall percentage AARD of the solid solubilities calculated using regressed parameters by the conventional VDW mixing rule and other complicated equations like HS VDW EOS (Dobbs and Johnston, 1987). The method can predict the temperature effects on solubilities quite well even though the adjustable parameter is temperature independent.

Earlier three correlations (Gangadhara Rao, 1990) were developed one each for the adjustable parameter needed by three approaches proposed to utilize the original CVD mixing rule, as given by Equation 2.21 for the solubility predictions of solid solutes using the PR EOS. The adjustable parameters, k_{ij}^1, k_{ij}, and m in the three approaches for the calculation of a_{ij} in the CVD mixing rule, are explained before presenting their correlations.

Approach I: With an indirect adjustable parameter, k_{ij}^1 (m = 1.5):

$$a_{ij} = 0.45724\,(RT_{c_{ij}})^2\,\alpha_{ij}/P_{c_{ij}} \tag{2.30}$$

$$b_{ij} = 0.07780\,RT_{c_{ij}}/P_{c_{ij}} \tag{2.31}$$

where

$$T_{c_{ij}} = (T_{c_i}\,T_{c_i})^{0.5}(1 - k_{ij}^1)$$

$$P_{c_{ij}} = 2T_{c_{ij}}/\left[\sum_k T_{c_k}/P_{c_k}\right]$$

$$\alpha_{ij}^{0.5} = 1 + k(1 - T_r^{0.5})$$

$$k = 0.37464 + 1.54226\omega_{ij} - 0.26992\omega_{ij}^2$$

$$T_r = T/T_{c_{ij}} \quad \text{and} \quad \omega_{ij} = (\omega_i + \omega_j)/2$$

Approach II: With a direct adjustable parameter, k_{ij} (m = 1.67):

$$a_{ij} = (a_{ii}\,a_{jj})^{0.5}\,(1 - k_{ij}) \tag{2.14}$$

$$b_{ij} = (b_{ii} + b_{jj})/2 \tag{2.15a}$$

$$a_{ii} = 0.45724\,(RT_{c_{ii}})^2\,\alpha_{ii}/P_{c_{ii}} \tag{2.30a}$$

where $\alpha_{ii}^{0.5} = 1 + k[1 - (T/T_{c_i})^{0.5}]$

Approach III: With "m" as the adjustable parameter (k_{ij} = 0 and k_{ij}^1 = 0)

$$a_{ij} = 0.45724\,(RT_{c_{ij}})^2\,\alpha_{ij}/P_{c_{ij}} \tag{2.30}$$

$$T_{c_{ij}} = (T_{c_{ii}}\,T_{c_{jj}})^{0.5} \tag{2.31}$$

or

$$a_{ii} = 0.45724\,(RT_{c_{ii}})^2\,\alpha_{ii}/P_{c_{ii}} \tag{2.30a}$$

$$a_{ij} = (a_{ii}\,a_{jj})^{0.5} \tag{2.31a}$$

In this approach, the intention is to distribute the co-volume dependency of the energy parameter symmetrically to both like and unlike molecules, i.e., $k_{ij}^1 = 0$, $k_{ij} = 0$.
The fugacity coefficient of solute i in the fluid phase is given as:

$$\ln\bar{\phi}_i = (Z-1)b_i/b - \ln(Z-B) - \frac{A}{2\sqrt{2}B} \tag{2.33}$$

$$[A_1 + (m-1)B_1 + 1 - 2m]\ln\left[\frac{Z+2.414B}{Z-0.414B}\right]$$

where $A_1 = 2\left[\sum_j y_j a_{ij}(b/b_{ij})^m\right]/a$, $A = aP/(RT)^2$ and $B_1 = 2\left[\sum_j y_j b_{ij}\right]/b$, $B = bP/(RT)$

The adjustable parameters listed in Table 2.4 were first regressed from experimental solubility data in three SCF solvents, CO_2, ethane and ethylene, and were correlated in terms of pure component properties. The overall percentage AARD of the calculated solubilities using both regressed parameters and the corresponding values calculated from their correlations are compared in Table 2.4. It is clearly seen that the pure component solubilities can be best calculated considering "m" as the adjustable parameters, i.e., using Approach III of the CVD mixing rule. The correlations are given below where their constants are listed in Table 2.5 for only carbon dioxide as the SCF solvent:

1. In the conventional VDW mixing rule (i.e., m = 0):

$$k_{12} = 1 + \alpha\log\left[1 + \frac{|\Delta\omega/\omega_2|^{\beta}|\Delta P_c/P_{c_2}|^{\gamma}}{1+|\Delta T_c/T|^{\delta}}\right] \tag{2.34}$$

2. In the CVD mixing rule (Approach I), with m = 1.5:

$$k_{12}^1 = 1 + \alpha\log\left[1 + \frac{|\Delta\omega/\omega_2|^{\beta}|\Delta P_c/P_{c_1}|^{\gamma}}{1+|\Delta T_c/T|^{\delta}}\right] \tag{2.35}$$

3. In the CVD mixing rule (Approach II), with m = 1.67:

$$k_{12} = 1 + \alpha\log\left[1 + \frac{|\Delta\omega/\omega_2|^{\beta}|\Delta P_c/P_{c_2}|^{\gamma}|\Delta T_c/T_{c_2}|^{\varepsilon}}{1+|\Delta T_c/T|^{\delta}}\right] \tag{2.36}$$

4. In the CVD mixing rule (Approach III), with $k_{12}^1 = k_{12} = 0$:

$$m = 1 - \alpha\log\left[1 + \frac{|\Delta\omega/\omega_j|^{\beta}|\Delta P_c/P_{c_j}|^{\gamma}|\Delta T_c/T_{c_j}|^{\varepsilon}}{1+|\Delta T_c/T|^{\delta}}\right] \tag{2.37}$$

TABLE 2.4
Performance of Correlations for Adjustable Parameters

Approach	Adjustable Parameter	m	% AARD in Solubilities: Regressed Parameters	% AARD in Solubilities: Correlated Parameters
VDW	k_{ij}	0.0	15.7	27.7
CVD-I	k_{ij}^{1}	1.5	13.9	20.0
CVD-II	k_{ij}	0.67	12.7	
		1.67	—	17.0
CVD-III	m	>0	13.3	16.0

TABLE 2.5
Empirical Constants in the Correlations for the Adjustable Parameters in SC CO_2 Solvent

Parameter	Equation No.	Constants: α	β	γ	δ	ε	% Dev.
k_{ij}(VDW)	2.34	−1.974	−0.060	−0.006	−0.400		9.2
k_{ij}^{1} (CVD-I)	2.35	−62.25	1.94	1.77	6.32		6.0
k_{ij}(CVD-II)	2.36	−47.26	1.14	1.20	0.23	4.25	14.2
m (CVD-III)	2.37	−5.52	0.30	−0.36	−0.72	1.65	8.0

where ΔP_c, ΔT_c, and $\Delta\omega$ represent, respectively, the differences in the critical pressure, critical temperature, and acentric factors of the pure components.

Another approach based on group contribution (GC) was developed (Gangadhara Rao and Mukhopadhyay, 1990) which utilizes a correlation, independent of temperature, for direct calculation of a_{12} from the pure component properties as:

$$a_{12} = q \times 10^7 (M_2/M_1)^r (b_1/b_{2,VDW})^s \tag{2.38}$$

where the constants q, r, and s are given in Table 2.6 for the three solvents and $b_{2,VDW}$ is the van der Waals volume which can be calculated using the group contribution method by Bondi (1968).

The solubilities of pure solids can thus be determined using the CVD mixing rule with m = 1.67, and with a_{12} from Equation 2.38, a_{22} from Equation 2.28, and b_2 (PR EOS) as $1.65b_{VDW}$. It is gratifying to note that this group contribution method for solubility predictions in three SCF solvents (CO_2, ethane, and ethylene) involving 25 systems and 70 isotherms yielded merely overall 22% AARD from the corresponding experimental solubility data, including six systems not included in the development of the correlation.

The group contribution approach by the CVD mixing rule, as given above, results in accurate predictions of solubilities of binary solid mixture (of components 2 and 3),

TABLE 2.6
Empirical Constants in the GC Correlation (Equation 2.38) for a_{12}

Solvent	q	r	s	% Dev.[a]
CO_2	1.0600	5.366	4.746	1.3
Ethylene	0.1165	3.062	1.678	2.1
Ethane	0.0474	3.940	2.420	3.1

[a] $[|a_{12,cal.} - a_{12,reg.}|/a_{12,reg.}] \times 100$.

as well as with the binary energy parameters a_{12} and a_{13} calculated from Equation 2.38, and a_{23} from the geometric mean of a_{22} and a_{33}. The success of the CVD mixing rule lies in its ability to reduce the sensitivity and temperature dependency of the adjustable parameters. This can be employed for solubility predictions of pure solids and their mixtures without having to use any separate solute–solute interaction parameter.

2.5.4 Liquid–Fluid Equilibrium Calculations

Analysis of liquid–fluid phase equilibrium behavior is more complicated due to the fact that the SCF solvent is dissolved in the liquid phase in significant amount. In recent years various approaches have been followed to model liquid–fluid or high pressure liquid–vapor phase equilibrium behavior using an equation of state which is capable of representing both of the equilibrium phases in question. However, all models are based on the thermodynamic criterion for equilibrium between the phases, namely, the fugacity of a component in all equibriated phases is the same, i.e.,

$$\bar{f}_i^v(T, P, y_i, v^v) = \bar{f}_i^L(T, P, x_i, v^L) \tag{2.39}$$

where $\bar{f}_i$ is the fugacity of component, i. The fugacity of each phase is dependent on its molar volume and composition in addition to its temperature and pressure. On expanding it further,

$$Px_i\bar{\phi}_i^L = Py_i\bar{\phi}_i^V \tag{2.40}$$

or

$$K_i = y_i/x_i = \bar{\phi}_i^L/\bar{\phi}_i^V \tag{2.41}$$

where K_i is the distribution coefficient between the vapor and liquid phases. It is already known that the fugacity coefficients in the liquid and vapor phases $\bar{\phi}_i^L$ and $\bar{\phi}_i^V$ can be calculated from the PR EOS and the conventional VDW mixing rules (Equation 2.5 to 2.15) as:

$$\ln\bar{\phi}_i^{\,L} = \frac{b_i}{b}\left(\frac{P_v^{\,L}}{RT} - 1\right) - \ln\left(\frac{P(v^L - b)}{RT}\right) \tag{2.42}$$

$$-\frac{a}{2\sqrt{2}bRT}\left[\frac{2\sum_k x_k a_{ik}}{a} - \frac{b_i}{b}\right]\ln\left[\frac{v^L + 2.414b}{v^L + 0.414b}\right]$$

A similar expression can be written for $\bar{\phi}_i^V$ by replacing v^L by v^v and x_i by y_i in Equation 2.42. It can be noticed that the expression for the fugacity coefficient is dependent on the mixing rule.

2.5.4.1 Mixing Rules

Johnston et al. (1989) critically reviewed the different EOS methods and prevailing mixing rules with a number of adjustable parameters in them, employed by various investigators for representing liquid–fluid and multiphase systems.

As the quadratic mixing rules were inaccurate for the polar and highly asymmetric systems, significant efforts were made in the development of new mixing rules. It is reported that the binary interaction parameter needs to be modified to include temperature dependency (Mohamed et al., 1987; McHugh et al., 1983), and/or pressure dependency (Mohamed, et al., 1987), and/or density dependency (Mohamed and Holder, 1987), and/or composition dependency (Panagiotopoulos and Reid, 1987). For natural molecules, in general, it is observed that the equilibrium data are better correlated if the mixing rules are modified to one or more of the following forms:

1. For the temperature dependency of the interaction parameters,

$$k_{ij} = \alpha_{ij} + \beta_{ij}T \quad \text{or} \quad k_{ij} = \alpha_{ij} + \beta_{ij}/T \tag{2.43}$$

2. For the pressure dependency of the interaction parameter,

$$k_{ij} = \alpha_{ij} + \beta_{ij}\, P \tag{2.44}$$

3. For the density dependency of the interaction parameter,

$$k_{ij} = \alpha_{ij} + \beta_{ij}\, \rho \tag{2.45}$$

4. For the composition dependency of the interaction parameter,

$$k_{ij} = \alpha_{ij} - (\alpha_{ij} - \alpha_{ji})\, x_i \tag{2.46}$$

where α_{ij} and β_{ij} are adjustable parameters.

TABLE 2.7
Correlation of High Pressure Fluid–Liquid Equilibria with New CVD Mixing Rule

System	T (°C)	m_{ij}	% AARD y_2	% AARD x_2	Ref.
CO_2 + methanol	37	1.09	0.3	15.9	Oghaki et al. (1989)
	40	1.10	0.1	15.7	Suzuki et al. (1990)
CO_2 + ethanol	40	1.15	0.1	15.9	Suzuki et al. (1990)
	60	1.07	0.5	14.9	Suzuki et al. (1990)
CO_2 + hexane	25	0.99	0.4	19.9	Oghaki et al. (1989)
CO_2 + benzene	25	1.04	0.1	15.3	Oghaki et al. (1989)

Mukhopadhyay and Raghuram Rao, 1993.

All these mixing rules improved the data correlation, but at the cost of an additional adjustable parameter. Mohamed et al. (1987), who tested these mixing rules for nine systems, opined that while the density- and pressure-dependent forms were superior for non-polar systems, the composition-dependent mixing rule was the best for polar systems. These mixing rules were used for perturbed-hard sphere (PHS) EOS to facilitate improved performance in the near critical region.

The perturbed hard chain theory (Cotterman et al., 1985; Radosz et al., 1987) and computer simulations are the few other techniques used for liquid-fluid modeling.

2.5.4.2 Regression of Binary Adjustable Parameters

Though it is observed that a larger number of adjustable parameters may represent the experimental data well, there is a greater uncertainty in their evaluations with more number of constants, as the uncertainty of one parameter gets distributed to the other parameters, making extrapolation of the experimental data unreliable. For highly asymmetric systems, like those encountered in SCF–liquid phase equilibria, a minimum of two parameters (k_{ij} and n_{ij}) are required to be evaluated if the VDW mixing rule is used. For hydrogen bonding or specific chemical interactions, the binary interaction constant, k_{ij}, may turn out to be negative. This often leads to apprehension about the sheer validity of the EOS method, since the concept of composition remaining uniform throughout the mixture and the uniform composition-dependent interactions between the components fails to give the correct picture in the case of hydrogen bonding and other chemical forces.

On the other hand, the CVD mixing rule takes into account the non-uniform distribution of the solute and solvent molecules in the form of co-volume dependency or size effects. However, the new CVD mixing rule (Equation 2.22) resulted in limited success when it was used with one adjustable parameter, m_{12}, i.e., the exponent of the co-volume dependency, for describing fluid–liquid equilibria (Mukhopadhyay and Raghuram Rao, 1993). The adjustable parameter, m_{12}, was regressed from the bubble point (P, y) pressure calculations at high pressures for four systems and their percentage AARD values are given in Table 2.7. It can be observed that y can be accurately calculated, since the CVD mixing rule is designed to represent molecular interactions in highly dilute supercritical mixtures. At lower pressures, the deviations of the liquid phase compositions are relatively large, since there is significant solubility of CO_2 in the liquid phase. A uniform co-volume dependency for all types of binary interactions, namely the original CVD mixing rule (Equation 2.21) with the adjustable parameter, m, may be a better choice for the calculation of liquid–fluid equilibrium compositions for systems with less asymmetry. However, the CVD mixing rule, in general, works better for systems having large molecular disparity between the components and thus for calculation of the vapor phase compositions of such systems.

2.5.4.3 Prediction of Multicomponent Data from Binary Interaction Constants

In recent years, SCF solvents have been extensively investigated as a solvent medium in simultaneous chemical reaction–separation schemes, owing to their unique thermodynamic and transport properties. Multicomponent solubility data and phase behavior in a reacting system are needed to ascertain the process conditions for achieving optimum selectivity and conversion of the reactants and to understand the reaction kinetics and pathways involved. For example, cyclohexane oxidation in SC CO_2 medium encounters phase separation as the products are formed even after starting with a homogeneous mixture (Srinivas and Mukhopadhyay, 1994a). In order to ascertain whether it is due to condensation of the reactants, namely oxygen and cyclohexane, or the products, namely, cyclohexanol, cyclohexanone, and water, solubility measurements were made for pure cyclohexanone, cyclohexanol, and their binary mixtures as well as their binary mixtures with cyclohexane, in SC CO_2 and N_2 (a homomorph of oxygen) (Mukhopadhyay and Srinivas, 1996). These equilibrium (P, T, x, y) data were utilized to regress for each binary pair, two adjustable interaction parameters, k_{ij} and n_{ij}, in the P-R EOS which are presented in Table 2.8a at three temperatures. The multicomponent data for three ternary systems, namely, (a) CO_2 (1)-cyclohexane(2)-cyclohexanone(4), (b) CO_2(1)-cyclohexane(2)-cyclohexanol(5), and (c) CO_2(1)-cyclohexanone(4)cyclohexanol(5), were calculated using these parameters at 150°C. They are compared in Figure 2.21 (a–c) with the corresponding experimentally measured ternary equilibrium data (Table 2.8b) at two pressures of 170 and 205 bar at 150°C (Srinivas and Mukhopadhyay, 1994b). The solubility calculations confirm that cyclohexanone is more soluble in SC CO_2 compared to cyclohexanol, which is the reverse in N_2. The solubilities of both products of oxidation, namely, cyclohexanone and cyclohexanol, are found to increase in the presence of cyclohexane. From the predicted multicomponent solubility, it can be visualized that the solubility of water in SC CO_2 is suppressed by the presence of all three organic compounds. With the progress of the reaction, water is facilitated to condense out of the solvent medium, resulting in a higher conversion of the reactants. Thus it can be seen that the operating conditions can be adjusted in the vicinity of the mixture critical point in order to condense out the products for circumventing side reactions. Further, it was demonstrated (Mukhopadhyay and Srinivas, 1997) that the proximity to the mixture critical point of the reaction mixture,

TABLE 2.8a
Regressed Binary Interaction Constants in P-R EOS

Binary Pair	T (°C)	k_{ij}	n_{ij}	% AARD (P)[a]
CO_2-$C_6H_{10}O$	137	0.1752	–0.0159	4.09
	150	0.1843	–0.0100	6.07
	160	0.2000	–0.0100	5.82
CO_2-$C_6H_{11}OH$	137	0.1475	–0.1485	1.59
	150	0.2018	–0.0900	6.71
	160	0.1851	–0.1174	4.94
N_2-$C_6H_{10}O$	150	0.2999		25.31
				12.55
C_6H_{12}-$C_6H_{10}O$	150	0.0153	–0.3188	0.16
C_6H_{12}-$C_6H_{11}OH$	150	0.0200	–0.0090	4.71
$C_6H_{12}O$-$C_6H_{11}OH$	150	0.0535	–0.0100	8.64
CO_2-H_2O[1]	150	0.0074		6.81
CO_2-C_6H_{12}[b]	150	0.099		
N_2-C_6H_{12}[c]	150	0.076		

[a] % AARD, $P = \Sigma(|P_{exp} - P_{cal}|/P_{exp})/NDP \times 100$.
[b] Interpolated from the regressed data of Shibata and Sandler (1989) and of Krichevski and Sorina (1960).
[c] Same as that from the data of Shibata and Sandler (1989).

Mukhopadhyay and Srinivas, 1996.

as represented on a ternary diagram, is an additional parameter for manipulating the reaction rate and conversion. It is important to note here that all of this information could be derived from the knowledge of the regressed binary interaction constants.

2.5.4.4 Prediction of Phase Boundaries

As can be seen from Figure 2.21b for the CO_2-cyclohexane-cyclohexanol system, the phase envelope predicted with two interaction constants is closer to the experimental data when compared to the one predicted with a single interaction constant, k_{ij} with $n_{ij} = 0$, which in turn is better than the one predicted with no interaction constant at all. This can be observed for other systems as well and at other temperatures (Mukhopadhyay and Srinivas, 1996). For ternary systems, inclusion of two interaction constants, in general, predicts shrinkage of the phase envelope when compared to one or no constant, resulting in higher predicted values of solubilities in SC CO_2.

Further, phase envelopes were calculated for the ternary system of cyclohexane-CO_2-N_2 at two pressures of 170 bar and 205 bar, each at three temperatures of 137, 150, and 160°C using the regressed interaction constants (Table 2.8a). It could be found that at a higher temperature the shrinkage of two-phase region with pressure is more pronounced (Srinivas, 1994). Thus, it was possible to have higher concentrations of the reactants in the homogeneous SCF phase at a higher pressure of 205 bar and at a higher temperature of 160°C, in order to retain the additional advantages of

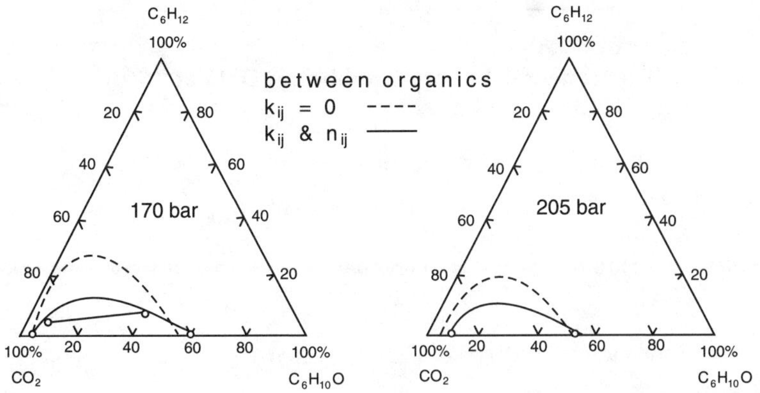

FIGURE 2.21a Phase diagrams for ternary system: CO_2-C_6H_{12}-cyclohexanone at 150°C and two pressures (Mukhopadhyay and Srinivas, 1996).

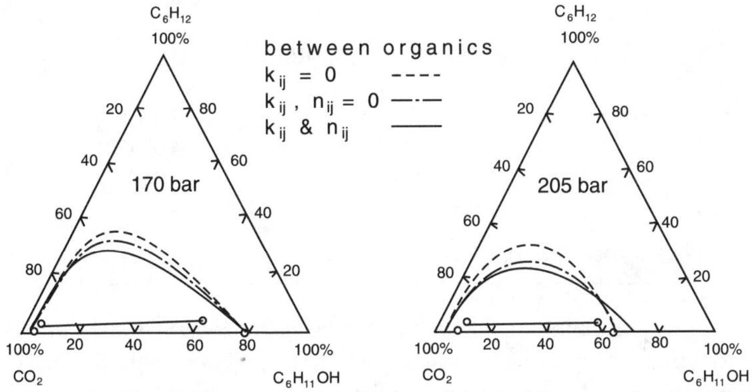

FIGURE 2.21b Phase diagrams for ternary system: CO_2-C_6H_{12}-cyclohexanol at 150°C and two pressures (Mukhopadhyay and Srinivas, 1996).

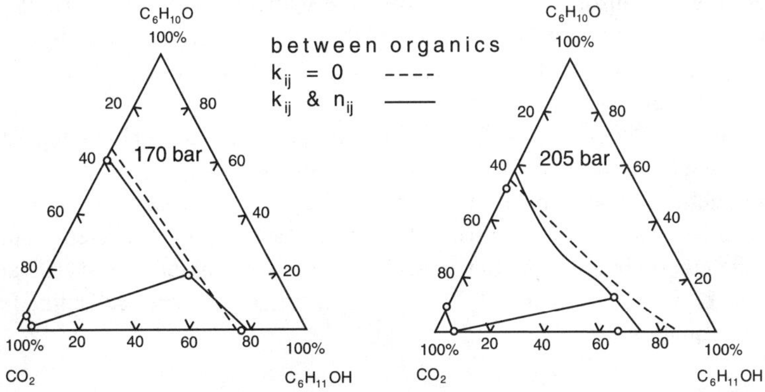

FIGURE 2.21c Phase diagrams for ternary system: CO_2-cyclohexanol-cyclohexanone at 150°C and two pressures (Mukhopadhyay and Srinivas, 1996).

TABLE 2.8b
Experimental Ternary Equilibrium Data at 150°C

P, bar	Mole Fraction Liq. Phase	Mole Fraction SC Phase
	$CO_2(1)$-$C_6H_{12}(2)$-$C_6H_{10}O(4)$	
170.0	1:0.5020	1:0.9003
	2:0.0791	2:0.0418
	$CO_2(1)$-$C_6H_{12}(2)$-$C_6H_{11}OH(5)$	
170.0	1:0.3289	1:0.9392
	2:0.0465	2:0.0157
204.9	1:0.3797	1:0.8792
	2:0.0304	2:0.0217
	$CO_2(1)$-$C_6H_{10}(4)$-$C_6H_{11}OH(5)$	
170.0	1:0.3008	1:0.9562
	4:0.2074	4:0.0127
204.9	1:0.2694	1:0.9287
	4:0.1322	4:0.0125

Mukhopadhyay and Srinivas, 1996.

the proximity to the plait point as well as the homogeneous SCF medium for cyclohexane oxidation. This example demonstrates how the knowledge of binary interaction parameters could be utilized for prediction of fluid–liquid phase boundaries.

2.5.5 Mixture Critical Point Calculations

Benmekki and Mansoori (1988) modeled the critical region of a ternary system (N_2-CO_2-CH_3OH) accurately considering ternary interactions in their conformal solution mixing rule for the PR EOS. The mixing rule required ten adjustable parameters for the ternary system.

Van der Haegen et al. (1988), Nitsche et al. (1984), and McHugh and Krukonis (1986) applied the mean-field lattice model with three adjustable parameters and could predict the upper critical end point (UCEP) of CO_2–water system accurately.

Palencher et al. (1986) compared the performance of five different cubic EOS and verified their ability to predict critical loci of binary mixtures. The critical points on the V–L boundary in the P–T space of a binary system are defined by the following equations:

$$\left(\frac{\partial^2 G}{\partial x_1^2}\right)_{P,T} = 0; \quad \left(\frac{\partial^3 G}{\partial x_1^3}\right)_{P,T} = 0 \qquad (2.47)$$

The Redlich-Kister form of the Gibbs free energy which depends on the measured variables, namely, P, T, y, x, gives the derivatives in the above equations as:

$$\left(\frac{\partial^2 G}{\partial x_1^2}\right)_{P,T} = \frac{RT}{x_1 x_2} + \int_v^{\infty}\left(\frac{\partial^2 P}{\partial x_1^2}\right)_{V,T} dv - \left(\frac{\partial P}{\partial x_1}\right)_{V,T}\left(\frac{\partial v}{\partial x_1}\right)_{P,T} = 0 \tag{2.48}$$

$$\left(\frac{\partial^3 G}{\partial x_1^3}\right)_{P,T} = \frac{RT(x_1 + x_2)}{x_1^2 x_2^2} - \int_v^{\infty}\left(\frac{\partial^3 P}{\partial x_1^3}\right)_{V,T} dv - \left(\frac{\partial P}{\partial x_1}\right)_{V,T}\left(\frac{\partial^2 v}{\partial x_1^2}\right)_{P,T} \tag{2.49}$$

$$-\left(\frac{\partial v}{\partial x_1}\right)_{P,T}\left(\frac{\partial^2 P}{\partial V \partial x_1}\right)_{P,T} - 2\left(\frac{\partial v}{\partial x_1}\right)_{P,T}\left(\frac{\partial^2 P}{\partial x_1^2}\right)_{V,T} = 0$$

The critical criteria, i.e., the above equations were solved using all of the five EOS, relating the variables P, v, T, x, and then the critical pressure was obtained from the EOS. Palencher et al. (1986) successfully predicted the binary critical loci of class I systems. For other classes, the performance of more complicated equations like the Teja EOS and the Adachi EOS are found to be better than the Redlich-Kwong (RK) and the Soave-Redlich-Kwong (SRK) EOS. These equations are listed in Table 2.9.

Billingsley and Lam (1986) formulated a rigorous procedure for the calculation of critical points of a mixture with non-zero interaction parameters in the EOS in the form of an eigen value of a 2 × 2 matrix. An initial guess for the volume as 4b and that for the temperature as $1.5\,\Sigma\, y_i\, T_{ci}$ are found suitable for solving the nonlinear equations by the ISML package (Srinivas, 1994).

2.5.6 Multiphase (LLV) Calculations

The phase rule provides the number of independent variables that must be fixed to define multiphase, multicomponent mixtures. For example, for a two-phase, binary mixture, the values of the compositions of the two equilibrium phases at fixed values of pressure and temperature are unique. These compositions can be calculated from the thermodynamic criterion of equilibria, namely, the fugacity of a component in the two equilibrium phases are equal. Similarly, when there are three equilibrium phases, two sets of equations can be written for each component as:

$$\bar{f}_i^{v}(T, P, y_i) = \bar{f}_i^{L_1}(T, P, x_i^{L_1}),\ i = 1,2,\ldots C \tag{2.50}$$

$$\bar{f}_i^{v}(T, P, y_i) = \bar{f}_i^{L_2}(T, P, x_i^{L_2}),\ i = 1,2,\ldots C \tag{2.51}$$

where C is the number of components, and the supercripts, L_1 and L_2, correspond to the two liquid phases. As the degree of freedom is one for a 3-phase mixture, in P–T space, LLV is represented by a line. In terms of the fugacity coefficients, the above equations can be written as:

TABLE 2.9
The Equations of State

Redlich-Kwong (RK) EOS		
$P = \dfrac{RT}{v-b} - \dfrac{a}{T^{0.5}v(v+b)}$		
$a = 0.4278\ R^2\ T_c^{2.5}/P_c$	$b = 0.0867\ R\ T_c/P_c$	
The Soave-Redlich-Kwong (SRK) EOS		
$P = \dfrac{RT}{v-b} - \dfrac{a(T)}{v(v+b)}$		
$a(T) = 0.42747\ \alpha(T)\ R^2\ T_c^2/P_c$	$b = 0.08664\ R\ T_c/P_c$	
$\alpha(t) = (1 + m(1 - Tr^{0.5}))^2$	$m = 0.480 + 1.574\ \omega - 0.176\ \omega^2$	
The Adachi EOS		
$P = \dfrac{RT}{v-b} - \dfrac{a(T)}{v(v+c)}$		
$a(T) = A_o\ Z_c\ (1 + \alpha(1 - Tr^{0.5})^2\ R^2\ T_c^2/P_c$	$b = B\ Z_c\ R\ T_c/P_c$	$c = C\ Z_c\ R\ T_c/P_c$
$\alpha = 0.479817 + 1.55553\omega - 0.287787\ \omega^2$	$A_o = B\ (1 + C)^3/(1 - B)^3$	
$B = 0.260796 - 0.0682692\omega - 0.0367338\ \omega^2$		
$C = [((4/B) - 3)^{0.5} - 3]/2$		
$Z_c = [1/(1 - B)] - A_o/(1 + C)$		
The Teja EOS		
$P = \dfrac{RT}{v-b} - \dfrac{a(T)}{v(v+b)+c(v-b)}$		
$a(T) = \Omega_a\ \alpha(T)\ R^2\ T_c^2/P_c$		
$b = \Omega_b\ RT_c/P_c$		
$c = \Omega_c\ RT_c/P_c$		
$\Omega_a = 3Z_c^2 + 3(1 - 2Z_c)\ \Omega_b + \Omega_b^2 + 1 - 3Z_c$		
Ω_b equals the smallest positive root of:		
$\Omega_b^3 + (2 - 3Z_c)\ \Omega_b^2 + 3Z_c^2\ \Omega_b - Z_c^3 = 0$		
$\Omega_C = 1 - 3Z_c$		
$Z_c = 0.329032 - 0.07699\ \omega + 0.0211947\ \omega^2$		
$a(T) = (1 + F(1 - Tr^{0.5}))^2$		
$F = 0.452413 + 1{,}30982\ \omega - 0.295937\ \omega^2$		

Palencher et al., 1986.

$$Py_i\bar{\phi}_i^v = P\bar{\phi}_i^{L_1}x_i^{L_1} \tag{2.52}$$

$$Py_i\bar{\phi}_i^v = P\bar{\phi}_i^{L_2}x_i^{L_2} \tag{2.53}$$

For a binary mixture, the location of the 3-phase (liquid–liquid–vapor) boundary in the P–T space is carried out by simultaneously solving the above equations, with the constraints that the summations of the mole fractions are unity.

The calculation of LLV phase behavior for a ternary mixture on the ternary diagram is more challenging than for a binary mixture, since the expression for the fugacity coefficient is highly nonlinear and uses the EOS model involving only binary interaction parameters. For a ternary mixture, six values or three sets of the binary constants are needed for the computation of equilibrium compositions.

2.5.7 Solubility Predictions Using the Solvent-Cluster Interaction Model

Most natural molecules have large molecular sizes and complex molecular structures with a variety of functional groups. When such a nonvolatile solid solute is dissolved in an SCF solvent, a dilute supercritical (SC) mixture is formed, representing a class by itself. The asymmetry between the large solute and the small solvent molecules and the strong solute–solvent interactions in such a dilute mixture, segregate the small solvent molecules, causing the population of small solvent molecules in the environment of a relatively large "attractive" solute molecule denser than that in the bulk. Consequently, a solvation model based on clustering of solvent molecules around the solute is visualized to be more appropriate for depicting the dilute SC mixtures, rather than the uniform molecular distribution, as in the case of conventional VDW-mixing-rule-based EOS model. Thus the Kirkwood-Buff approach (Pfund et al., 1988) based on molecular theories required an accurate EOS for correlating the concentration derivatives of chemical potential and molecular distribution functions of solute–solvent, solvent–solvent, and solute–solute interactions. However, in order to overcome the mathematical complexities of these models, several approximations were considered which enabled step-wise regression of a number of molecular parameters from experimental solubility data. Chrastil's model (1982), on the other hand, was based on a concept of chemical association or formation of a stoichiometric complex between the solute and the solvent molecules, but it did not account for the pressure and temperature effects on solubilities separately. Rather, it described solubility to be solely density-dependent. It postulated a constant number of solvent molecules participating in the complex formation, disregarding their variations with temperature and to some extent with pressure, as is the case of any complexation process. Lemert and Johnston (1991) combined the density-dependent, local-composition (DDLC) mixing-rule-based PR EOS with the principle of chemical reaction equilibrium for formation of a stoichiometric complex with a solute and a cosolvent, both being polar. It assumed that some solute molecules preferentially took part in the formation of a chemical complex whereas the rest of the solute molecules preferred to remain "free."

In another approach to tackle non-uniform, non-random distribution of solvent molecules and asymmetry between the nonvolatile solute and SCF solvent molecules, Sastry and Mukhopadhyay (1993) described the solid–SCF phase equilibrium in terms of long-range interactions between the solvent cluster around the solute molecules and the residual "free" solvent molecules. The short range interactions between the solute and the solvent molecules were considered within the cluster and were utilized to characterize the energy and size parameters of the cluster in such a way that the assymetry between the interacting species could be greatly reduced. According to this model, the formation of a cluster D out of a system-specific number of molecules of solvent B surrounding molecules of solute A is described as:

$$R_oA + R_oKB \rightarrow D \qquad (2.55)$$

where K is the ratio of the solvent to solute populations in a cluster and R_o is the stoichiometric ratio of the solute to cluster molecules. Thus the model incorporates the solute–solute aggregation as well as solvent–solute aggregation within a cluster. The dilute SC mixture is visualized to be constituted of cluster and solvent molecules as the interactive species. The average mole fraction of solute, y_A, is related to the true cluster mole fraction z_D in the SC mixture by material balance as:

$$y_A = z_D/(1 + Kz_D) \qquad (2.56)$$

The molar volume of the cluster D in the frozen (solid) state is expressed as:

$$v_D = v_A + Kv_B \qquad (2.57)$$

where v_A and v_B are the molar volumes of A and B in their solid states.

The mole fraction of cluster z_D in the SC phase is related to the vapor pressure of the solute and the fugacity coefficient, $\bar{\phi}_D$, of cluster D based on the concept of a pseudoequilibrium between the SC-phase and the closely packed clusters in a hypothetical condensed phase as:

$$z_D = \frac{P_A^S}{\bar{\phi}_D P} \exp[v_D(P - P_A^S)/RT] \qquad (2.58)$$

The fugacity coefficient, $\bar{\phi}_D$, is calculated from the P-R EOS with the mixture parameters a and b as:

$$a = z_D^2 a_D + 2z_D z_B \sqrt{a_B a_D} + z_B^2 a_B \qquad (2.59)$$

$$b = z_D b_D + z_B b_B \qquad (2.60)$$

where $z_B = 1 - z_D$.

The above mixing rules are justified in view of the highly reduced asymmetry of B and D as a_B and a_D are of the same order of magnitude as are also b_D and b_B. The ratio of the energy parameters of solute to solvent, namely, a_{AA}/a_{BB}, ranges from 16 to 50, whereas, as per the present model, a_{DD}/a_{BB} is reduced to the range from 1.03 to 1.25. Similarly, the ratio of the co-volumes of the interacting species in the conventional VDW mixing rule, i.e., b_A/b_B ranges from 3.53 to 10.34, whereas in the present model the size disparity could be greatly reduced such that b_D/b_B varies between 1.03 to 1.58 for a large number of solutes. This justifies the reason why the conventional VDW mixing rules are applicable for evaluating the PR EOS mixture parameters a and b in the present model.

The cluster parameters, b_D and a_D, are calculated based on the local mole fractions within the cluster as:

$$b_D = X_A b_A + X_B b_B \tag{2.61}$$

and

$$a_D/b_D = X_A^2 (a_A/b_A) + X_B^2 (a_B/b_B) + 4X_A X_B (a_A a_B)^{1/2}/(b_A + b_B) \tag{2.62}$$

where $X_A = 1/(K + 1)$.

It can be noticed that the energy parameter of the cluster, a_D, is calculated from the quadratic mixing rule of the co-volume dependent energy parameters, i.e., a_{ii}/b_{ii} of the constituents, A and B, without any interaction constant. The only regressed parameter in this model is the solvent population, K in the cluster, whereas R_o is taken as 1 for pure solid solutes. K is regressed from experimental solubility data of pure solid solutes in an SCF solvent at each condition of pressure and temperature. The regressed values of K for 32 solids from 101 isotherms reported elsewhere (Sastry and Mukhopadhyay, 1993) reveal that K decreases with increasing temperature under isobaric condition, in conformity with the effect of temperature on entropy. At isothermal condition, K initially increases with pressure, passes through a maximum, and then decreases with pressure beyond the crossover pressure. Typically K ranges from 15 (at 400 bar) to 110 (at 95 bar) for naphthalene in SC CO_2 at 35°C. It is gratifying to note that the maximum value of K and the lowest partial molar volume, $\bar{v}_2^\infty$, are observed at the same pressure at a particular temperature. K asymptotically approaches a constant value at higher pressures, indicating that the difference in the solvent population within and outside the cluster decreases with pressure. At higher pressures, the solvent distribution becomes more compact and uniform, leading to smaller clusters. It is to be noted that an increase in solubility with pressure has nothing to do with the formation of large size clusters, since K decreases with pressure. For substances which have very low solubilities in SC CO_2, K values are very low. For example, β-carotene accommodates a relatively much smaller number of solvent molecules as compared to a relatively more soluble solid, such as naphthalene. The validity of the cluster–solvent interaction model was justified by its ability to predict partial molar volumes at infinite dilution (Sastry, 1994).

In order to make the method of solubility calculation a predictive one, the regressed values of K were correlated in terms of pure component properties, such as critical temperatures and critical pressures of the solute and solvent, and the isothermal compressibility, K_T, of the solvent as:

$$K = 68.13 - K_T^{-0.52} \left\{ \left[\frac{T - T_{c_B}}{T - T_{c_A}} \right]^{2.13} + \left[\frac{P - P_{c_B}}{P - P_{c_A}} \right]^{0.40} \right\} \tag{2.63}$$

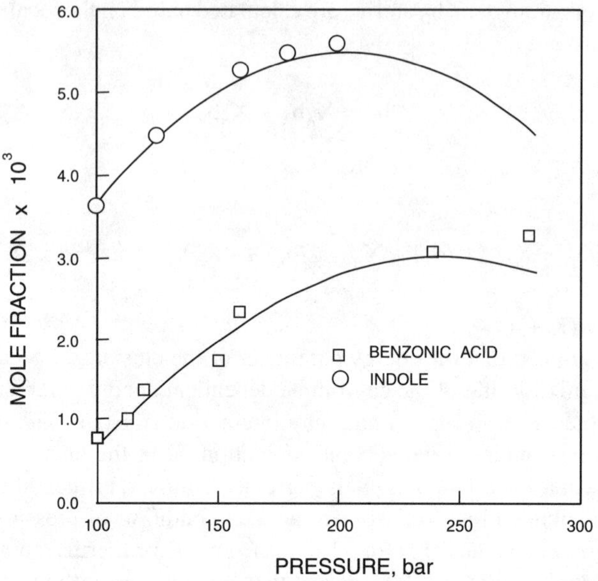

FIGURE 2.22a Comparison of the model predicted solubilities with experimental values for pure solids in SC CO_2 at 35°C.

Using the above correlation, the solid solubilities could be predicted in terms of the pure component properties within percentage AARD of 22.5% for 32 solid solutes and 101 isotherms as demonstrated in Figure 2.22a. Further, this correlation allows estimation of solvent population in the mixed solvent clusters for which the mole fraction averaged properties of mixed solvent are used in place of pure solvent properties (Mukhopadhyay and Sastry, 1997).

The model also predicts the solubility of mixture of solids only from the knowledge of the solvent populations for the individual solid solutes, considering the interactive species to be the composite cluster D^1 and the free solvent molecules. The composite cluster D^1 is visualized to be formed as:

$$R_1A_1 + R_1K_1B + R_2A_2 + R_2K_2B \rightarrow D^1 \quad (2.64)$$

where K_1 and K_2 are the solvent populations for the individual solutes A_1 and A_2, respectively, and R_1 and R_2 are the relative proportions of A_1 and A_2 in the composite cluster. The mole fraction of cluster D^1 in the SC phase is given as:

$$Z_{D^1} = \frac{P_D^{\,s}}{\bar{\phi}_{D^1} P} \exp\left[\frac{v_{D^1}^{\,s}}{RT} (P - P_{D^1}^{\,s}) \right] \quad (2.65)$$

where $P_{D^1}^s = X_1 P_{A_1}^s + X_2 P_{A_2}^s$ and

$$v_{D^1}{}^s = X_1 v_{A_1}{}^s + X_2 v_{A_2}{}^s + K' v_B \quad (2.66)$$

where X_1 and X_2 are the molecular fractions of A_1 and A_2, respectively, in the cluster on solvent-free basis and are expressed as:

$$X_1 = \frac{R_1}{R_1 + R_2} \text{ and } X_2 = \frac{R_2}{R_1 + R_2} \text{ and } K' = X_1K_1 + X_2K_2$$

The mixture parameters a and b in P-R EOS for the mixture of composite cluster and free solvent molecules can be obtained as before as:

$$a = Z_{D^1}^2 a_{D^1} + Z_B^2 a_B + 2Z_B Z_{D^1}\sqrt{a_B a_{D^1}} \quad (2.67)$$

and

$$b = z_{D^1} b_{D^1} + z_B b_B \quad (2.68)$$

where

$$b_{D^1} = X_{A_1} b_{A_1} + X_{A_2} b_{A_2} + x_B b_B \quad (2.69)$$

and

$$\frac{a_{D^1}}{b_{D^1}} = X_{A_1}^2 a_{A_1}/b_{A_1} + X_{A_2}^2 a_{A_2}/b_{A_2} + X_B^2 a_B/b_B \quad (2.70)$$

$$+ 4\left\{ X_{A_1} X_{A_2} \frac{\sqrt{a_{A_1} a_{A_2}}}{b_{A_1} + b_{A_2}} + X_{A_1} X_B \frac{\sqrt{a_{A_1} a_B}}{b_{A_1} + b_B} + X_B X_{A_2} \frac{\sqrt{a_B a_{A_2}}}{b_{A_2} + b_B} \right\}$$

where X_A, X_{A_2}, and X_B are the local mole fractions in the composite cluster, D^1, given as:

$$X_{A_1} = \frac{X_1}{X_1 + X_2 + K'}$$

$$X_{A_2} = \frac{X_2}{X_1 + X_2 + K'}$$

$$X_B = 1 - X_{A_1} + X_{A_2}$$

The average solubilities y_{A_1} and y_{A_2} are transformed to cluster mole fractions, $Z_D{}^1$, as:

$$y_{A_1} = \frac{X_1 Z_{D^1}}{1 + K' Z_{D^1} + Z_B[1/(R_1 + R_2) - 1]} \quad (2.71)$$

$$y_{A_2} = \frac{X_2 Z_{D^1}}{1 + K' Z_{D^1} + Z_B[1/(R_1 + R_2) - 1]} \quad (2.72)$$

which satisfy the constraint

$$y_{A_1}/y_{A_2} = R_1/R_2 \quad (2.73)$$

The following correlations are used for evaluation of R_1 and R_2:

$$R_1 + R_2 = -1.68\left[\frac{R_1 a_{D_1} + R_2 a_{D_2}}{a_B}\right] + 2.64\left[\frac{R_1 b_{D_1} + R_2 b_{D_2}}{b_B}\right] \quad (2.74)$$

and

$$(R_1R_2)^{0.05} = 0.016(a_{D_1}/a_B)(P_{A_1} + P_{A_2})/P_{A_1} + 0.75(a_{D_2}/a_B)(P_{A_1} + P_{A_2})/P_{A_2} \quad (2.75)$$

The solubilities of mixed solids could be predicted with a maximum deviation of 21% from the reported experimental data for six systems (Mukhopadhyay and Sastry, 1997) using this correlation and the validity of this method of prediction is demonstrated in Figure 2.22b for mixed solid solute solubilities in SC CO_2.

For prediction of solid solubilities in a cosolvent-modified SC solvent, the mixed solvent (M) is initially treated as the solvent having mole-fraction averaged properties. However the preferential interaction between the solute and the cosolvent leads to an enrichment or depletion of cosolvent molecules resulting in a supercritical mixture of the modified mixed solvent (MM) and the modified clusters (D_M). In view of this, the first step is to regress the mixed solvent population in the intermediate cluster (D_i), and subsequently in the second step, the enrichment/depletion of the cosolvent, C_e, in the cluster per one solute molecule is regressed, considering the following sequence:

$$RA + RK_M M \rightarrow D_i \quad (2.76)$$

$$D_i + R\,C_e C \rightarrow D_M \quad (2.77)$$

where D_i is the intermediate cluster and D_M is the modified mixed solvent cluster. M represents the mixed solvent having the mole fraction-averaged properties, K_M represents the mixed solvent population ratio, and C represents the cosolvent.

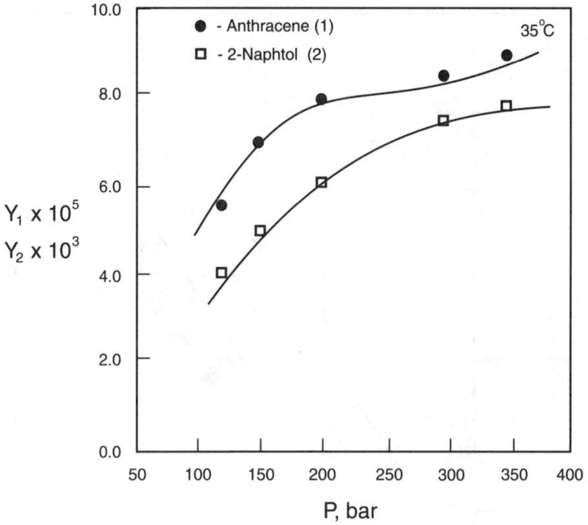

FIGURE 2.22b Comparison of the predicted and experimental solubilities of mixed solid components in SC CO_2 at 35°C.

The average solubility is related to the cluster mole fraction by material balance as:

$$y_A = \frac{z_{D_M}}{1 + (K_M + C_e)z_{D_M} + (1/R - 1)} \tag{2.78}$$

where z_{D_M} is the mole fraction of the modified mixed solvent cluster and $z_{MM} = 1 - z_{D_M}$.

The mixed solvent population, K_M, can also be predicted using the correlation (Equation 2.63) given earlier for a pure SCF solvent. With the help of a correlation for C_e (reported elsewhere), one can predict the solid solubilities in mixed SCF solvent from the knowledge of the pure component properties. An approach similar to that for solubility predictions of mixed solids in pure SCF solvent can be employed for solubility predictions of mixed solids in cosolvent modified mixed solvent (Mukhopadhyay and Sastry, 1997).

2.5.8 Solubility Calculations from Correlations

Sometimes the EOS method cannot be applied for calculation of solubilities of nonvolatile solutes due to non-availability of the necessary parameters, such as critical constants, molar volume, or vapor pressure of the solute. In such a situation, it is often convenient to use a correlation for the solubility calculations. Chrastil (1982) proposed a simple correlation for solubility prediction in terms of the solvent density, which is based on the concept of complex formation between the solvent and the solute molecules. Although the reliability of this correlation is questionable (Nilsson et al., 1991; Sastry and Mukhopadhyay, 1993), solubilities of many compounds,

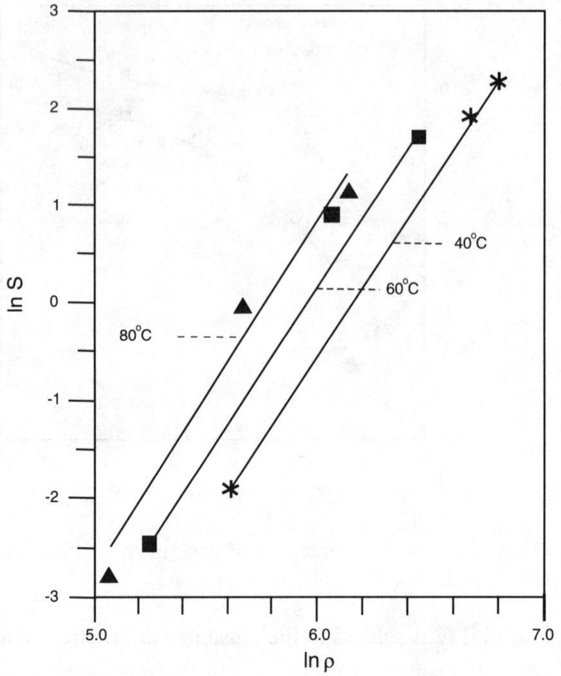

FIGURE 2.23a Chamomile flower extract solubility (S) with density (ρ), (Tolic et al., 1996).

particularly lipids in CO_2, can be correlated by means of this logarithmic expression, which is given as:

$$s = \rho^{A_1} \exp(A_2/T + A_3) \tag{2.79}$$

where s is the solubility (g/l), ρ is the solvent density (g/ml), T is the temperature (K), and A_1, A_2, and A_3 are the empirical constants for the particular component.

de Valle and Aguilera (1988) presented a modified expression of the Chrastil's equation in the form:

$$\ln s = C_o + C_1/T + C_{11}/T^2 + C_2 \ln \rho \tag{2.80}$$

where C_o, C_1, C_{11}, and C_2 are regressed constants and the correlation can be used in the range of measurement. Figure 2.23 (a–c) illustrates the linear behavior of logarithmic solubility with logarithmic density of the solvent for different substances.

The enhancement factor, E, is the ratio of the partial vapor pressure of the solute in the supercritical phase to the vapor pressure of the pure solute at the same temperature as given by Equation 2.81. It is also found to be dependent on density. It is found that the logarithmic value of E is linearly dependent on density as:

$$E = Py_2/P_2^s \tag{2.81}$$

$$\ln E = a + b\rho \tag{2.82}$$

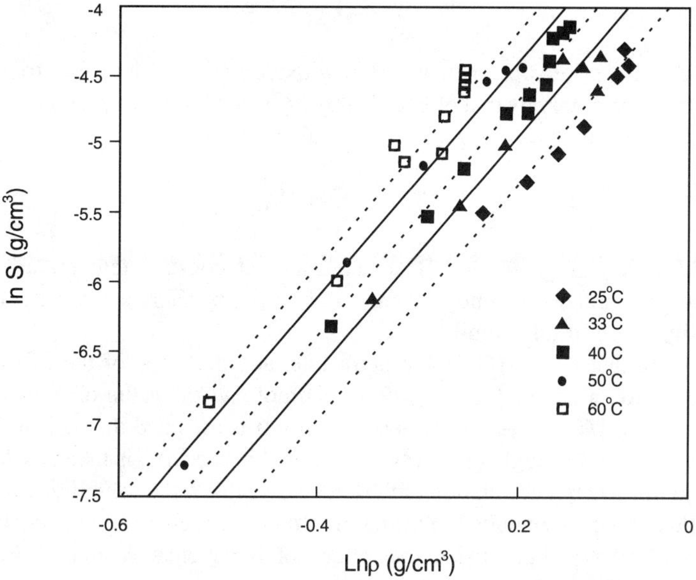

FIGURE 2.23b Log-log plots for the solubility (S) of α-tocopherol vs. solvent density, ρ (da Ponte et al., 1997).

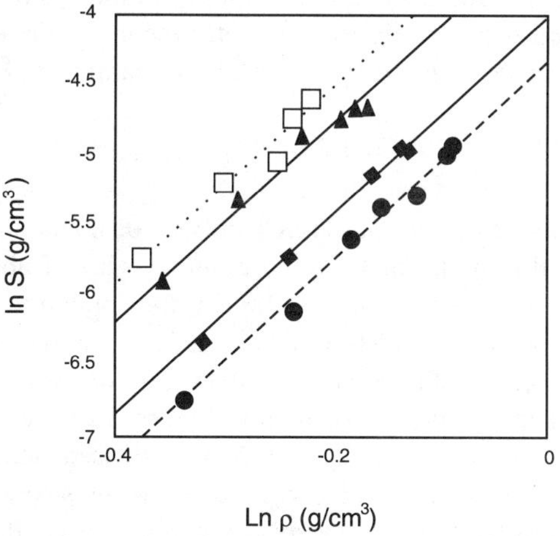

FIGURE 2.23c Log-log plots for the solubility (S) of δ-tocopherol vs. solvent density, ρ, at (●) 33°C; (♦) 40°C; (▲) 50°C; (□) 60°C (da Ponte et al., 1997).

where a and b are constants at a constant temperature. However, vapor pressures of nonvolatile solutes may not be always available. Accordingly, Bartle et al. (1991) proposed a method for estimation of solubility of nonvolatile solutes in SC CO_2 using the constants A and B suggested and tabulated by them using the following simple relation:

$$\ln (Py_2/P_{ref}) = A + B\rho \tag{2.83}$$

The density, ρ, of supercritical carbon dioxide may be calculated from an equation of state, preferably by that of Ely (1986). P_{ref} is the reference pressure, preferably 1 bar. Alternatively, the solubility y_2 is calculated from:

$$\ln (Py_2/P_{ref}) = A^1 + B(\rho - \rho_{ref}) \tag{2.84}$$

where $A^1 = A + B\rho_{ref}$. A_1 and B are constants at constant temperature, P_{ref} is a reference pressure conveniently taken as 1 bar; ρ_{ref} is reference density, central to the density range of all solubilities.

It is seen that A^1 vs. 1/T is a straight line, and the slope of the line $(-\Delta H_v/R)$ corresponds to an enthalpy change of 77 KJ/mol and the value of A^1 is read from the plot (A^1 vs. 1/T) at the desired temperature. It is assumed that B is independent of temperature and a single or average value of B is used. ρ_{ref} is taken as 700 kg/m^3. It is found that a better correlation of the data is possible if the solubilities at pressures above 100 bar are not included in fitting the straight line correlation (Equation 2.84). Bartle et al. (1991) have listed the values of parameters A^1 and B for various substances in CO_2 at different temperatures.

Giddings et al. (1968) suggested that the solvent power of a compressed gas depends on its state relative to its critical condition and its chemical forces. Assuming the equivalence of a liquid solvent and an SCF solvent at a common density with some manipulation for liquid at normal boiling point, Giddings and co-workers (1968, 1969) extended the solubility parameter theory to the solvent power of an SCF solvent, and characterized it by the solubility parameter, δ, expressed as:

$$\delta = 1.25\ P_c^{0.5}\ (\rho_r/\rho_{r,L}) \tag{2.85}$$

where, ρ_r is the reduced density of the SCF solvent or liquid and $\rho_{r,L}$ is the reduced density of the liquid at its normal boiling point, a value of about 2.66, with P_c in bar and δ in (cal/cc)$^{0.5}$. For example, at 400 bar, the solubility parameter of carbon dioxide is about 7.3 at 40°C and 6.0 for ethane at 37°C. At low pressures, the density of a gas is much lower than that at near-critical conditions where the density rapidly increases to that of the liquid. As seen from Figure 2.24, the solubility parameter varies from zero at normal pressures to a rapidly increased value beyond the critical pressure (73.8 bar). The solubility of a solute in a solvent is believed to be maximum when the respective solubility parameters are the same or the difference is small, as can be deduced from the solubility prediction equation based on the Hildebrand theory for a non-polar solute in a non-polar solvent (Prausnitz et al., 1986):

$$\ln x_2 = \frac{\Delta H_f}{R}\left(\frac{1}{T_f} - \frac{1}{T}\right) - \frac{v_2\phi_1^2(\delta_2 - \delta_1)^2}{RT} \tag{2.86}$$

where T_f is the freezing point of the solute, ΔH_f is the heat of fusion, and ϕ_1 is the volume fraction of the solvent component (1). The rule of thumb is that miscibility

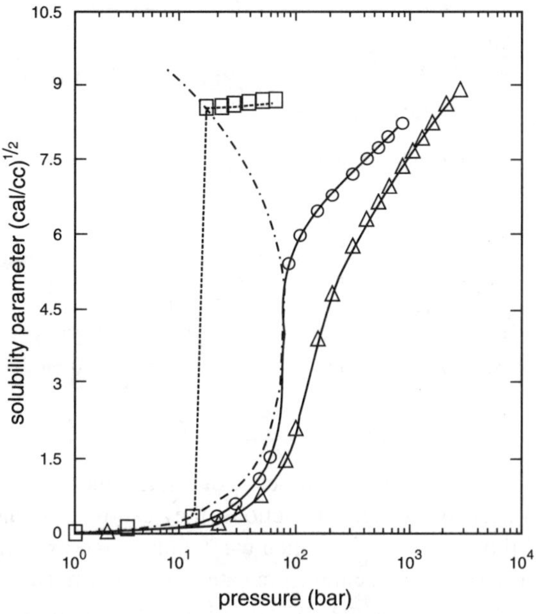

FIGURE 2.24 Hildebrandt solubility parameter for CO_2 (□) 30°C; (○) 31°C; (△) 70°C. (Rizvi et al., 1994).

will occur if $(\delta_1 - \delta_2)$ is about 1 and is valid for solid–liquid and liquid–liquid behavior, but the rule of thumb is not generally applicable to solid solubilities in an SCF solvent. The regular solution theory has the underlying assumption that the volume change upon mixing is zero, which cannot be true for dissolution of a solid or liquid component in the SCF solvent. For this reason, Giddings considered the state effect in terms of reduced density and visualized a liquid-like solvent power from the liquid-like density of the SCF solvent. Giddings et al. (1969), however, cautioned against applying the solubility parameters for esters, ketones, alcohols, and other polar liquids. CO_2 has zero dipole moment but a large quadruple moment which can interact with other polar molecules. It is therefore doubtful whether solubility parameter concept is valid for supercritical CO_2, although many investigators (Prausnstz, 1958; Giddings et al., 1968; Allada, 1984) calculated solubility parameters of the SCF solvent.

The Kirkwood-Buff solution theory (Cochran et al., 1987; Pfund et al., 1988), Monte-Carlo simulations (Shing and Chung, 1987), and Mean-Field theory (Jonah et al., 1983) are some of the other techniques which were used to represent the solubilities in an SCF solvent.

2.5.9 Selectivity of Natural Molecules from Pure Component Solubilities

Supercritical CO_2 can dissolve natural molecules depending on their molecular size, polarity, and functional groups. Development of an SCFE process often requires

TABLE 2.10a
Physical Properties of some close Mol wt Compounds Studied

Compounds	MW	t_b (°C)	t_f (°C)	ρ (g·cm^{-3})
Tetralin	132	207	–35	0.973
Decalin	138	190	–43	0.870
p-anisaldehyde	136	248		1.119
p-anisic acid	152		180	
Thymol	150	232	50	0.965
Menthol	156	212	35	0.890

Mukhopadhyay, and De, 1995.

knowledge of selectivities and in the absence of reliable thermodynamic models and other parameters needed for their predictions, they can be estimated from the pure component solubilities in SC CO_2, which can be measured experimentally.

Separation of aromatic molecules from nature is considered commercially important as they may be the starting raw materials for several high-value, fine chemicals and are often required in the purest forms. Some of these naturally occurring chemicals are in a very close range of molecular weights or boiling points and hence their separation, in general, poses a serious problem. A supercritical fluid solvent, such as CO_2, is known to have the capability of separating such difficult mixtures by virtue of its adaptable molecular association or interactions.

There is no substitute for experimental measurements of solubility data of natural molecules in SC CO_2. Accordingly, fluid–liquid equilibrium measurements were carried out for three binaries having close molecular weights, namely, (1) decalin-tetralin, (2) anisaldehyde-anisic acid, and (3) menthol-thymol in SC CO_2 over a pressure range of 55 to 160 bar and at temperatures in the range of 50 to 100°C (Mukhopadhyay and De, 1995). The physical properties listed in Table 2.10a indicate their closeness in molecular weights and boiling points, yet their solubilities are quite different as listed in Table 2.10b for (1) decalin at 50 and 70°C, (2) tetralin at 50°C, and (3) anisaldehyde at 50 and 100°C. The distribution coefficient, K_i, and selectivity of separation, S_{ij}, are calculated from these experimental phase equilibrium data, which are defined as:

$$K_i = \frac{y_i}{x_i} \qquad (2.87)$$

$$S_{ij} = K_i/K_j = \frac{y_i/x_i}{y_i/x_j} \qquad (2.88)$$

The fluid–liquid equilibrium data for decalin + tetralin + CO_2 are presented in Table 2.10c.

TABLE 2.10b
Solubility of Various Pure Aromatic Compounds in Supercritical CO_2[a]

P (bar)	x_1	y_1	10^3K_2	K_1
		Carbon dioxide (1) + decalin (2)		
		T = 50°C		
100	0.791	0.9889	53.10	1.25
90	0.689	0.9940	19.20	1.44
80	0.500	0.9973	5.40	1.99
70	0.387	0.9990	1.63	2.58
60	0.281	0.9992	1.11	3.55
		T = 70°C		
130	0.728	0.9741	95.20	1.36
110	0.579	0.9889	26.30	1.70
100	0.521	0.9908	19.20	1.90
90	0.441	0.9950	8.94	2.26
75	0.312	0.9971	4.21	3.20
55	0.258	0.9977	3.10	3.87
		Carbon dioxide (1) + tetralin (2)		
		T = 50°C		
110	0.740	0.9916	32.30	1.34
95	0.620	0.9975	4.31	1.61
90	0.597	0.9986	3.48	1.67
85	0.520	0.9990	2.08	1.92
80	0.494	0.9993	1.38	2.02
52	0.322	0.9997	0.44	3.10
		Carbon dioxide (1) + anisaldehyde (2)		
		T = 50°C		
115	0.7949	0.9970	58.00	1.25
106	0.7815	0.9916	38.40	1.28
100	0.7661	0.9945	23.20	1.30
90	0.726	0.9980	7.30	1.37
80	0.6802	0.9993	2.19	1.49
72	0.5987	0.9997	0.75	1.67
55	0.3022	0.9998	2.29	3.31
		T = 100°C		
137	0.6420	0.9860	39.10	1.53
125	0.6300	0.9918	22.10	1.57
104	0.5500	0.9956	9.77	2.11
90	0.4705	0.9980	3.77	2.63
69	0.3307	0.9994	0.89	3.02
55	0.1827	0.9996	0.49	5.47

[a] x_1 is the mole fraction of carbon dioxide in the liquid phase and y_1 is the mole fraction in the vapor phase.

Mukhopadhyay and De, 1995.

TABLE 2.10c
Fluid–Liquid Equilibria for Decalin + Tetralin + CO_2

P (bar)	x_2	x_3	10^3y_2	10^3y_3	10^3K_2	10^3K_3
			T = 50°C			
72	0.2394	0.2503	0.66	0.41	2.75	1.63
84	0.1861	0.1945	1.08	2.50	5.79	1.29
90	0.1652	0.1727	1.93	0.40	11.69	2.34
100	0.1327	0.1696	4.80	1.10	36.17	6.49
110	0.0951	0.1215	6.00	1.70	60.90	13.99
			T = 70°C			
85	0.0955	0.0955	1.00	0.65	10.47	6.78
110	0.0808	0.1286	1.28	0.82	15.85	7.27
120	0.0751	0.1177	4.69	2.31	62.45	19.58
130	0.061	0.0958	7.40	6.33	121.14	66.07
135	0.0586	0.0919	9.09	8.77	161.49	95.42
140	0.0556	0.0871	11.98	11.53	199.47	132.37

Mukhopadhyay and De, 1995.

The solubility of anisic acid, a solid, is presented in Table 2.11a at 50, 70, and 100°C, and those of menthol and thymol are presented in Table 2.11b at 50°C and 70°C.

It is interesting to note that the solubility of anisic acid in supercritical CO_2 increases with temperature beyond 100 bar, the crossover pressure, whereas the crossover pressure for decalin solubility is 85 bar, for anisaldehyde solubility it is 96 bar, and for tetralin solubility it is 95 bar (Mukhopadhyay and De, 1995). Apparently these are the upper level crossover pressures. However, a reverse trend was observed for thymol as the solubility was found to decrease with temperature beyond 65 bar and no upper level crossover pressure was observed for menthol in the range of pressures studied. From the trend of isothermal solubility variation with pressure for menthol, it is expected that the lower crossover pressure would occur at a much lower pressure, beyond which the reverse trend of the decrease in solubility with an increase in temperature is observed as in the case of thymol (Mukhopadhyay and De, 1995). The solubility of menthol at 50°C is five times higher than that of thymol in the pressure range of 25 to 120 bar and four times higher at 70°C in the pressure range of 90 to 135 bar, as can be seen from Figure 2.25.

The distribution coefficient and solubility in SC CO_2 exhibit a similar behavior. However, the selectivity of separation truly represents the molecular association among the components. In the absence of reliable experimental equilibrium data on multicomponent mixtures or a rigorous thermodynamic model, there is a tendency to approximate them from the distribution coefficients of the respective binary systems, which are termed as the apparent distribution coefficients. A comparison of the apparent selectivity calculated from binary data with the corresponding experimental

TABLE 2.11a
Solubility of Anisic Acid in Supercritical Carbon Dioxide

P (bar)	10^6y_2	P (bar)	10^6y_2
	T = 50°C		
110	9.64	71	5.78
100	9.21	55	4.61
80	7.35		
	T = 70°C		
140	32.00	75	4.58
120	21.02	55	3.57
100	10.20		
	T = 100°C		
151	137.30	92	5.56
130	59.03	75	3.65
112	22.67	50	2.87

Mukhopadhyay and De, 1995.

TABLE 2.11b
Solubility of Menthol and Thymol in Supercritical CO_2

Menthol P (bar)	10^3y_2	Thymol P (bar)	10^3y_2
	T = 50°C		
120	57.83	127	12.56
112	42.31	113	7.21
95	8.127	104	4.14
88	5.331	98	2.44
78	2.890	78	0.83
65	1.720	65	0.47
	T = 70°C		
135	22.08	140	4.49
115	7.29	119	1.71
100	3.40	90	0.44
82	1.02	70	0.51
75	0.88	57	0.54
65	0.72		

Mukhopadhyay and De, 1995.

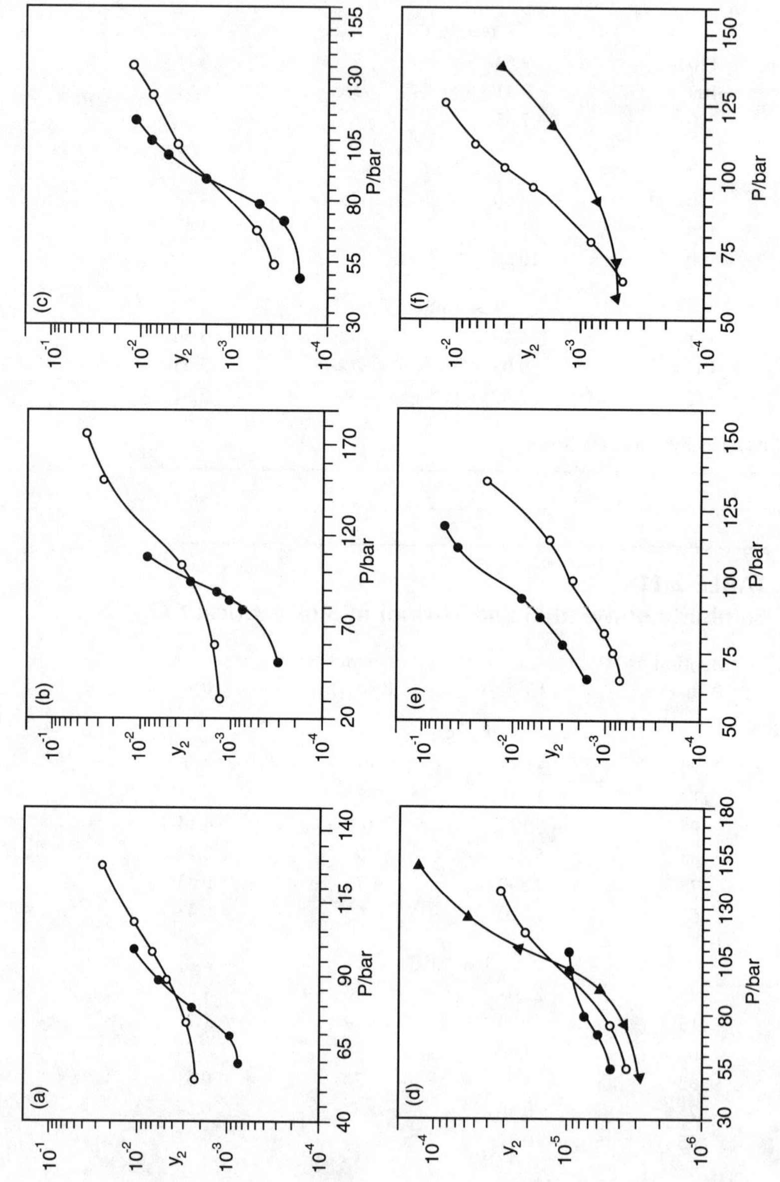

FIGURE 2.25 Solubility-Pressure diagrams for (a): decalin (2) at (●) 50°C, (○) 70°C; (b): tetralin (2) at (●) 50°C, (○) 70°C; (c): anisaldehyde (2) at (●) 50°C, (○) 70°C; (d): anisic acid (2) at (●) 50°C, (○) 70°C, (▲) 100°C; (e): menthol (2) at (●) 50°C, (○) 70°C; (f): thymol (2) at (○) 50°C, (▲) 70°C in SC CO_2(1) (Mukhopadhyay and De, 1995).

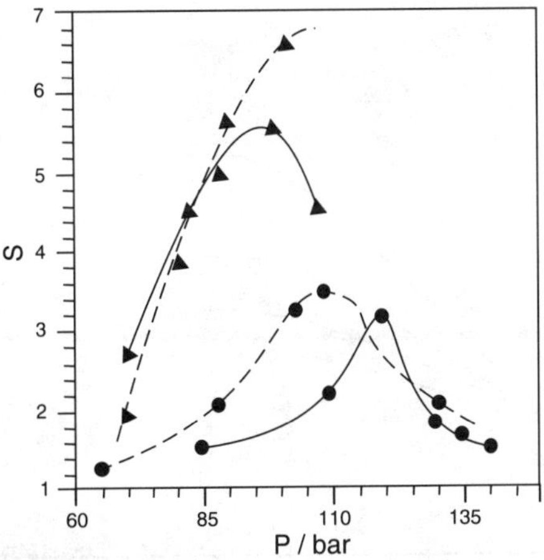

FIGURE 2.26 Selectivity (S_{23}) of decalin in $CO_2(1)$ + decalin (2) + tetralin (3) mixture as a function of pressure: (—) from ternary data, (- - -) from binary data; (▲) 50°C; (●) 70°C (Mukhopadhyay and De, 1995).

value from the ternary system (decalin + tetralin + CO_2) at 50°C and 70°C is presented in Figure 2.26. It can be observed that the trends of variation with pressure and temperature are very much similar for the components in the close range of molecular weights and physical properties, and that the maximum selectivity of separation is very close to the corresponding apparent selectivity calculated from the binary data. Hence the apparent selectivity can be reliably considered for the selection of SCFE process conditions in the absence of elaborate phase equilibrium data of multicomponent natural systems.

2.6 THERMOPHYSICAL PROPERTIES OF CO_2

The most important thermophysical property is the density and its knowledge is essential for the process design calculations involving SCFE. While experimentally measured values are preferable, theoretical models for its prediction are considered very useful. Ely et al. (1989) measured densities of CO_2 at temperatures from –23 to 63°C and pressures up to 350 bar covering a wide range of densities from gaseous state to the supercritical fluid state of CO_2. A wider range of experimental densities were reported earlier by Holste et al. (1987) for temperatures in the range from triple point temperature, –56°C, to 175°C. For the supercritical region the P-R EOS gives good representation of the experimental data of CO_2 densities. For density calculations of various states, the multiparameter Bender equation may be considered useful, which is given as:

$$P = T\rho[R + B\rho + C\rho^2 + D\rho^3 + E\rho^4 + F\rho^5 + (G + H\rho^2)\rho^2 \exp(-a_{20}\rho^2)] \quad (2.89)$$

The constants, B, C, D, E, F, G, H, and a_{20}, are reported in the literature or evaluated from experimental data (Brunner, 1994).

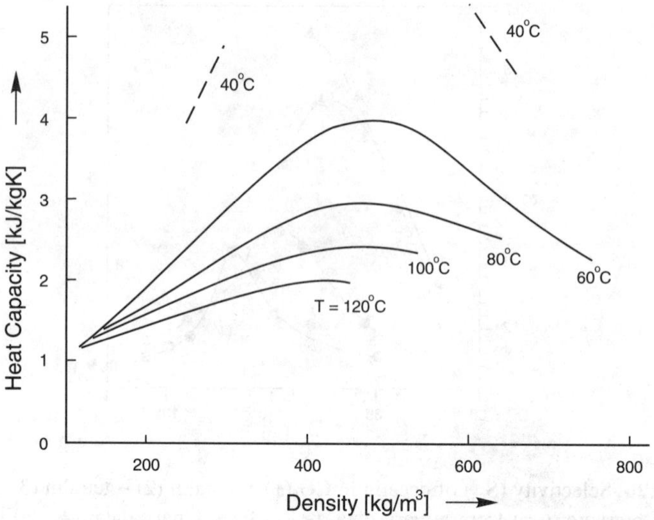

FIGURE 2.27a Heat capacity of CO_2 as a function of density (Brunner, 1994).

Thermodynamic properties of CO_2 like enthalpy, entropy, and density can be obtained from the thermodynamic charts compiled by Canjar et al., 1966. The specific heat capacity at constant pressure, or simply heat capacity, Cp, of CO_2 rapidly increases as the critical point (31.1°C, 73.72 bar, and 467.7 g/l) is approached. Like enthalpy and entropy, heat capacity is also dependent on temperature, pressure, and composition and can be calculated from P-v-T behavior or an equation of state. Ernst et al. (1989) reported accurate experimental data on Cp at pressures up to 900 bar and temperature in the range of 30 to 120°C. In general, both enthalpy and entropy of a gas decrease with pressure at a constant temperature and increase with temperature at a constant pressure. Enthalpy of a liquid at a constant temperature remains more or less constant with pressure at a moderate pressure level, but the pressure effect is substantial at a very high pressure (>1000 bar). Heat capacity, Cp, of a gas decreases with temperature but increases with pressure and so it depends on density, as shown in Figure 2.27a. Cp increases with pressure from a very low pressure, attains a maximum at a moderate pressure, and then decreases with pressure, as can be seen from Figure 2.27b. Heat capacity of a liquid increases with temperature at a constant pressure near its boiling temperature and the increase is nearer the critical temperature. At the critical point, Cp is infinite. At $P > P_c$, Cp passes through a maximum with pressure at a constant temperature.

NOMENCLATURE

A	A dimensionless EOS parameter, aP/R^2T^2
a	Attractive energy parameter of the mixture
a_{ij}	Interaction energy between components i and j
b	Covolume or size parameter, cm^3/mol

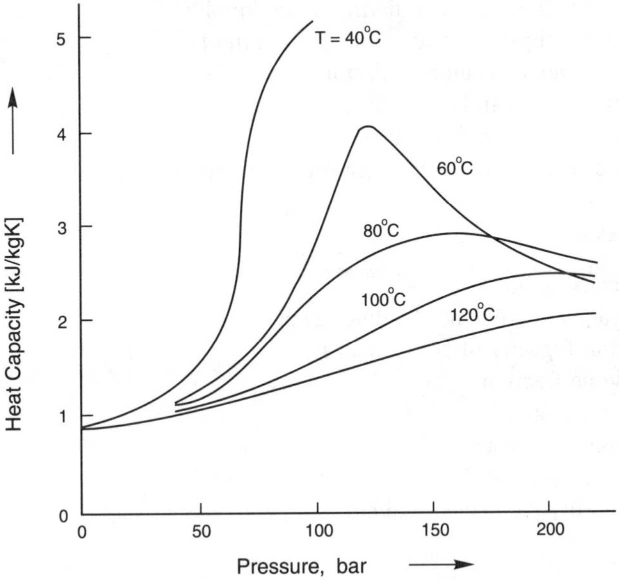

FIGURE 2.27b Heat capacity of CO_2 as a function of pressure (Brunner, 1994).

b_{ij}	Size parameter or covolume for representing interaction between the molecules i and j, cm^3/mol
b_{VDW}	van der Wall's volume, cm^3/mol
B	A dimensionless EOS parameter, bP/RT
$\bar{f}_i$	Partial fugacity of component i, bar
G	Gibbs free energy, cal/mol
g^E	Excess Gibb's free energy, cal/mol
ΔH_2^s	Heat of sublimation of solute 2, cal/mol
$\bar{h}_2$	Partial molar ethalpy, cal/mol
$\bar{h}_2^s$	Molar enthalpy of solid solute (2), cal/mol
K	Ratio of the SCF solvent to solute molecules
K_i	Distribution coefficient, y_i/x_i
K_T	Isothermal compressibility, bar^{-1}
k_{ij}, k_{ij}'	Interaction constants
M	Molecular weight
m	Exponent in CVD mixing rule
P	Pressure, bar
P_c	Critical pressure, bar
P_r	Reduced pressure
P_2^s	Saturation pressure of solid solute (2), bar
R	Universal gas constant, cal/g mole K
R_o	Ratio of solute to cluster molecules
T	Temperature, K
T_c	Critical temperature, K,
T_r	Reduced temperature

v	Molar volume of the fluid/liquid, cm^3/mol
v_2^s	Molar volume of solid solute (2), cm^3/mol
$\bar{v}_2$	Partial molar volume, cm^3/mol
x	Mole fraction in liquid phase
y	Mole fraction in fluid phase
z_D	Mole fraction of cluster D in dilute SC mixture

Greek Letters

ρ	Density, g/cm^3
φ_2^s	Fugacity coefficient of solute (2) at P_2^s
$\bar{\phi}_i$	Partial fugacity of component i
ϕ_1	Volume fraction
ω	Acentric factor
α, β, γ	Empirical constants
μ	Dipole moment, D
δ	Solubility parameter, $(cal/cm^3)^{1/2}$

REFERENCES

Allada, S. R., Solubility parameters for supercritical fluids, *I. E. C. Proc. Des. Dev.*, 23, 244, 1984.

Bamberger, T., Erickson, I. C., Cooney, C. L., and Kumar, S., Measurement and model prediction of solubilities of pure fatty acids, pure TG and mixtures in SC CO_2, *J. Chem. Eng. Data*, 33, 320, 1988.

Bartle, K. D., Clifford, A. A., Jafar, S. A., and Shilstone, G. F., Solubilities of solids and liquids of low volatility in supercritical CO_2, *J. Phys. Chem. Ref. Data*, 20(4), 713–755, 1991.

Benmekki, E. J. and Mansoori, G. A., The role of mixing rules and three-body forces in the phase behavior of mixtures; simultaneous VLE and VLLE calculations, *Fluid Phase Equilibria*, 41, 43, 1988.

Billingsley, D. S. and Lam, S., Critical point calculation with non zero interaction parameters, *A. I. Ch. E. J.*, 32, 1396, 1986.

Bondi, H., *Physical Properties of Molecular Crystals, Liquids, and Glasses*, John Wiley & Sons, New York, 1968.

Brennecke, T. F. and Eckert, C. A., Phase equilibria for supercritical fluid process design, *A. I. Ch. E. J.*, 35, 1409, 1989.

Brunner, G. and Peter, S., State of the art extraction with compressed gases, *Ger. Chem. Eng.*, No. 5, 181, 1982.

Brunner, G., *Gas Extraction*, Steinkopff Darmstadt, Springer, New York, 1994.

Canjar, L. N., Pollock, E. K., and Lee, W. E., *Hydrocarbon Process*, 45, 139, 1966.

Chimowitz, E. H. and Pennisi, K. J., Process synthesis concepts for supercritical gas extraction in the cross over region, *A. I. Ch. E. J.*, 32, 1655, 1986.

Chrastil, J., Solubility of solids and liquids in supercritical gases, *J. Phys. Chem.*, 86, 3016, 1982.

Cocero, M. J. and Calvo, L., Supercritical fluid extraction of sunflower seed oil with CO_2-ethanol mixtures, *J. Am. Oil Chem. Soc.*, 73(11), 1573, 1996.

Cochran, H. D., Lee, L. L., and Pfund, D. M., Application of Kirkwood-Buff theory of solutions to dilute supercritical mixtures, *Fluid Phase Equilibria*, 34, 219, 1987.

Cotterman, R. L., Demitrelis, D., and Prausnitiz, J. M., Design of supercritical fluid extraction processes using continuous thermodynamics, in *Supercritical Fluid Technology*, Penninger et al., Eds., Elsevier, New York, 1985, 107.

da Ponte, M. N., Menduina, C., and Pereira, P. J., Solubilities of tocopherols in supercritical CO_2 correlated by the Chrastil equation, Proc. 4th Intl. Symp. Supercritical Fluids, Sendai, Japan, May 1997, 303.

Debenedetti, P. G. and Kumar, S. K., The molecular basis of temperature effects in supercritical fluid extraction, *A. I. Ch. E. J.*, 34, 645, 1988.

de Valle, J. M. and Aguilera, J. M., An improved equation for predicting the solubility of vegetable oils in supercritical CO_2, *Ind. Eng. Chem. Res.*, 27, 1551, 1988.

Dobbs, J. M., Modification of SCF Equilibrium and Selectivity Using Polar and Non-Polar Co-Solvents, Ph.D. thesis, University of Texas, Austin, 1986.

Dobbs, J. M. and Johnston, K. P., Selectivities in pure and mixed supercritical fluid solvents, *Ind. Eng. Chem. Res.*, 26, 1476, 1987.

Dobbs, J. M., Wong, J. M., Lahiere, R. J., and Johnston, R. P., Modification of supercritical fluid phase behavior using polar cosolvents, *Ind. Eng. Chem. Res.*, 26, 56, 1987.

Eckert, C. A., Ziger, D. H., Johnston, K. P., and Ellison, T. S., The use of partial molar volume data to evaluate EOS for SCF mixtures, *Fluid Phase Equilibria*, 14, 167, 1983.

Eckert, C. A., Ziger, D. H., Johnston, K. P., and Kim, S., Solute partial molar volume in supercritical fluids, *J. Phys. Chem.*, 90, 2738, 1986.

Ekart, M. P., Brennecke, J. F., and Eckert, C. A., Molecular analysis of phase equilibria in supercritical fluids, in *Supercritical Fluid Technology*, Bruno, T. J. and Ely, J. F., Eds., 1991, 163.

Ely, J. F., Proc. 65th Annu. Conv. Gas Processors Assoc., San Antonio, TX, 1986, 185–192.

Ely, J. F., Haynes, W. M., and Bain, B. C., Isochoric (P,V,T) measurements on CO_2 and (CO_2 + N_2) from 250 to 330K at pressure to 35 MPa, *J. Chem. Thermodyn.*, 21, 879, 1989.

Ernst, G., Maurer, G., and Wideruh, E., Flow calorimeter for the accurate determination of the isobaric heat capacity at high pressures; results for CO_2, *J. Chem. Thermodyn.*, 21, 53, 1989.

Holste, J. C., Hall, K. R., Eubank, P. T., Esper, G., Watson, M. Q., Warowny, W., Bailey, D. M., Young, J. G., and Ballomy, M. T., Experimental (P,V,T) for pure CO_2 between 220 and 450K, *J. Chem. Thermodyn.*, 19, 1233, 1987.

Gangadhara Rao, V. S. and Mukhopadhyay, M., A covolume-dependent mixing rule for prediction of supercritical fluid solid equilibria, in *Proc. 1st Intl. Symp. SCFs* (Nice, France), 1, 161, 1988.

Gangadhara Rao, V. S. and Mukhopadhyay, M., The influence of the binary interaction parameter on the prediction of SCF phase equilibrium data, *Ind. Chem. Eng. Trans.*, 31(3), 27, July–September 1989a.

Gangadhara Rao, G. V. and Mukhopadhyay, M., The effect of covolume dependency of the energy parameter on predictability of SCFE data using PR EOS, *J. Supercrit. Fluids*, 2, 22, 1989b.

Gangadhara Rao, V. S., Studies on Supercritical Fluid Extraction, Ph.D. dissertation, Indian Institute of Technology Bombay, India, 1990.

Gangadhara Rao, V. S. and Mukhopadhyay, M., Solid solubilities in supercritical fluids from group contributions, *J. Supercrit. Fluids*, 3(2), 66, June 1990.

Giddings, J. C., Meyers, M. N., McLaren, S., and Keller, R. A., Higher pressure gas chromatography on nonvolatile species, *Science*, 162, 67, 1968.

Giddings, J. C., Meyers, M. N., and King, J. W., Dense gas chromatography at pressures to 2000 atm, *J. Chromatogr. Sci.*, 7, 276, 1969.

Hannay, J. B. and Hogarth, J., On the solubility of solids in gases, *Proc. R. Soc. London*, 29, 324, 1879.

Hannay, J. B. and Hogarth, J., On the solubility of solids in gases, *Proc. R. Soc. London*, 30, 178, 1880.

Holste, J. C., Hall, K. R., Eubank, P. T., Esper, G., Watson, M. Q., Warowny, W., Bailey, D. M., Young, J. G., and Bellomy, M. T., Experimental (P,v,T) for pure CO_2 between 220 and 450K, *J. Chem. Thermodyn.*, 19, 1233, 1987.

Hong, G. T., Model, M., and Tester, J. W., Binary phase diagrams from a cubic EOS, in *Chemical Engineering at Supercritical Fluid Conditions*, Paulaitis, M. E. et al., Eds., Ann Arbor Science, Ann Arbor, MI, 1983, 263.

Hubert, P. and Vitzthum, O. G., Fluid extraction with hops, spices and tobacco with supercritical gases, in *Extraction with Supercritical Gases*, Schneider, G. M., Stahl, E., and Wilke, G., Eds., Verlag Chemie, Weinheim, Gemany, 1980, 25.

Irani, C. A., Cozewith, C., Lower critical solution temperature behavior of ethylene-propylene copolymers in muticomponent solvents, *J. Appl. Polym. Sci.*, 31, 1879–1899, 1986.

Johnston, K. P., Ziger, D. G., and Eckert, C. A., Solubilities of hydrocarbon solids in supercritical fluids, the augmented van der Waal's treatment, Ind. Eng. Chem. Fundam., 21, 191, 1982.

Johnston, K. P., Perk, D. G., and Kim, S., Modelling supercritical mixtures; how predictive is it, *Ind. Eng. Chem. Res.*, 28, 1115, 1989.

Jonah, D. A., Shing, K. S., and Venkatasubramanian, V., Molecular thermodynamics of dilute solids in supercritical solvents, in *Chemical Engineering at Supercritical Fluid Conditions*, Paulaitis, M. E. et al., Eds., Ann Arbor Science, Ann Arbor, MI, 1983, 221.

Kelly, F. D. and Chimowitz, E. H., Experimental data for the cross over process in a model supercritical system, *A. I. Ch. E. J.*, 35, 981, 1989.

Kim, S., Wong, J. H., and Johnston, K. P., Theory of the pressure effect in dense gas extraction, in *Supercritical Fluid Technology*, Elsevier Science Publishers, New York, 1985, 45.

Klesper, E., Chromatography with supercritical fluids, in *Extraction with Supercritical Gases*, Schneider, G. M., Stahl, E., and Wilke, G., Eds., Verlag Chemie, Weinheim, Germany, 1980, 115.

Kosal, E. and Holder, G. D., Solubilities of anthracene and phenanthrene mixtures in supercritical CO_2, *J. Chem. Eng. Data*, 32, 148, 1987.

Krichevskii, I. R. and Sorina, G. A., Liquid-gas phase equilibria in the cyclohexane carbon dioxide and cylcohexane nitrous oxide systems, *Russ. J. Phys. Chem.*, 34, 679, 1960.

Kurnik, R. T. and Reid, R. C., Solubility extrema in solid fluid equilibria, *A. I. Ch. E. J.*, 27, 861, 1981.

Kurnic, R. T. and Reid, R. C., Solubility of solid mixtures in supercritical fluids, *J. Fluid Phase Equilibria*, 8, 93, 1982.

Kwiatkowski, J., Lisicki, Z., and Majewski, W., An experimental method for measuring solubilities of solids in supercritical fluids, *Ber. Bunsen-Ges. Phys. Chem.*, 88, 865, 1984.

Lemert, R. M. and Johnston, K. P., *Ind. Eng. Chem. Res.*, 30, 1222, 1991.

McHugh, M. A. and Paulaitis, M. E., Solid solubilities of naphthalene and biphenyl in SC CO_2, *J. Chem. Eng. Data*, 25, 326, 1980.

McHugh, M. A., Mallet, M. W., and Kohn, J. P., High pressure fluid phase equilibria of alcohol-water-supercritical fluid mixtures, in *Chemical Engineering at Supercritical Conditions*, Paulaitis, M. E. et al., Eds., Ann Arbor Science, Ann Arbor, MI, 1983, 221.

McHugh, M. A. and Krukonis, V. J., Supercritical fluid extraction, *Principles and Practice*, 2nd ed., Butterworth-Heinemann, Boston, MA, 1994.

Mohamed, R. S., Enick, R. M., Bendale, P. G., and Holder, G. D., Empirical two-parameter mixing rules for a cubic equation of state, *Chem. Eng. Commun.*, 59, 259, 1987.

Mohamed, R. S. and Holder, G. D., High pressure phase behavior in systems containing CO_2 and heavier compounds with similar vapor pressures, 32, 295, 1987.

Mukhopadhyay, M. and Rao Raghuram, G. V. R., Thermodynamic modelling for supercritical fluid process design, *Ind. Eng. Chem. Res.*, 32, 922–930, 1993.

Mukhopadhyay, M. and De, S. K., Fluid phase behavior of close molecular weight fine chemicals with supercritical carbon dioxide, *J. Chem. Eng. Data*, 40, 909–913, 1995.

Mukhopadhyay, M. and Srinivas, P., Multicomponent solubilities of reactants and products of cyclohexane oxidation in SC CO_2 medium, *Ind. Eng. Chem. Res.*, 35, 4713–4717, 1996.

Mukhopadhyay, M. and Srinivas, P., Influence of the thermodynamic state of cylcohexane oxidation kinetics in carbon dioxide medium, *Ind. Eng. Chem. Res.*, 36, 2066–2074, 1997.

Mukhopadhyay, M. and Sastry, S. V. G. K., Modelling cosolvent induced solubilities using solvent-cluster interactions, Proc. 4th Intl. Symp. SCFs, Sendai, Japan, May 1997, 343.

Nilsson, W. B., Gauglitz, E. J., and Hudson J. K., Solubilities of methyl oleate, oleic acid, oleyl glycerols and oleyl glycerol mixtures in supercritical carbon dioxide, *J. Am. Chem. Oil. Soc.*, 68, 87, 1991.

Nitsche, J. M., Teletzke, G. F., Scriven, L. E., and Davis, H. T., Phase behavior of binary mixtures of water, carbon dioxide and decane predicted with a lattice gas model, *Fluid Phase Equilibria*, 17, 243, 1984.

Oghaki, K., Tsukahara, I., Semba, K., and Katayama, T. A., A fundamental study of extraction with supercritical fluid solubilities of α-tocopherol, palmitic acid and tripalmitin in compressed CO_2 at 25°C and 45°C, *Intl. Chem. Eng.*, 29, 302, 1989.

Palencher, R. M., Erickson, D. D., and Leland, T. W., Prediction of binary critical loci by cubic equations of state, in *Equations of State, Theories and Applications*, Chao, K. C. and Robinson, R. L., Eds., *Am. Chem. Soc. Symp. Ser.*, 1986.

Panagiotopoulos, A. Z. and Reid, R. C., High pressure phase equilibria in ternary fluid mixtures with a supercritical component, *Am. Chem. Soc. Symp. Ser.*, No. 406, 1987.

Peng, D. Y. and Robinson, D. B., A new two-constant equation of state, *Ind. Eng. Chem. Fundam.*, 15, 59, 1976.

Pfund, D. M., Lee, L. L., and Cochran, H. D., Application of Kirkwood-Buff theory of solutions to dilute supercritical mixtures. II. the excluded volume and local composition models, *Fluid Phase Equilibria*, 39, 161, 1988.

Prausnitz, J. M., Regular solution theory for gas liquid solution, *A. I. Ch. E. J.*, 4, 269, 1958.

Prausnitz, J. M., Lichtenthaler, R. N., and de Azevedo, E. G., *Molecular Thermodynamics of Fluid Phase Equilibria*, Prentice Hall, Englewood Cliffs, NJ, 1986.

Radosz, M., Ho-Mu, L., and Chao, K. C., High pressure VLE in asymmetric mixtures using new mixing rules, *Ind. Eng. Chem. Proc. Des. Dev.*, 21, 653–658, 1982.

Radosz, M., Cotterman, R. L., and Prausntiz, J. M., Phase equilibrium in supercritical propane systems for separation of continuous oil mixtures, *Ind. Eng. Chem. Res.*, 26, 731, 1987.

Raghuram Rao, G. V., Thermodynamic Modelling of Supercritical Fluid Phase Behavior, M. Tech. dissertation, Indian Institute of Technology, Bombay, India, 1992.

Raghuram Rao, G. V., Srinivas, P., Sastry, S. V. G. K., and Mukhopadhyay, M., Modelling solute-cosolvent interactions for supercritical fluid extraction of fragrances, *J. Supercrit. Fluids*, 5(1), 19, 1992.

Reverchon, E. and Osseo, L. S., Comparison of processes for the SC CO_2 extraction of oil from soybean seeds, *J. Am. Oil Chem. Soc.*, 71(9), 1007, September 1994.

Rizvi, S. S. H., Yu, Z. R., Bhaskar, A. R., and Raj, C. B. C., Fundamentals of processing with supercritical fluids, in *Supercritical Fluid Processing of Food and Biomaterials*, Blackie Academic Professional, an imprint of Chapman & Hall, Glasgow, 1994, 1.

Sastry, S. V. G. K. and Mukhopadhyay, M., Modelling dilute supercritical mixtures utilising solvent-cluster interactions, *J. Supercrit. Fluids*, 6, 21–30, 1993.

Sastry, S. V. G. K., Supercritical Fluid Extraction of Fragrances from Jasmine Flowers, Ph.D. dissertation, Indian Institute of Technology, Bombay, India, 1994.

Schneider, G. M., Physico-chemical principles of extraction with supercritical gases, in *Extraction with Supercritical Gases*, Schneider, G. M., Stahl, E., and Wilke, G., Eds., Verlag Chemie, Weinheim, Germany, 1980, 46.

Scott, R. L. and van Konynenberg, P. B., Static properties of solutions: van der Waals and related models for hydrocarbon mixtures, *Discuss Faraday Soc.*, 49, 87, 1970.

Shibata, S. K. and Sandler, S. I., Higher pressure vapor-liquid equilibria of mixture of nitrogen, carbon dioxide and cyclohexane, *J. Chem. Eng. Data*, 34, 419, 1989.

Shim, J. J. and Johnston, K. P., Phase equilibria, partial molar enthalpies, and partial molar volumes determined by supercritical fluid chromatography, *J. Phys. Chem.*, 95, 353, 1991.

Shing, K. S. and Chung, S. T., Computer simulation methods for the calculation of solubility in supercritical extraction systems, *J. Phys. Chem.*, 91, 1674, 1987.

Soave, G., Equilibrium constants from a modified R-K equation of state, *Chem. Eng. Sci.*, 27, 1197-1203, 1972.

Smith, W. R., Perturbation theory and one fluid corresponding states theories for fluid mixtures, *Can. J. Chem. Eng.*, 50, 271, 1972.

Srinivas, P., Oxidation of Cyclohexane in Supercritical Carbon Dioxide Medium, Ph.D. disseration, Indian Institute of Technology, Bombay, India, 1994.

Srinivas, P. and Mukhopadhyay, M., Oxidation of cyclohexane in supercritical carbon dioxide medium, *Ind. Eng. Chem. Res.*, 33, 3118–3124, 1994a.

Srinivas, P. and Mukhopadhyay, M., Supercritical fluid-liquid equilibria of binary and ternary mixtures of cyclohexanone and cyclohexanol with CO_2 and N_2, presented at 1994 AIChE National Spring Meeting, Georgia, April 1994b.

Streett, W. B., Phase equilibria in fluid and solid mixtures at high pressures, in *Chemical Engineering at Supercritical Conditions*, Paulaitis, M. E. et al. Eds., Ann Arbor Science, MI, 1983.

Suzuki, K., Sue, H., Itou, M., Smith, R. L., Inomata, H., Arai, K., and Saito, S., Isothermal vapor-liquid equilibrium data at high pressures, *J. Chem. Eng. Data*, 35, 63, 1990.

Tolic, A., Zekovic, Z., and Pekic, B., Dependence of camomile flower solubility on CO_2 density at SC conditions, *Sep. Sci. Technol.*, 31(13), 1889–1892, 1996.

Van der Haegen, R., Koningsveld, R., and Kleintjens, L. A., Solubility of solids in supercritical solvents. IV. Mean-field lattice gas description for the P-T-X diagram of the system ethylene-naphthalene, *Fluid Phase Equilibria*, 43, 1, 1988.

Van Leer, R. A. and Paulaitis, M. E., Solubilities of phenol and chlorinated phenols in supercritical CO_2, *J. Chem. Eng. Data*, 25, 257, 1980.

Vidal, J., Mixing rules and excess properties in cubic equations of state, *Chem. Eng. Sci.*, 33, 787, 1978.

Wong, J. M., Pearlman, R. S., and Johnston, K. P., Supercritical fluid mixtures prediction of phase behavior, *J. Phys. Chem.*, 89, 2671, 1985.

Wong, J. M. and Johnston, K. P., Solubilization of biomolecules in CO_2 based SCFs, *Biotechnol. Prog.*, 2(1), 29–38, March 1986.

3 Fundamental Transport Processes in Supercritical Fluid Extraction

The purpose of this chapter is to focus attention on the fundamental transport processes such as heat and mass transfer involved in extraction from natural products using a supercritical fluid (SCF) solvent. Thermodynamic and phase equilibrium properties dictate the feasibility of such a process and the conditions for maximum possible separation. Knowledge of transport properties of the SCF solvent and resistance to transport processes are required for deciding the time required for extraction and the sizes of the critical components of the process plant. This chapter is useful for process design, simulation, and optimization purposes. It summarizes the relevant procedures for generating the required quantitative information.

3.1 TRANSPORT PROPERTIES

Successful design and development of a supercritical fluid extraction (SCFE) process plant on the commercial scale while retaining its economic feasibility rely not only on the knowledge of the unique solvent characteristics, but also on the understanding of excellent transport properties of supercritical carbon dioxide. Three principal transport properties of industrial interest are the viscosity, η, the diffusivity, D, and the thermal conductivity, λ, which characterize the dynamics of the SCFE process, involving the momentum transport related to the pressure difference, the mass transport related to the concentration difference, and the heat transport related to the temperature difference, respectively. The behavior of these transport properties in the vicinity of the critical point is not fully understood due to very rapid changes in their values with respect to a small change in any one of the thermodynamic state variables. An attempt will be made in this section to present their behaviorial trends and simple prediction methods, excluding the ones based on rigorous theoretical foundation.

The measurement of transport properties poses a serious problem, since the thermodynamic state at which measurements are needed is required to be maintained as close to equilibrium as possible. This implies that a minimal disturbance required to this effect, should cause sufficient transient change in the system which can be measured with an instrument having its sensitivity limit less than this change. For reliable experimental values of transport properties, it is thus essential to maintain a small transient change from equilibrium, to secure the instrument with the highest possible resolution for its detection, and to correlate the measured transport property as a function of temperature and pressure (or density) to be able to interpolate the data within the range of measurements.

3.1.1 Viscosity

The coefficient of the shear stress of a flowing fluid or viscosity is defined by the linear relation:

$$\tau = -\eta \frac{\partial u}{\partial z} \tag{3.1}$$

where, τ is the shear stress resulting from the applied velocity gradient $\partial u/\partial z$, normal to the direction of flow. The proportionality constant, η, simply called viscosity, depends on the thermodynamic state variables, such as its temperature and pressure (or density). In general, the isothermal viscosity of an SCF solvent increases with increasing pressure. It decreases with an increase in temperature at a constant pressure up to a minimum as in the case of a liquid (McHugh and Krukonis, 1994), then increases with temperature as in the case of a gas at high reduced temperatures (Brunner, 1994). Figure 3.1a–b show the variation of viscosity with pressure at different temperatures and with temperature at different pressures, respectively. Although the viscosity increases rapidly in the critical region, its order of magnitude is an order less than those of liquid organic solvents even at very high pressures (300–400 bar) (de Fillipi et al., 1980). Viscosity of CO_2 was measured (Assale et al., 1980) very accurately using a torsional piezo electric crystal method with an uncertainty of better than 2%. Table 3.1 compares viscosity data for different SCF solvents at two levels of temperature higher than the corresponding critical temperatures (Vesovic and Wakeham, 1991). The viscosity data for CO_2 as a function of temperature at different pressures are also given in Appendix A (Figure A.2).

Vesovic et al. (1990) developed viscosity correlations for supercritical CO_2 which are preferred to any prediction method. The viscosity of the fluid, at any density and temperature, is calculated with reference to its dilute gas limit, η^o, in terms of the excess viscosity, $\Delta\eta$, and the critical enhancement, $\Delta\eta^c$, near the critical region as:

$$\eta(\rho,T) = \eta^o(T) + \Delta\eta(\rho,T) + \Delta\eta^c(\rho,T) \tag{3.2}$$

Lucas (1987) gave an empirical expression for the zero-density viscosity, $\eta^o(T)$, with ±3% accuracy as:

$$\eta^o(T) = \frac{F_p}{\xi}[0.807T_r^{0.618} - 0.357\exp(-0.449T_r) \tag{3.3}$$

$$+ 0.340\exp(-4.058T_r) + 0.018]$$

where

$$T_r = T/T_c \text{ and } \xi = 1.76\,(T_c/M^3\,P_c^4)^{1/6}$$

$\eta^o(T)$ is in micropascal seconds, P_c is in bar, and T in °K. M is molecular weight. F_p is the correction factor to account for the polarity, where

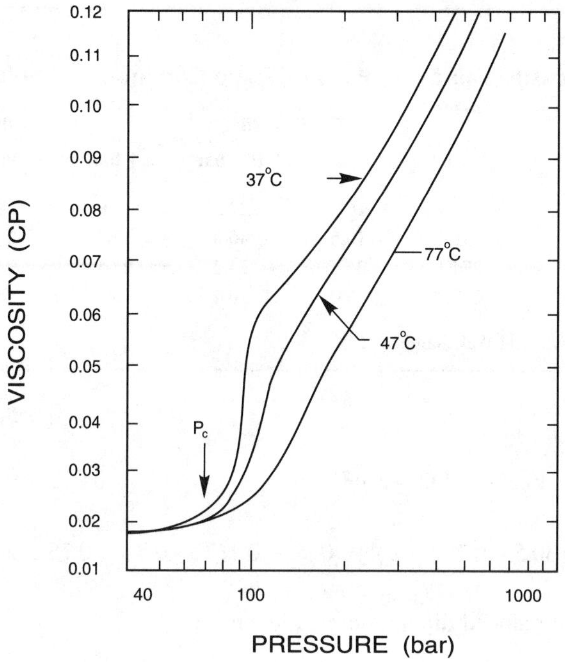

FIGURE 3.1a Viscosity behavior of carbon dioxide (McHugh and Krukonis, 1994).

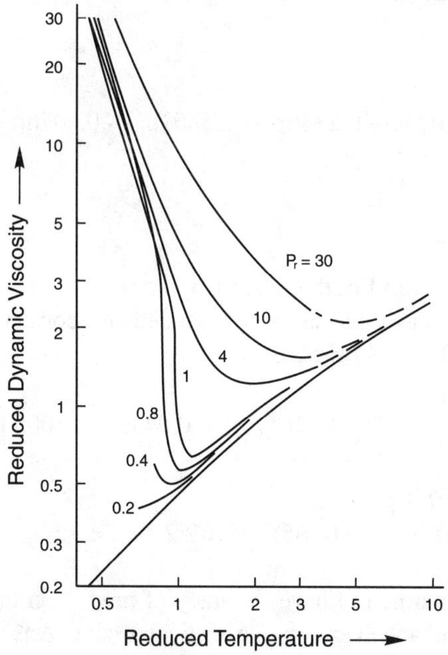

FIGURE 3.1b Generalized viscosity behavior (Brunner, 1994).

TABLE 3.1
Viscosity Values (η, Pa·s) of Some Common SCF Solvents

		$T = T_c + 20°C$		$T = T_c + 100°C$	
SCF	**T_c (°C)**	**1 bar**	**100 bar**	**1 bar**	**100 bar**
CO_2	31.1	16.2	27.5	19.9	22.3
C_3H_8	96.7	10.6	46.6	12.7	20.8
H_2O	374.1	24.2	24.2	27.5	27.8
NH_3	232.4	15.0	16.4	18.0	18.7

Vesovic and Wakeham, 1991.

$$F_p = 1.0 \qquad 0 \le \mu_r < 0.022 \quad (3.4)$$

$$F_p = 1.0 + 30.55\,(0.292 - z_c)^{1.72} \qquad 0.022 \le \mu_r < 0.075 \quad (3.5)$$

$$F_p = 1.0 + 30.55\,(0.292 - z_c)^{1.72}\,\left|0.96 + 0.1(T_r - 0.7)\right| \quad 0.75 \le \mu_r \quad (3.6)$$

where μ_r is the reduced dipole moment given as:

$$\mu_r = 52.46\,(\mu^2 P_c/T_c^2)$$

Jossi et al. (1962) gave the following correlation for excess viscosity, $\Delta\eta$, of nonpolar gases:

$$\left(\frac{\xi\Delta\eta}{0.176} + 1\right)^{0.25} = 1.0230 + 0.2336\rho_r + 0.5853\rho_r^2 - 0.4076\rho_r^3 + 0.09332\rho_r^4 \quad (3.7)$$

where $\rho_r = \rho/\rho_c$.

The correlation is valid in the range $0.1 \le \rho_r < 3$.

For polar gases, the same authors extended the correlation for estimation of viscosity in the range $0.9 \le \rho_r < 2.6$ as:

$$\log(4 - \log(\xi\Delta\eta/0.176)) = 0.6439 - 0.1005\,\rho_r - s \quad (3.8)$$

where $s = 0$ for $0.9 \le \rho_r < 2.2$
$s = 4.75 \times 10^{-4}\,(\rho_r^3 - 10.65)^2$ for $2.2 \le \rho_r < 2.6$

This method is claimed to have accuracy of prediction in the range of 5–10%.

The viscosity enhancement in the critical region $\Delta\eta^c$ is found to rise to a maximum as the critical point is approached (Vesovic et al., 1991), as can be seen from Figure 3.2. At very high densities, this contribution to viscosity may be

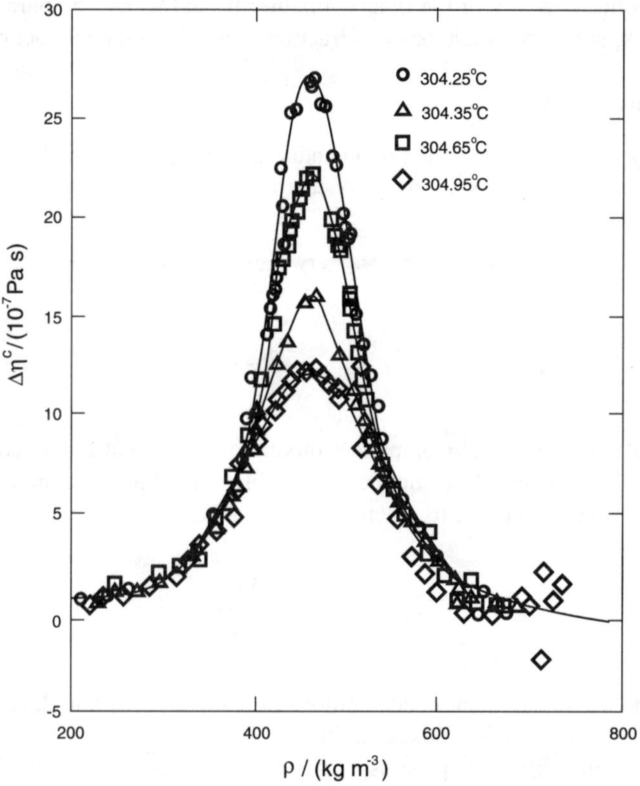

FIGURE 3.2 The viscosity enhancement, $\Delta\eta^c$ of SC CO_2 calculated from a model due to Vesovic et al. (1990) (Magee, 1991).

neglected. The principle of corresponding states (CS) was employed by Ely and Hanley (1981) for estimation of viscosity of non-polar fluids and their mixtures at high pressures and at reduced temperatures between 0.5 and 2.0. The method requires knowledge of critical constants and acentric factors of the pure components. It relates viscosity of the SCF of interest with respect to a reference fluid. The method was tested for a number of compounds and their mixtures including hydrocarbon and CO_2 and the viscosity could be predicted within an accuracy of ±8%.

Viscosities of liquids are much higher than those of gases and decrease with temperature but increase with pressure at moderate pressures. When an SCF is dissolved in the liquid phase, its viscosity decreases and the viscosity of the saturated liquid phase decreases with increasing pressure due to the increasing amount of SCF dissolved in it. The viscosity of the saturated fluid phase also increases with pressure corresponding to the increasing amount of liquid component dissolved in the SCF phase. The viscosities of saturated phases in equilibrium with each other can be calculated using the Grunberg equation (Brunner, 1994) as:

$$\eta_m = \eta_1^{x_1}\eta_2^{x_2}\exp[G_{12}x_1x_2] \tag{3.9}$$

where η_m is the viscosity of the binary mixture; η_1 and η_2 are the pure component viscosities; x_1 and x_2 are mole or mass fractions; and G_{12} is the interaction constant.

3.1.2 Diffusivity

For a binary mixture at a constant temperature and pressure, the molecular diffusion flux of a component with respect to its concentration gradient is given by Fick's law as:

$$\bar{J}_1 = -D_{12}\frac{\partial c_1}{\partial z} \tag{3.10a}$$

$$\bar{J}_2 = -D_{21}\frac{\partial c_2}{\partial z} \tag{3.10b}$$

The molar flux in a multicomponent mixture is dependent on the concentration gradients of all species. If component 1 is diffusing through a multicomponent mixture at a small value of flux, then

$$\bar{J}_1 = -\left(\sum_{j=2}^{N}\frac{x_{1j}}{D_{1j}}\right)\frac{\partial c_1}{\partial z} \tag{3.11}$$

where c_1 and c_2 are the molar concentrations of components 1 and 2, and $\bar{J}_1$ and $\bar{J}_2$ are the corresponding molar fluxes, and $D_{12} = D_{21}$.

Although the diffusion process is a transient phenomenon involving small changes in temperature, pressure, and concentration in the system, their effects are usually very small and an average value is considered for the thermodynamic state of the system at which the diffusivity is measured. Figure 3.3 shows the self diffusivity of CO_2 as a function of temperature over a wide pressure range, which is approximately the same as the diffusivity of a molecule having a similar size diffusing through CO_2. It can be seen that diffusivities of solutes in organic liquids are of the order of 10^{-5} cm^2/sec, which are much lower (by one or two orders of magnitude) than the self-diffusivity of CO_2. The diffusivity in supercritical CO_2, in general, increases with temperature and decreases with pressure. At a low pressure, the diffusivity is nearly independent of composition, whereas at higher densities, the composition dependence becomes more significant.

Table 3.2 presents binary diffusivity of solutes present in traces in supercritical CO_2 at 40°C and compares them with the values of the self-diffusivity at low pressures in the gas and liquid phases (Vesovic and Wakeham, 1991).

There are several methods available for measuring binary diffusivities in an SCF solvent, but not all of them are reliable, and some are very tedious. Diffusivity is inversely proportional to pressure and accordingly, if atmospheric diffusivity measurement takes 5 h involving a low pressure method, the diffusivity measurement at a pressure of 100 bar might require 20 days using the same technique. However, a gas chromatographic technique in conjunction with a high pressure UV detector has proved to be more reliable and faster than the other methods (Feist and Schneider, 1980).

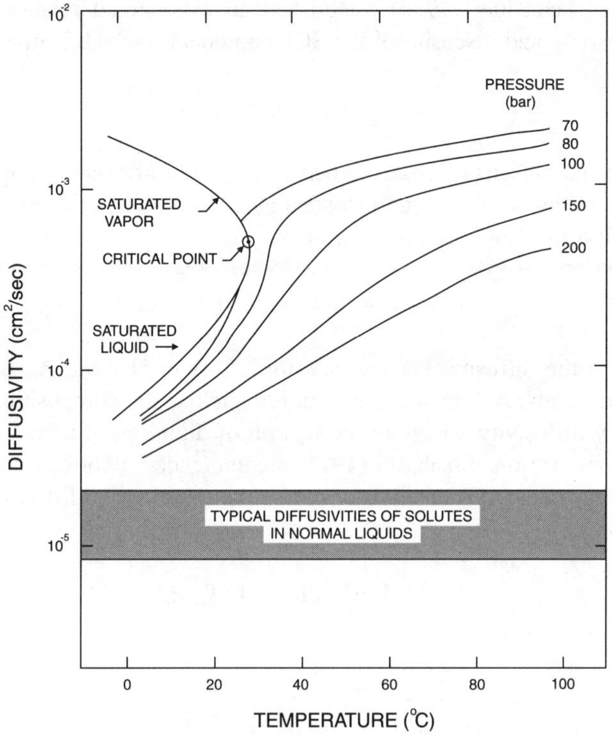

FIGURE 3.3 Diffusivity behavior of carbon dioxide (McHugh and Krukonis, 1994).

TABLE 3.2
Binary Diffusivities in CO_2 at 40°C

Component	P (bar)	$D_{12} \times 10^4$ (cm²/s)
Benzene	100	1.67
Propyl benzene	100	1.39
Naphthalene	100	1.52
Dilute Gases	1.0	1000–4000
Liquids		0.1–0.4

Vesovic and Wakeham, 1991.

Diffusivity in dense fluids may be related to viscosity by the Stokes-Einstein equation (Gubbins, 1973) as:

$$D_{12} = kT/c\pi\eta\ (v_b/N_A)^{1/3} \tag{3.12}$$

where k is Boltzmann's constant and c is a parameter having a value between 3 and 6; v_b is the molar volume of the solute at the normal boiling point; η is the viscosity of the solvent; N_A is Avogadro's number.

Fiest and Schneider (1980) established an exponential relation between the binary diffusivity and viscosity of the SCF component, which is given as:

$$D_{12} \approx \eta^{0.66} \tag{3.13}$$

Marrero and Mason (1972) developed a correlation, after compiling experimental diffusivities at low densities, and extended its validity over a range of temperatures by making it a function of temperature as:

$$\ln P\, D_{12}^{o} = \ln A_1 + A_2 \ln T - A_3/T \tag{3.14}$$

where D_{12}^{o} is the diffusivity at low densities; T is in °K; and A_1, A_2, and A_3 are empirical constants. A separate correlation depicting the composition dependence of the binary diffusivity allows for correction of dilute gas diffusion coefficient to any other composition. Takahashi (1974) recommended a correlation based on the corresponding states (CS) method for predicting the binary diffusion coefficient of a dense gas as:

$$(D_{12}P)/(D_{12}^{o}P^{o}) = f\,(T_r, P_r) \tag{3.15}$$

with

$$T_r = T/(x_1 T_{c_1} + x_2 T_{c_2})$$

and

$$P_r = T/(x_1 P_{c_1} + x_2 P_{c_2})$$

where T_{c_1}, P_{c_1}, T_{c_2}, P_{c_2} are the critical temperatures and critical pressures of the components 1 and 2, respectively. P^o is the low pressure at which the value of D_{12}^{o} is available. The function $f(T_r, P_r)$ is as given in Figure 3.4.

Dawson et al. (1970) suggested a correlation for diffusivity in terms of the ratio of the products of diffusivity and density at high pressures to those at low pressures as a function of reduced density. The correlation is valid up to very high pressures.

$$\frac{D\rho}{(D\rho)^{o}} = 1 + 0.053432\rho_r - 0.030182\rho^2{}_r - 0.02975\rho^3{}_r \tag{3.16}$$

where $0.8 < T_r < 1.9$ $0.3 < P_r < 7.4$

The diffusivity from natural systems encountered in SCFE can be calculated from the correlation for the ratio of Schmidt numbers at the system pressure to the ambient pressure, as proposed by Funazukuri et al. (1992):

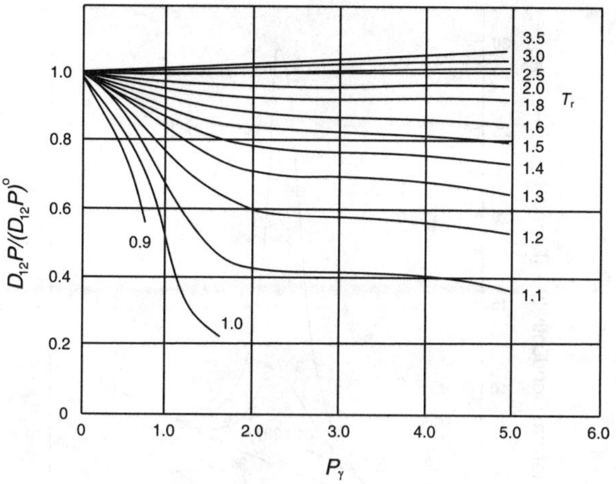

FIGURE 3.4 Takahashi correlation for evaluating binary diffusivity for dense gases by the CS method (Vesovic and Wakeham, 1991).

$$SC/SC^o = 1 + 2.45(M_1/M_2)^{-0.089} F_v^{1.12} \tag{3.17}$$

where $SC = \eta/\rho D_{12}$; M_1 and M_2 are molecular weights of the solvent and the solute molecules. Sc^o is the Schmidt number at the ambient pressure. The other parameters in the equation are

$F_v = x/(x - 1)^2$

$x = v_1/[1.384 (\bar{v}_1)_o]$

$(\bar{v}_1)_o = N_A \sigma_1^3/\sqrt{2}$ m^3/mol

N_A = Avogadro's Number

σ_1 = hard sphere diameter of solvent

v_1 = molar volume of the SCF component (1)

The diffusion coefficient of a component in an SCF solvent may reduce when an entrainer is added to it due to interactions between solute and cosolvent and between solute and solvent.

Diffusivities in CO_2 are also calculated by using the Fuller's Method as given by Reid et al. (1989). The method was tested for twelve nonvolatile components up to a pressure of 250 bar and was found to be satisfactory, though in some regions the deviation was 20%.

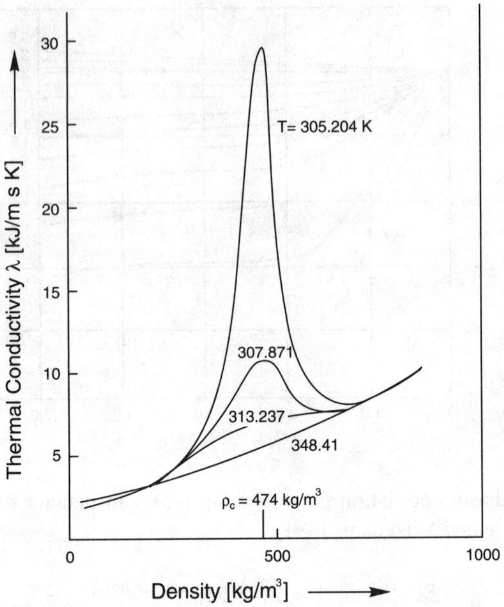

FIGURE 3.5a Thermal conductivity of CO_2 in the vicinity of the critical point (Brunner, 1994).

3.1.3 Thermal Conductivity

Thermal conductivity is defined as the proportionality constant of the linear relationship of heat flux with respect to temperature gradient as:

$$q = -\lambda \frac{dT}{dz} \tag{3.18}$$

where q is the heat flux and dT/dz is the temperature gradient. Thermal conductivity, λ, depends on temperature, pressure, or density of the fluid. In general, thermal conductivity decreases with increasing temperature at the supercritical condition, passes through a minimum value at any pressure, and then increases with increasing temperature and increasing density for most SCF, as represented in Figure 3.5a. Some typical values are given in Table 3.3. The minimum is shifted to a higher temperature with pressure. The influence of pressure is less after a minimum value of λ is observed at each pressure when temperature is increased (Figure 3.5b). Thermal conductivities of CO_2 as a function of temperature at different pressures are presented in Appendix A (Figure A.3). Thermal conductivity of a gas in the supercritical region may be calculated from three contributions, namely, (1) the dilute gas or zero density thermal conductivity, λ^o, (2) the excess thermal conductivity Δλ, and (3) the critical enhancement thermal conductivity, $\Delta\lambda^c$, as:

$$\lambda(\rho, T) = \lambda^o(T) + \Delta\lambda(\rho, T) + \Delta\lambda^c(\rho, T) \tag{3.19}$$

TABLE 3.3
Thermal Conductivity of Some SCF Solvents

	λ (mW/m/K)			
	$T = T_c + 20°C$		$T = T_c + 100°C$	
	P = 1 bar	P = 100 bar	P = 1 bar	P = 100 bar
CO_2	18.8	51.1	25.5	31.9
H_2O	54.0	67.8	63.7	73.0
C_3H_8	29.0	69.9	39.6	54.0

Vesovic and Wakeham, 1991.

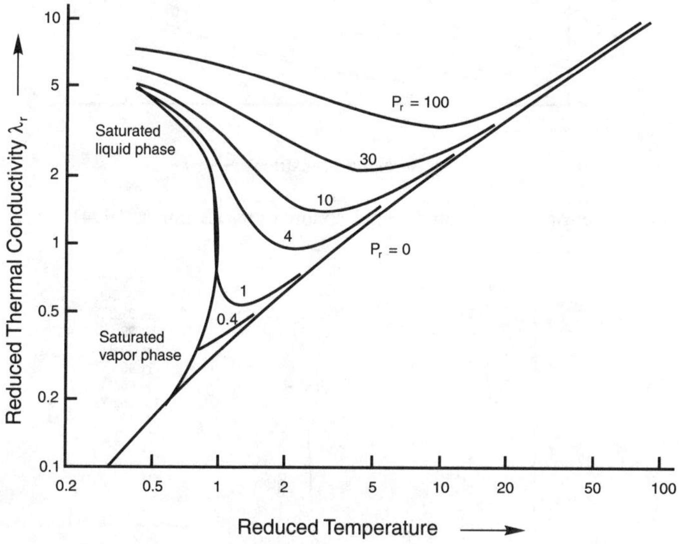

FIGURE 3.5b Reduced thermal conductivity for bi-atomic gases (Brunner, 1994).

The zero density or dilute gas concentration, λ^o, is a function of temperature only, and the polynomial correlation is available elsewhere (Nieto de Castro, 1991), $\Delta\lambda$ corrects for the pressure effect over λ^o and are correlated in terms of temperature and density.

The pressure dependency of thermal conductivity can be calculated within ±20% accuracy using a generalized CS correlation developed by Lenoir et al. (1953), as illustrated in Figure 3.6a. For nonpolar gases, Stiel and Thodos (1964) developed a prediction method based on departures with respect to low density gas in terms of reduced density. Figure 3.6b shows the behavior of the critical enhancement contribution, $\Delta\lambda^c$ with density.

3.1.4 Interfacial Tension

Interfacial tension between two phases decides the stability of the phase boundary and may even break a continuous liquid film into spherical droplets depending on

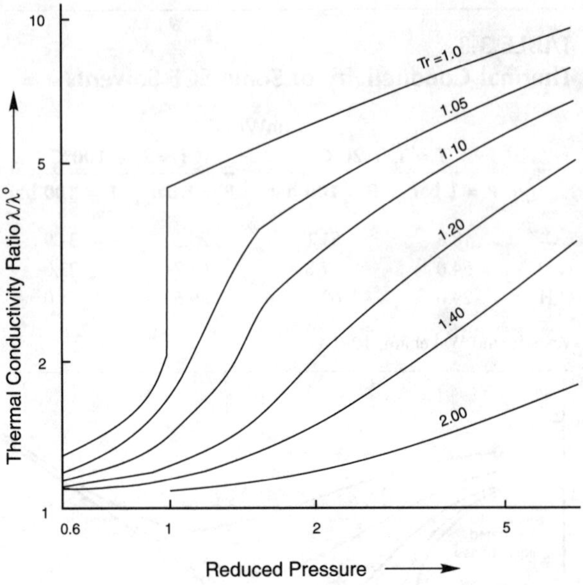

FIGURE 3.6a Effect of pressure on thermal conductivity (Brunner, 1994).

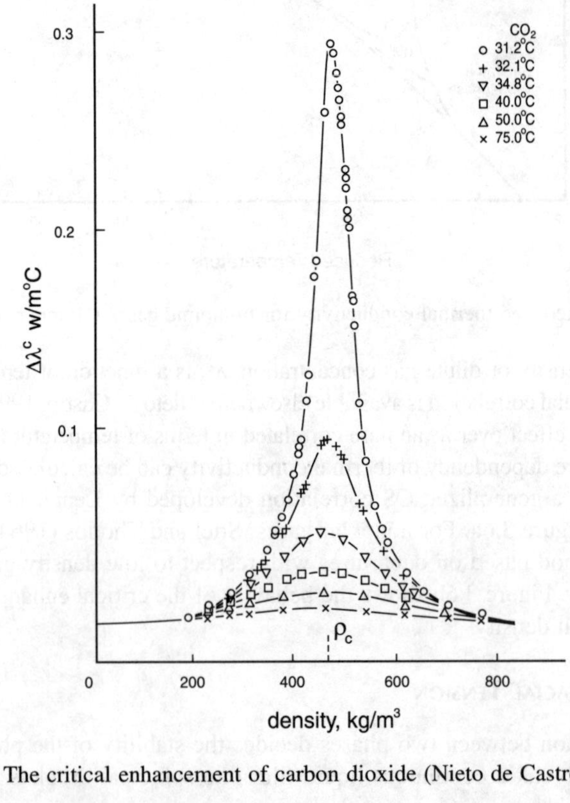

FIGURE 3.6b The critical enhancement of carbon dioxide (Nieto de Castro, 1991).

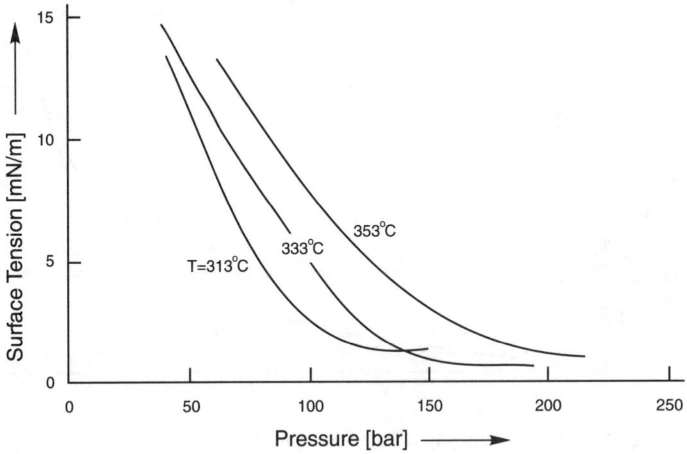

FIGURE 3.7a Surface tension for the system squalane-CO_2 (Brunner, 1994).

the relative magnitudes of strong attraction on the liquid side over the weak attraction on the gaseous side. The surface tension of a pure liquid varies between 20 and 40 dynes/cm (with the exception of water for which σ = 72.8 dynes/cm at 20°C). The interfacial tension can be calculated from the following equation:

$$\sigma^{1/4} = [P]\,(\rho_L - \rho_v) \tag{3.20}$$

where ρ_L and ρ_v are the densities of liquid and vapor phases and [P] is a proportionality constant, called Parachor, which can be determined from regression of experimental data. For mixtures, the following equation may be used:

$$\sigma_m^{\;1/4} = \sum_{i=1}^{n} \{[P_i](\rho_L x_i - \rho_v y_i)\} \tag{3.21}$$

where x_i and y_i are the liquid and vapor phase compositions. Surface tension of a liquid decreases with pressure and increases with temperature, as can be seen in Figure 3.7a. Surface tension of a gas, however, increases with pressures, and near the critical point it was suggested to use $^3/_{11}$ for the exponent instead of $^1/_4$ (Reid et al. 1988). Surface tension approaches zero at the critical point, as the density difference between the phases is zero, as can be seen from Figure 3.7b.

3.2 MASS TRANSFER BEHAVIOR

Process design and simulation of supercritical fluid extraction (SCFE) of natural products requires both qualitative and quantitative understanding of mass transfer kinetics. When the feed is in the form of a solid, the SCFE process is carried out from a fixed bed of solids with a continuous flow of the SCF solvent, since solids are difficult to handle continuously in pressurized vessels. When the feed is in the

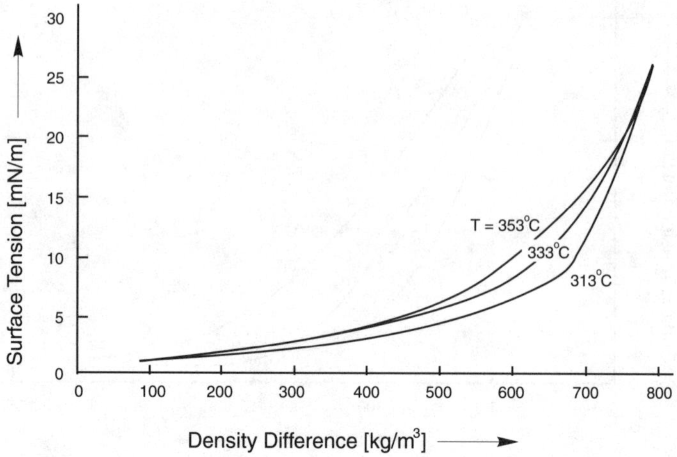

FIGURE 3.7b Surface tension for the system squalane-CO_2 with density difference between coexisting phases (Brunner, 1994).

form of a liquid, the extraction process is carried out continuously by flowing the SCF solvent and the liquid mixture counter-currently in a multistage contacting device, such as a column. When a liquid feed needs to be separated into a number of fractions and the final products are desired in pure form, then chromatographic separation is carried out on preparative scale with an SCF solvent as the mobile phase.

Extraction from a natural material by an SCF solvent may require much less time for the initial 50% extraction, compared to the later part of the extraction, and accordingly the extraction curve invariably ends up in a tail. The mechanism of mass transfer may alter depending on the time of extraction and the nature of the botanical substrates other than the flow characteristics of the SCF solvent.

3.2.1 SCFE from Solid Feed

For extraction from a fixed bed of solid particles loaded in a batch extractor, the SCF solvent is continuously passed through the extractor and evenly, either in downward or in upward direction, with a fixed flow rate of solvent to solid feed ratio. The pressure and temperature of the extractor are maintained constant and these parameters should be decided such that the SCF solvent has sufficient solvent capacity for the solutes to be extracted, or, in other words, the solutes should be sufficiently soluble. The solutes must be rapidly transported from the substrate to the bulk of the SCF solvent or the diffusivity should be sufficiently high for faster completion of extraction. For this purpose the particle size of the solid feed must be sufficiently small in order to provide a large surface area for the solute to transport from the matrix to the bulk of the solvent. The solid particles are depleted of the solutes in the direction of flow. The concentration of the solutes in the flowing SCF solvent, which is also known as loading, at the exit of the bed is at the highest in the beginning of extraction and may remain steady for some time before the concentration starts falling. Also, the flowing solvent may get saturated with the solutes after traversing a short bed height initially, beyond which it does not retain any solvent capacity for further extraction while it traverses the rest of the bed height. This saturation line moves in the direction of flow and reaches the exit of the bed at the end of the constant rate period. The constant rate period depends on the flow rate of the solvent and the characteristics of the solid substrate. However, for certain natural materials containing low extractables, this may not exist at all. Depending on the mechanism of mass transfer, there may be up to three different regimes of extraction, namely (1) constant rate (solubility controlled) regime, (2) falling rate — phase I (diffusion-controlled) regime, and (3) falling rate — phase II (desorption controlled) regime.

The nature of the cellulose matrix, in general, decides the mass transfer characteristics in the dynamic extraction process, where the extract is continuously separated and the regenerated SCF solvent is continuously recycled to the same bed of solids. In addition to temperature, pressure, and cosolvent concentration of the SCF solvent, which decide the solubility of the extractables in the SCF solvent, the other process variables are flow rate, flow mode, particle size, porosity, void fraction, and extractor configurations, such as, height and diameter.

Sometimes an entrainer (e.g., water or alcohol) is directly added to the solid matrix prior to the extraction for pretreatment of the solid substrate with the purpose of its morphological modification (i.e., dilation of pores) or surface modification (i.e., reduction of its affinity for the solutes). There are other methods of pretreatment of solids, such as sudden expansion of the solids, "cryogrinding" or grinding at low temperature, and controlled-humidity drying at ambient condition. These pretreatment methods allow the solid substrate to reduce the hindrance to the release of the solutes. As a result the behavior of extraction yield with pressure tends to be similar to that of isothermal neat solubility of the extract with pressure. The ratio of the neat solubility to the maximum loading of the extractables in the SCF solvent in the presence of the solid matrix is called the "hindrance factor" (HF). HF is made to reduce with pretreatment (Sastry and Mukhopadhyay, 1994). It also depends on temperature, pressure, flow rate, and particle size, shape, and distribution. The success of the extraction process depends on the reduction of this lumped parameter, implying an improvement of desorption from the solid surface and reduction of resistance to diffusion through pores.

3.2.2 Mechanism of Transport from Solids

When a fixed bed of solid is contacted with flowing CO_2 at a selected supercritical condition, the mass transport mechanism involves diffusion and adsorption of SCF solvent followed by solute desorption, diffusion through pores, and the convective transport along with the flowing SCF solvent across the bed height. But the crucial factor is the initial distribution of the extractable substance within the solid substrate which may exist in the adsorbed state either on the outer surface or on the surface of the pores, or may exist in the dissolved state in the cytoplasm or the vacuoles within the plant cells (Brunner, 1994). The resistance to mass transfer across the plant cell membranes is immaterial if the natural material is crushed and cell structures are broken.

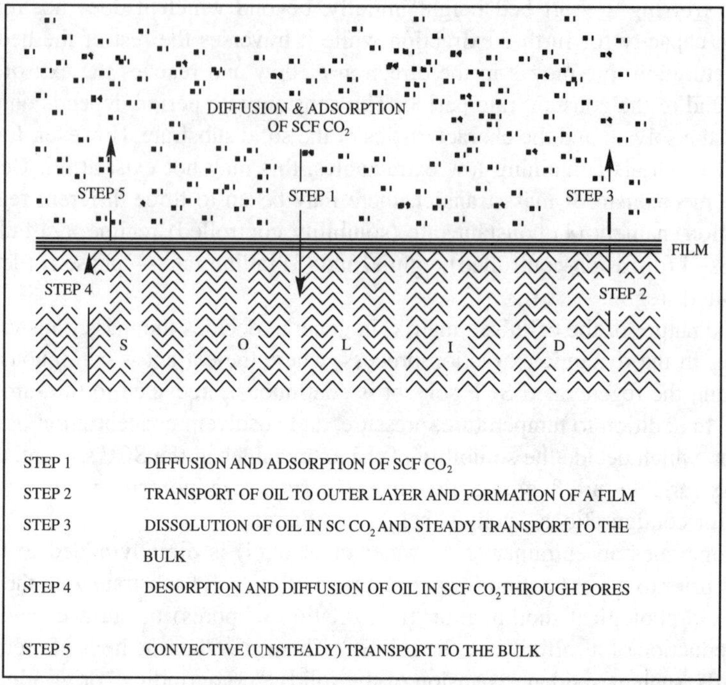

FIGURE 3.8 Schematic description of the transport mechanisms of SCFE from solids.

As schematically described in Figure 3.8, the extraction process entails the following sequential and parallel steps at steady mode of extraction at the beginning of extraction:

1. Diffusion of CO_2 into the pores and adsorption of CO_2 on the solid surface
2. Transport of oil to the outer layer and formation of a thin liquid film around the solid particles
3. Dissolution of oil in SC CO_2
4. Convective transport of the solute to the bulk of the fluid

Subsequently, at the unsteady mode of extraction, the SCFE process entails:

5. Desorption of solute from the solid or pore surface followed by
6. Dissolution of the solute in SC CO_2
7. Diffusion of the solute in the pores
8. Convective unsteady state transport of the solute to the bulk of the fluid

From Figure 3.8, it can be visualized that step 4 is the slowest one when there is a liquid film of oil present on the outer surface of the solid particle and the mass transfer takes place at a constant rate depending on the solubility of oil in the solvent, and the mass transfer rate is controlled by the external film resistance. On the other hand, if there is no liquid film present at all on the outer layer of the solid particle, then there is no initial constant rate period. Then the rate of mass transfer takes place at the first falling rate regime and is controlled by internal diffusional resistance, i.e., by step 7. If step 5 controls the mass transfer rate, then mass transfer takes place at the second falling rate. However, there is no such sharp demarcation observed between the two unsteady modes of extraction, as there is between the steady and the unsteady modes.

3.2.3 Stages of Extraction for Different Natural Materials

Supercritical CO_2 extraction from natural products, such as rosemary, basil, and marjoram leaves (Reverchon et al., 1993) clearly revealed that the extraction of cuticular waxes in the initial period of extraction occurred according to the constant-rate mass transfer regime, while the subsequent extraction of essential oil at a later period exhibited the falling-rate mass transfer regime. This variation in the extraction behavior of two types of products from the same botanical source is attributed to the fact that waxes are located at the surface of the leaves and accordingly their extraction is not hindered by internal diffusion within the solid substrate, resulting in a constant-rate mass transfer behavior. On the contrary, the essential oils are uniformly distributed in the plant cells of the leaves and their extraction is hindered by the intraparticle resistance due to diffusion within the solid substrate resulting in a falling-rate regime. This was further substantiated by the increase in diffusional resistance with an increase in the particle size, as reported in the case of supercritical CO_2 extraction of basil leaves (Reverchon et al., 1993) at 100 bar and 40°C with a fixed solvent flow rate of 1.2 kg/h.

For supercritical CO_2 extraction of evening primrose oil from its seeds, which contains 24.3 wt% of extractable oil, it was observed that up to 95% recovery of oil, a constant-rate mass transfer behavior was observed from a bed of crushed particles having sizes less than 0.355 mm and no falling-rate regime was observed, implying that most of the oil was available as a liquid film on the outer surface of the particle, and that no diffusional resistance (Favati et al., 1991) hindered the mass transfer of oil. However, Lee et al. (1994) observed that the loading of SC CO_2 at the end of extractor remained substantially constant, up to about 50% recovery of evening primrose oil at 300 bar and 50°C for a larger particle size of 0.635 mm, and beyond 50% recovery, there was a sharp decline in the loadings. This is due to a progressive increase in the diffusional path by which the oil reaches the outer surface of the seed particles. For a larger diameter particle, there is an increase in the length of the diffusional path and consequently the falling-rate mass transfer regime sets in earlier than for a smaller size particle.

In some cases, two falling-rate regimes are observed, as demonstrated by Peker et al. (1992) for the extraction of caffeine from water-soaked coffee beans by supercritical CO_2 at 138 bar and 64°C with a flow rate of CO_2 of 1.51 g/min. The two falling-rate regimes are attributed to the differences in the resistances due to affinity of the solute to the substrate and transport within the solid. However, neat solubility of caffeine in SC CO_2 at 138 bar and 64°C is very low and accordingly the commercial process of decaffeination is invariably carried out at much higher pressures of 300 bar or above. The loading of caffeine in SC CO_2 in the decaffeination process is always much less than the neat solubility of caffeine at the same condition due to high affinity of caffeine to the solid substrate and very low (1 to 5 wt%) caffeine content in the beans. As a result, the solvent requirement for decaffeination is much higher than that given by the solubility (McHugh and Krukonis, 1994).

For supercritical CO_2 extraction of clove oil from a bed of ground clove buds containing 20.9% oil and having an average particle size of 0.297 mm (Mukhopadhyay and Rajeev, 1998), it was observed that at 40°C and 200 bar, 85% of the clove oil could be recovered at a steady state within 45 min of extraction, requiring only 50% of CO_2 consumed for the complete extraction, and the complete extraction required 180 min. At up to 85% of the extraction, the mass transfer rate was found to be strongly influenced by the flow rate, indicating that the mass transfer is controlled by the solubility-limited driving force and external film resistance. It is also observed that the intraparticle diffusional resistance can be made negligible by reducing the particle size to less than 0.2 to 0.3 mm.

The transition from a constant-rate regime to a falling-rate regime also depends on the initial oil content in the solid substrate as well as the cell structure of the plant material. For example, for extraction of floral fragrance from unground jasmine flowers containing 0.44% of essential oil using supercritical CO_2 at 120 bar and 40°C, it was observed (Sastry, 1994) that the entire mass transfer regime was controlled by diffusional resistance. On the other hand, for a very high initial oil content, as in the case of finely ground evening primrose seeds, the mass transfer kinetics was found to be predominantly at a constant rate for which extraction was controlled by the solubility-limited driving force. The nature of mass transfer kinetics thus depends on how much, where, and how the solute is bound to the cellulosic matrix, the mechanism of its release from substrate, and the nature of solute transport within the solid matrix, besides other parameters like pressure, temperature, flow rate, particle size, etc.

3.2.4 SCFF of Liquid Feed

For continuous supercritical fluid extraction and fractionation of a multicomponent liquid feed, the mass transfer resistances associated with the morphology of the external surface and internal structure of the feed material are absent. Among the three modes of contact between the liquid feed and the supercritical fluid solvent, namely, by counter-current, cross-current, and cocurrent schemes, it is the counter-current multistage contacting device for supercritical fluid fractionation (SCFF) which has been mostly practiced (Siebert and Moosberg, 1988; Lahiere and Fair, 1987; Brunner, 1994; de Haans et al., 1990; Brunner et al., 1991). The continuous counter-current fractionation at steady state is the most efficient operation as it reduces the solvent requirement, increases the throughput, and facilitates high extraction recovery and high extract purity. However, the batchwise operation of counter-current SCFF is also possible, as in the case of rectification or distillation with reflux.

For multistage counter-current separation with an SCF solvent, the components distribute between the SCF solvent (extract) phase and the liquid (raffinate) phase which are contacted countercurrently in a separation column, as schematically represented in Figure 3.9. The extract phase is taken out of the top of the column (1)

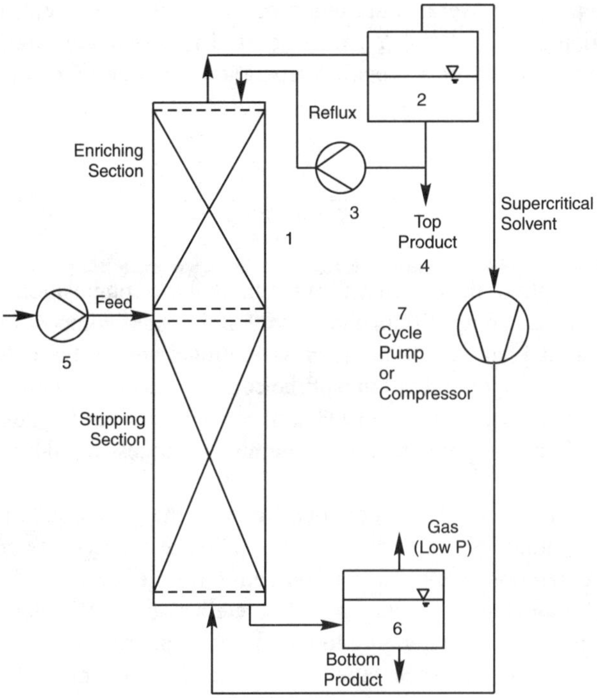

FIGURE 3.9 Process scheme of countercurrent supercritical fluid fractionation of liquid mixtures (Brunner, 1994).

and depressurized in a separation drum (2) as the extract is separated from the solvent. A part of the extract is recycled (3) to the top of the column as reflux while the rest is the top product (4). The solvent liberated from the separator is conditioned and recycled (7) along with the make-up solvent to the bottom of the column. The feed (5) is introduced to the column somewhere near the middle of the column. The raffinate phase is taken out from the bottom of the column and depressurized in a separator drum (6) to collect the bottom product. If the bottom product is also desired at high purity, a part of the raffinate may be returned to the column. In order to fractionate a multicomponent liquid feed having C number of components, it is necessary to employ a cascade of (C – 1) columns.

The continuous countercurrent supercritical fluid fractionation is a separation process which is performed close to the phase equilibrium condition due to low flow velocities, low viscosity, and high diffusivity, irrespective of whether a plate column or a packed column or a spray column is employed as the contacting device. Phase equilibrium calculations using an equation-of-state method are utilized first, most importantly to characterize the two-phase region, the distribution of the components in the SCF and the liquid phases, and their separation factors at various compositions from the top to the bottom of the column. In packed columns, the compositions in both SCF and liquid phases undergo continuous change, whereas in plate columns, the compositions change in steps.

The concept of the overall fluid-phase mass transfer coefficient, K_{OF}, based on the overall driving force, $(Y^* - Y)$, may be used for computing the height of the packed column required for a specified separation using an SCF solvent (McCabe et al., 1989) as:

$$Z = \frac{G}{AK_{OF}a} \int_{bottom}^{top} \frac{dY}{Y^* - Y} \tag{3.22}$$

where Z is the height of the column, Y^* is the fluid phase composition in equilibrium with the liquid phase composition, and Y is the fluid phase composition; G is molar flow rate of the SCF solvent, A is the cross-sectional area of the column, and a is the mass transfer area per unit volume of the contactor. K_{OF} is usually obtained from experimental data. Bhaskar et al. (1993) utilized this concept for designing a continuous supercritical CO_2 fractionation column for processing anhydrous milk fat (AMF).

For plate columns, the concept of equilibrium or theoretical stages is considered for designing continuous supercritical fractionation at steady-state conditions and subsequently corrections are made for non-equilibrium effects.

An equilibrium stage is defined as a discrete section of the plate/tray column, such that the flow streams belonging to two different phases leaving the section are in phase equilibrium with each other, while the flow streams belonging to two different phases entering the section are not in equilibrium. Enrichment or depletion achievable in one discrete equilibrium stage depends on the flow rates and concentrations of the entering streams and the equilibrium compositions of the leaving streams, in other words, on the mass balance and equilibrium relations of the components. For example, for a particular nth equilibrium stage (Guarise et al., 1994) the component (i) balance gives

$$L_{n-1}\, x_{n-1,i} - L_n x_{n,i} + V_{n+1} K_{n+1,i}\, x_{n+1,i} - V_n K_{n,i} x_{n,i} = 0 \text{ (C – 1) equations} \tag{3.23}$$

and the overall mass balance gives

$$L_{n-1} - L_n + V_{n+1} - V_n = 0 \tag{3.24}$$

with the equilibrium relation for the last component (c):

$$1 - \sum_{C-1} K_{n,i} x_{n,i} - K_{n,c}\left(1 - \sum_{C-1} x_{n,i}\right) = 0 \tag{3.25}$$

where C is the number of components and the stages are numbered from the top.

In the absence of a rigorous multicomponent equilibrium model or data, the multicomponent mixtures are simplified to be equivalent to binary or ternary systems and then the separation calculations are simplified. The method, like the

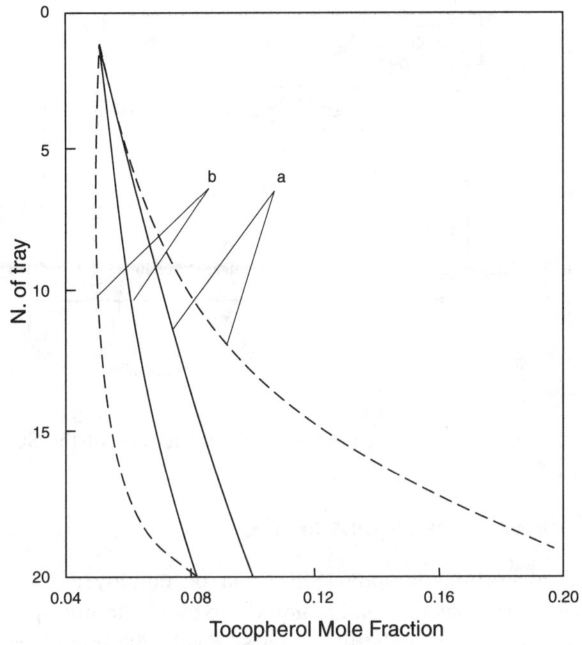

FIGURE 3.10 Simulated solvent-free liquid composition profiles in a continuous supercritical extraction column for the system α-tocopherol + oleic acid + CO_2: discrete equilibrium stage model (dashed lines) and continuous mass transfer model (solid lines) for (a) V = 640 mol/hr, and (b) V = 468 mol/h (Guarise et al., 1994).

McCabe-Thiele method or Ponchon Savarit method, may be utilized for estimating the number of stages required for the desired separation. Two key components are selected to represent a multicomponent mixture by a binary mixture on a solvent-free basis. If the volumetric flow rates of the countercurrently flowing phases are not constant due to changes in concentration and solubility along the height, then the presence of solvent is also considered by treating the multicomponent mixture as a ternary system. If an entrainer is added to the SCF solvent, the distribution coefficient may change with concentration along the height, then the multicomponent system is treated as a quarternary system comprising the two key components, the solvent and the cosolvent (Brunner, 1994). The difference between the approximated solvent-free liquid mole fraction profiles in the discrete equilibrium stage concept and the continuous mass transfer concept is illustrated in Figure 3.10 for continuous supercritical CO_2 fractionation of the system α-tocopherol-oleic acid at two different flow rates. The fluxes across the fluid–liquid interface and the stage-wise composition profiles were calculated and reported by Bartolomeo et al. (1990). For continuous supercritical CO_2 fractionation of deodorizer condensates for separation of tocopherol and fatty acid methyl esters (FAME), Brunner et al. (1991) reported detailed methods of analysis of the problems and compared the results obtained by different methods elsewhere (Brunner, 1998; Brunner, 1994).

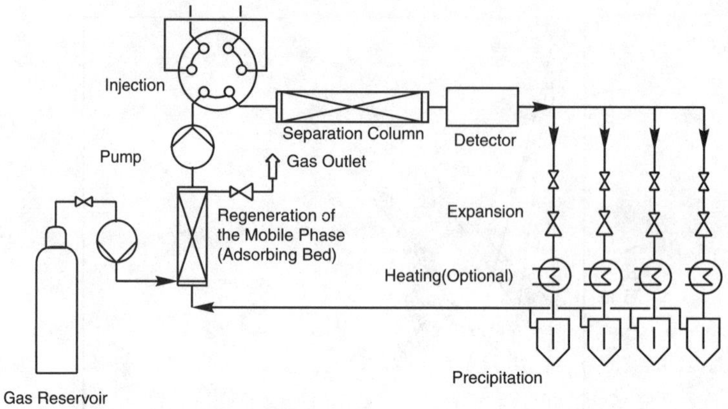

FIGURE 3.11 Flow scheme of a preparative SFC with recycle of the SCF solvent mobile phase (Brunner, 1994).

3.2.5 Fractionation of Liquids by SFC

Supercritical fluid chromatography (SFC) can be employed for preparative and production scale operations to separate liquid mixtures into high-purity components in order of their polarity and volatility in an SCF solvent which is characteristically used as the mobile phase. SFC is a relatively new separation technique with a high potential for isolation and purification of natural molecules to 95 to 99% purity which find uses as high value pharmaceuticals or drug intermediates. It operates in an intermediate region between those of gas chromatography (GC) and liquid chromatography (LC), as the density and solvent power of an SCF solvent are comparable to those of liquids, and the transport coefficients of an SCF solvent are closer to those of a gas. The SFC utilizes the high density and high diffusivity of the SCF solvent. SFC has the advantage of selectively dissolving the compounds of interest and separating them through a stationary phase, after which they can be easily separated from the mobile phase. The conditions of pressure, temperature, and cosolvent concentrations can be suitably selected for large differences in the retention times of several compounds which cannot be separated by a GC column. The addition of cosolvent not only allows the operation of the SFC at a lower pressure (than without it), but it also gives long retention times and broad peaks. This happens because some of the polar sites of the stationary phase get occupied by the polar cosolvent, thereby allowing the polar components of the mixture to compete with each other. They are thus made to separate without merging with each other. Pressure and density programming to enable elute and separate larger molecules are the additional features of SFC which are not possible with GC. However, for SFC operation on a preparative scale, the pressure and density of the SCF solvent (mobile phase) are kept constant because the feed mixtures are injected several times and the elution of the components takes place continuously. A schematic description of the preparative SFC with recycle is shown in Figure 3.11 (Brunner, 1994).

The preparative SFC can be carried out either batchwise for which feed is injected intermittently or in continuous operation for which the feed mixture is continuously fed counter-current to the stationary phase. The components are transported through the column in order of their strength of interaction with the stationary phase. The component with the weakest interaction is eluted first. In a counter-current continuous chromatography column, either the column is rotated in order to achieve a moving stationary phase but fixed points of feed injection and removal of separated components, or the SFC column is kept stationary with moving injection and removal points which are operated by means of a switching valve (Brunner, 1994).

SFC columns may be open tubular or packed. While the packed columns are advantageous due to higher loadability, wider choice of selectivities, higher efficiency of separation, and easier flow controls, yet the open tubular columns have the advantage of allowing a higher number of stages for the same pressure drop. For chromatographic separations of polar components, the packed columns have an additional advantage of effective utilization of a cosolvent for the benefit of selective separation of chemically similar components. Berger and Perrut (1990) presented a detailed review of both large-scale and small-scale preparative SFC.

The throughput may vary from a few grams to a few kilograms of the substance with the diameter of the SFC column increasing from 0.01 to 0.1 m. The particle size of the packings for larger columns is about 0.05 mm, i.e., one order magnitude higher than that for analytical SFC.

When the feed mixture is injected at a low concentration in the mobile phase (SCF solvent), the components get distributed between the stationary phase and the mobile phase linearly, that is, the concentration in the two phases are directly proportional as given by the linear adsorption isotherm and the ratio is known as the partition coefficient, K'. However, at higher concentrations, the stationary phase may get saturated for some of the components and can not adsorb more, consequently the SCF solvent phase may get saturated limited by its solvent capacity for the components. As a result, the adsorption isotherm becomes nonlinear and the peaks of the chromatograph appear at lesser retention times. The other important process parameter affecting SFC is the mobile phase flow rate as the retention time is inversely proportional and the throughput is directly proportional to the mobile phase flow rate. The pressure drop is important in SFC if the separation is carried out in the vicinity of the critical point, where density is very sensitive to the change of pressure. The retention time increases with an increase in temperature in the vicinity of the critical point, as density decreases with increasing temperature and solvent power of the mobile phase decreases. At higher temperatures the rentention times again become shorter. A longer column is not always better, since both pressure drop and retention times are enhanced.

Most preparative separation columns for SFC are packed with silica gel (both unmodified and modified with C_8- or C_{18}-ODS) in the range of 20 to 100 μm particles. The length of the column varies between 0.6 to 0.9 m and is selected based on pressure drop considerations, which is of the order of 10 to 40 bar for a length of 0.25 m (Brunner, 1994). Recently chiral stationary phases (CSP) derived from small and stable molecules are being used for separation of stereoisomers on preparative scale for higher efficiency, durability, sample capacity, and reproducibility of the SFC columns. The small chiral selectors may be incorporated into linear polysiloxane which is then coated and bonded to silica support. The robust nature of CSP allows the usage of higher flow rate and increase in pressure or temperature to increase the throughput. A wide variety of additives may be used in the mobile phase without damaging the column (Pirkle et al., 1996).

Mostly SCFE of natural materials and preparative SFC are combined to get the best results. Extensive investigations have been carried out to fractionate supercritically extracted products like lemon peel oil for deterpenation, wheat germ oil for recovering tocopherols, fish oils for separation of methyl esters of unsaturated fatty acids, prostaglandin isomers for their isolation, etc. (Brunner, 1994; Alkio et al., 1988).

For example, Bruns et al. (1988) investigated preparative SFC of tocopherol isomers using a 0.4 m long and 0.1 m diameter silica column from an enriched tocopherol fraction containing about 55 wt% tocopherols. The distribution of the isomers of tocopherols were 5 to 6 wt% α-, negligible β-, 35 to 44 wt% of γ-, and 12 to 14 wt% of δ-tocopherols, apart from 10 to 12 wt% squalance, 1 to 3 wt% sterols and unidentified components. These components were eluted in an SFC column among the different tocopherol isomers to produce pure tocopherol isomers from such a fraction. The partition coefficient and selectivity of separation with methanol as the cosolvent, as shown in Figure 3.12, reveals that both partition coefficient and selectivity of separation decrease with an increase in the concentration of methanol (cosolvent) in the SCF solvent (CO_2) with maximum selectivity at 0.5 wt% methanol. It also decreases with density as shown in Figure 3.13 for hydrocarbon mixtures. Pure α-tocopherol could be separated from a 660 g of tocopherol per liter of feed mixture when SFC column was operated at an average pressure of 230 bar at 40°C. Bruns et al (1988) obtained a throughput of 15 g of tocopherol in 1 h using a 0.1 m diameter column from 24 g of the tocopherol mixture processed in 1 h.

Prostaglandins are unsaturated hydroxy (keto)-carboxylic acids with 20 carbon atoms containing a ring with five carbon atoms. The active substance is the α-cis isomer which is separated from other isomers, such as, B isomer, β-cis isomer, α-trans isomer, and β-trans isomer. Prostaglandins are dissolved in ethyl acetate for injection to preparative SFC column with SC CO_2 mixed with a polar cosolvent as the mobile phase. Using a silica gel (Li Chrosorb Si 60) column of 25 cm length and 1 cm diameter and particle diameter of 7 μm, α-cis isomer with a purity of 80% could be isolated in the first fraction with SC CO_2 mixed with 10% methanol at 200 bar and 19°C from a feed mixture containing 18% α-cis isomer, 61% B-isomer, 17% β-cis isomer and the rest trans isomers (Brunner, 1994).

A fully automated SFC pilot plant equipped with a 30 l column with internal diameter of 0.3 m is in operation in Germany, used for isolation of high-purity EPA ethyl ester (>90%) from esterified fish oil. The largest SFC plant has an annual production capacity of 30 MTPA of high purity EPA and has a 400 l column with 0.54 m internal diameter and is being operated at 200 bar. It is reported that the cost of such a plant is one third of the traditional separation plant based on preparative HPLC technology (Lembke, 1996).

3.3 MASS TRANSFER MODELING FOR SCFE FROM SOLIDS

As mentioned earlier, there are predominantly two types of mass transfer kinetics related to the extraction from naturally occurring solid substrate using supercritical

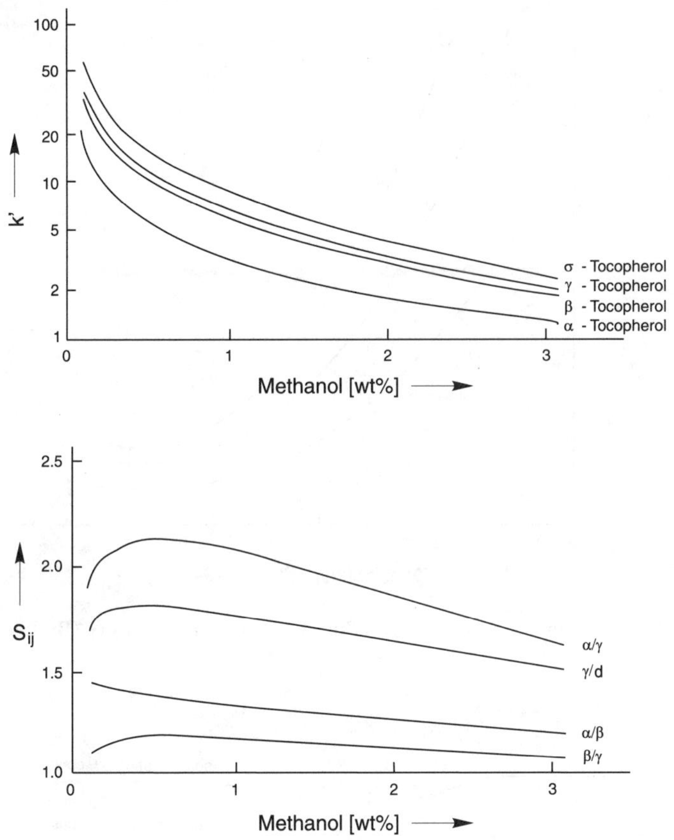

FIGURE 3.12 Separation of tocopherols with SFC at analytical conditions with mobile phase: CO_2 + methanol at average pressure of 150 bar; column, 125 × 4 mm; 5 μm Li Chrosorb Si 60 (Brunner, 1994).

CO_2. For a natural material with a high initial content of extractables such as an oil seed, the rate of extraction from a fixed bed of solids at the initial period of extraction remains constant at a constant process condition after the lapse of the residence time of the flowing solvent. However, after a certain time of extraction, the loading of the extract in the flowing SCF solvent stream declines, indicating a falling rate of extraction. On the other hand, for a natural material containing relatively much less amount of extractables, such as coffee beans, the loading of the extract in the SCF solvent from the extractor starts declining from the beginning of the extraction process, even though the operating conditions are maintained constant. These two types of extraction rate curves are illustrated in Figure 3.14.

The declining of loading in the SCF solvent may be due to two reasons, namely, (1) the outer surface of the solid is depleted after a certain time of extraction and/or (2) the extractables are distributed within the solid particle requiring more time to reach the fluid–solid interface. However, the saturation or attainment of maximum loading of the SCF solvent at the exit of the extractor, depends on whether the length

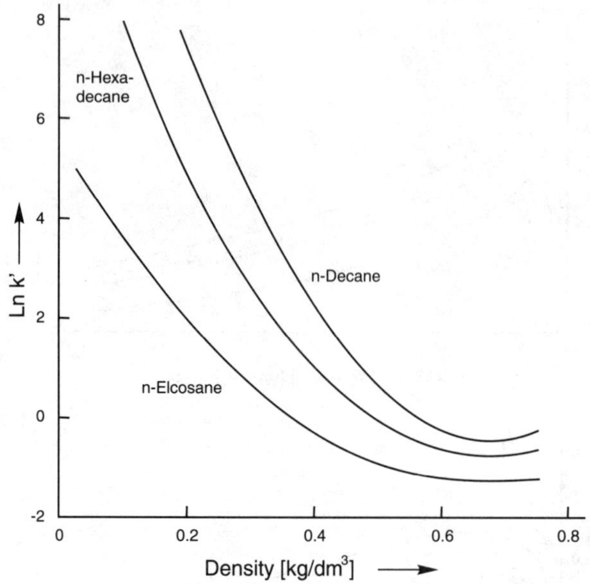

FIGURE 3.13 Partition coefficients of paraffins as a function of density (Brunner, 1994).

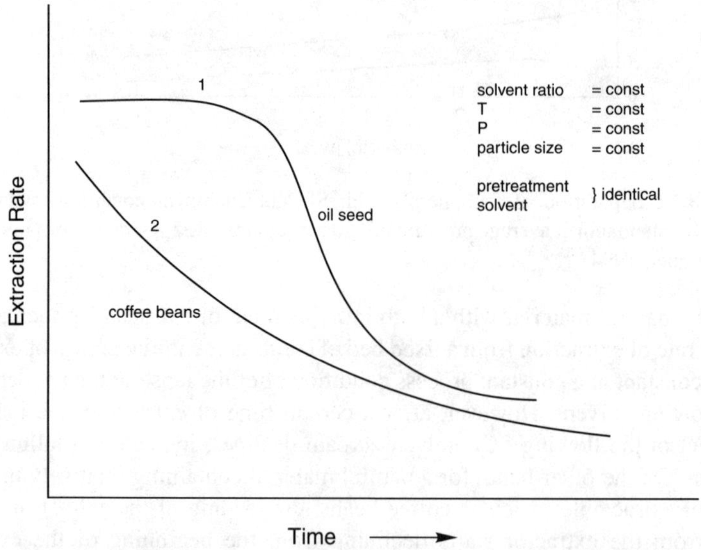

FIGURE 3.14 Two types of extraction rate curves: for natural materials with high and low initial content of extractables.

of the bed of the solids in the extractor is sufficiently long. In other words, a constant value of loading of solutes in the SCF solvent at the exit of the extractor due to the constant rate of extraction does not necessarily correspond to the neat solubility nor the maximum loading possible at the equilibrium condition. The two types of extraction behavior are observed irrespective of the length of the bed. In addition, a rising trend of loading of the SCF solvent is observed at the starting point from almost zero concentration to the maximum loading at any condition up to a time equal or close to the residence time of the solvent. Though this period is almost negligible in most cases, it may continue for a while for very low solvent flow rates.

3.3.1 Process Parameters

The thermodynamic state of the SCF solvent influences the extraction rate in a manner very much similar to its effects on the solubility. This is particularly true if the mass transfer rate is controlled by the external fluid film resistance, in which case the difference in the solubility and loading of the solvent at any instant of time is the driving force for the steady rate of extraction. In general, the time required for extraction decreases with an increase in pressure and with a decrease in temperature at relatively lower pressures. The rate of extraction may however increase with temperature at higher pressures. Extraction rate also increases with density at a constant temperature, but different observations may be expected at the same density at different temperatures. The rate of extraction is enhanced by the addition of a cosolvent in small amounts to the SCF solvent and to the solid substrate (Mukhopadhyay and Sastry, 1994). After the thermodynamic state parameters, the next most significant process parameter is the solvent to the solid feed ratio. A high solvent to solid ratio increases the rate of extraction and reduces the extraction time, but the loading of the SCF solvent is lower due to a lower residence time of the solvent. This implies less capacity utilization of the solvent. It also increases the throughput requirement of the pump and increases the cost of the plant, but decreases the unit cost of production due to a faster rate or higher recovery and larger capacity utilization to process the solid feed. The optimum solvent to solid feed ratio is however decided by many other factors in addition to economic considerations. For example, there may be restrictions on the shelf life of the raw material and the quality of the finished product. As a consequence it may be required to reduce the time of extraction and thus to select a high solvent to solid feed ratio.

The size of the solid particles is a crucial factor in deciding the nature of extraction kinetics, particularly for solids containing less solutes. In general, extraction rate as well as yield increase with decreasing particle size, by reducing the path of transport within the solid and by breaking larger number of oil sacs and vacuoles to release more extractables. However, the smaller particle size may result in a higher pressure drop and uneven distribution of particles, rendering a lowering of mass transfer rate.

3.3.2 Mass Transfer Coefficients

In order to design a commercial plant, it is necessary to quantitatively evaluate the parametric and mass transfer characteristics of SCFE from natural products for its optimal performance. It has already been realized that extraction behavior from natural materials is immensely complicated since they do not have uniform physical and chemical characteristics within the solid particles which are of a cellular nature.

Further the partial destruction of cell structures occurs in the process of crushing and other pretreatment methods. It is futile to consider any rigorous mathematical model for accurate representation of mass transfer phenomena around and within the solid material. Consequently, a simplified model based on the linear driving force approach is considered reasonable.

An unsteady-state, packed-bed mass transfer model based on mass conservation concept (Gangadhara Rao, 1990) can be utilized to represent the concentration profile of the SCF solvent phase in the extractor with respect to time and length of the bed for continuous supercritical CO_2 extraction from a fixed bed of solids. All constituents in the extract are usually clubbed together as the solute, as it is believed that they would have similar mass transfer characteristics. After the solute-free SCF solvent CO_2 enters the bed of ground solid particles, its loading of the solute increases with time, t, and length of the bed, h, in the direction of flow. Assuming plug flow and neglecting axial dispersion of the solutes, the accumulation of the extract in the SCF phase over a differential length, δ_h, is given as:

$$\frac{\partial}{\partial t}(\varepsilon \rho y \delta h) = -\rho u \partial y + A_p K(y^* - y) \tag{3.26}$$

where ε is the void fraction, ρ is the density of the SCF phase, y is the mass fraction of the solute or loading in the SCF phase, u is the superficial velocity, y^* is the equilibrium mass fraction or loading at equilibrium with residual solute in the solid bed at any instant of time and height. A_pK is the overall effective volumetric mass transfer coefficient which lumps all probable modes (series or parallel) of mass transfer from the fluid–solid interface to the bulk of the fluid. Assuming that the porosity, density, and superficial velocity of the SCF solvent do not significantly change across the length of the bed, in view of the slow rate of mass transfer, Equation 3.26 gets simplified to:

$$\varepsilon\rho(\partial y/\partial t) = -\rho u\,(\partial y/\partial h) + A_p K\,(y^* - y) \tag{3.27}$$

A similar equation can be written for the depletion of solute in the solid bed as:

$$-(1-\varepsilon)\,\rho_s(\partial x/\partial t) = A_p K\,(y^* - y) \tag{3.28}$$

where x is the residual oil content of the solid at any instant of time and height. The boundary conditions are

$$t = 0 \qquad 0 \le h \le H \qquad x = x_o \tag{3.29}$$

$$t > 0 \qquad h = 0 \qquad y = 0 \tag{3.30}$$

where H is the total length of the bed of ground solids and x_o is the initial oil content of the solids. Equations 3.27 and 3.28 can be solved provided y^* is known as a function of x, which gets depleted with time (Gangadhara Rao and Mukhopadhyay, 1988).

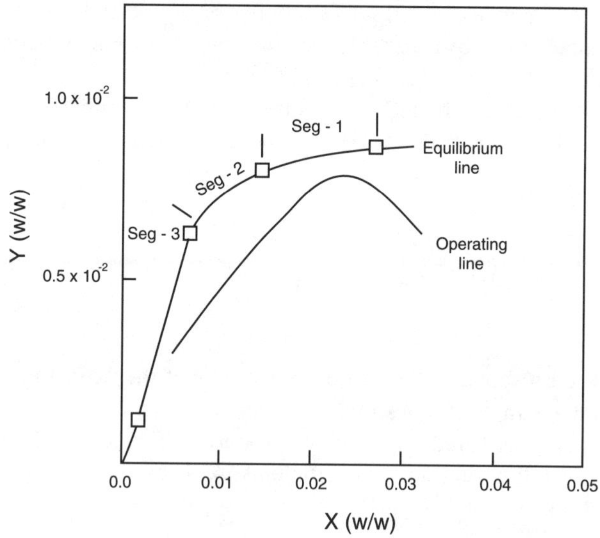

FIGURE 3.15 Comparison of equilibrium and operating lines at 40°C and 120 bar for cumin oil extraction.

Accordingly, it is necessary to regress y^* vs. x data which are obtained by the static equilibrium method (Gangadhara Rao, 1990). In this method, the solid bed is left for equilibration for a period of 2 h before allowing the extract-loaded SCF phase from the bottom of the equilibrium cell to expand to the atmospheric pressure for a while at a very low flow rate. The pressure of the equilibrium cell is maintained constant by allowing fresh supercritical CO_2 solvent to enter the cell from the top without disturbing the equilibrium. The extract collected is quantified, and depressurized CO_2 is measured by a calibrated wet test meter. The residual solute content in the solids, x, can be calculated from the knowledge of the net amount of extract collected and the total (initial) amount of extractable solute from solid, x_o. The value of the corresponding y^* is obtained from the extract collected and the corresponding amount of depressurized CO_2 as per the definitions:

$$y^* = \frac{\text{Amount of extract in the SCF Phase at equilibrium}}{\text{Amount of extract} + \text{amount of SCF } CO_2} \quad \text{and}$$

$$x = \frac{\text{Amount of residual extract in the solid mass}}{\text{Amount of solid mass}}$$

The experimental (y^* vs. x) data over the entire region of extraction is divided into three segments as shown in Figure 3.15 and each segment is fitted to a third degree polynomial as:

$$y^* = c_1 + c_2x + c_3x^2 + c_4x^3 \tag{3.31}$$

and the empirical constants, c_1, c_2, c_3, and c_4 are regressed from the experimental data.

The solid bed may be divided into a number of segments for solving these two equations simultaneously by a suitable method, such as by the numerical method of lines. The mass fraction of solutes in the SCF phase, y(t,h), and the solid bed, x(t,h), at various points of length, h, and time, t, are determined so that the time averaged values at the exit of the bed, y(t,H), can be calculated at different intervals of extraction as:

$$y_{avg}(H) = \int_{t_1}^{t_2} y(t, H)dT \tag{3.32}$$

where t_1 and t_2 correspond to the subsequent values of the extraction time between which the extract sample in question is collected. These calculated time-averaged values of y(H) are compared with the corresponding experimental values in order to regress the overall mass transfer coefficients, A_pK, by minimizing the following objective function:

$$F_H = \sum_{NP} [|y_{\exp}(H) - y_{cal}(H)| / y_{\exp}(H)] \tag{3.33}$$

The overall mass transfer coefficients evaluated for supercritical CO_2 extraction from cumin seeds are found to increase with pressure at isothermal conditions and decrease with temperature at isobaric conditions (Gangadhara Rao, 1990). It was found to increase with superficial velocity of the SCF solvent.

It is not surprising that the unsteady-state, packed-bed model with a lumped overall mass transfer coefficient for the entire regime of extraction is able to provide good representation of the extraction profiles. This is because the loading of the solute in the SCF phase at the gas–solid interface, i.e., y^* is considered to be dependent on the residual solute content of the solid at any instant of time and height, thus recognizing the actual characteristics of the solid bed during the extraction process.

On the other hand, if y^* is considered to be constant throughout the extraction as that at the starting point, a constant overall mass transfer coefficient, A_pK, is not sufficient to represent the entire extraction curve (Mendes et al., 1994). To obviate the problem, A_pK is made dependent on the residual solute content of the solid as:

$$A_pK = A_pK_o \exp [\ln 0.01 (x_o - x)/(x_o - x_{shift})] \tag{3.34}$$

where x_{shift} is the solute mass fraction in the solid bed after which there is a transition from the steady-state extraction regime to diffusion-controlled regime. However, the shift to the diffusion-controlled regime is found to be dependent on the actual characteristics of the solid bed during extraction and needs to be evaluated experimentally. According to this model, at $x = x_{shift}$, A_pK is taken to be lower than its initial value, A_pK_o, by two orders of magnitude. The initial overall mass transfer coefficient, A_pK_o, in this model is also found to increase with pressure and gave good representation of the extraction profiles of lipids from algae (Cygnarowicz-Provost et al., 1992).

The extraction behavior of clove oil from ground clove buds was analyzed at pressures in the range of 150 to 250 bar at two temperatures: 40°C and 60°C, with different upward flow rates (Mukhopadhyay and Rajeev, 1998). It was observed that at 150 bar, 68% extraction was completed in 1 h whereas it was possible to have 98% extraction completed at 200 bar and 40°C in the same period of time. At any pressure in this range of pressures, it was distinctly observed that at the beginning of the extraction, the loading of clove oil in the effluent SCF CO_2 from the extractor remained constant, which later decreased with time. For these two regimes of extraction, two different equations (3.27 and (3.28 were applied sequentially, namely, a modified form of Equation 3.27 at the beginning of extraction and a modified form of Equation 3.28 toward the end of extraction, using two different types of mass transfer coefficients, respectively (Mukhopadhyay and Rajeev, 1998).

For the initial period of extraction, the steady-state approximation of Equation 3.27 leads to:

$$\rho u(\partial y/\partial h) = A_pK_e\,(y^* - y) \tag{3.29}$$

where K_e is the external mass transfer coefficient and A_p is the surface area of the particle per unit volume. Equation 3.29 can be solved with the boundary conditions given as:

$$y = 0 \text{ at } h = 0 \tag{3.30}$$

$$y = c^*y^* \text{ at } h = H \tag{3.31}$$

where c^* is a constant determined experimentally.

At a constant condition of temperature, pressure, and flow rate of supercritical CO_2, it is expected that the initial loading of the solutes in the bulk SCF phase for the steady rate of extraction will effectively remain constant at the exit of the extractor. Equation 3.29 on integration gives:

$$y/y^* \equiv 1 - \exp[-A_pK_eH/\rho u] \tag{3.32}$$

For a sufficiently long extractor, the value of c^*, which is the ratio of the loadings of the bulk SCF phase to that at the gas–solid interface under equilibrium, is close to unity. For supercritical CO_2 extraction of clove oil from ground clove buds (with $x_o = 21\%$), the value of c^* was experimentally determined and found to be 0.85. The external mass transfer coefficient, A_pK_e, was evaluated at various conditions of pressure, temperature, and flow rates when SC CO_2 was flowed in the upward direction. It was observed that the external mass transfer coefficient was more at 60°C than that at 40°C due to an increase in diffusivity in SC CO_2 with temperature. It decreased with pressure at 40°C due to the pressure effect on diffusivity. The external mass transfer coefficient was found to be distinctly enhanced by an increase in the superficial velocity and was related by the following correlations:

$$A_pK_e = 0.008\ (u/\varepsilon)^{0.54} \tag{3.33}$$

$$Sh\ = 0.003\ (Re)^{0.45}\ (Sc)^{1/3} \tag{3.34}$$

in the range of $4.8 < Sc < 12.8$ and $0.4 < Re < 1.9$.

However, the very small value of Sherwood number (of the order of 10^{-3}) in the case of clove oil extraction with upward flow of CO_2 is attributed to the dominance of free convection over forced convection due to very low velocity of SC CO_2 (of the order of 10^{-4} m/s). The corresponding Reynolds number in terms of particle size is also very low (in the range of 0.4 to 1.9) for a particle diameter of 2.97×10^{-4}m. It is believed that for upward flow of the solvent, the two Sherwood numbers (Brunner, 1994) oppose each other and natural convection is more pronounced for small interstitial velocity and large voidage. The downward flow of CO_2 causes extraction to take place as very close to that in a plug flow, whereas upward flow causes channeling and less efficient extraction. It is therefore advantageous to operate the SCFE process in the down-flow mode rather than in the up-flow mode for low flow rates to get the effect of gravity-assisted extraction. Similar observations were reported by Sovova et al. (1993) and Lim et al. (1994).

For the later part of the extraction, an overall internal mass transfer coefficient, K_i, is assumed to be the controlling parameter for the unsteady state falling rate period of extraction and the Equation 3.28 is modified as:

$$-(1-\varepsilon)\ \rho_s\ dx/dt = A_pK_i(x - x^*) \tag{3.35}$$

where x^* is the solute content at the solid surface corresponding to the isothermal adsorption equilibrium with the bulk fluid phase; x is the solute content in the solid at any time. On integration, Equation 3.35 gives:

$$x(t)/x_r = \exp(-K^*_i\ \Delta t) \tag{3.36}$$

where $K_i^* = \dfrac{A_pK_i}{(1-\varepsilon)\rho_s}$ and Δt = time elapsed since the onset of unsteady state, when $x = x_r$. The value of x_r depends on the pressure, temperature, particle size, voidage, solid density, solid diffusivity, and surface characteristics. For clove oil extraction, its value ranges from 0.012 to 0.18 (Mukhopadhyay and Rajeev, 1998).

It can be noted that x_r is very much similar to x_{shift} mentioned earlier. The only difference in the present model is that the external and internal mass transfer coefficients are calculated independently and are not made dependent on the residual solute content as in the case of the earlier model. The internal mass transfer coefficient, K_i^*, is found to increase with temperature and is not dependent on flow rate. The value of K_i for clove oil extraction is found to be of the same order of magnitude as those obtained by Hong et al. (1990) for soybean oil extraction.

Thus the entire extraction regime can be simply divided into two sectors, namely, (1) the constant rate period in which the external mass transfer resistance is the controlling factor, and (2) the declining rate period, in which the internal mass transfer resistance within the solid phase is the controlling factor. By reducing the particle size x_r may be made to reduce to a very small value for natural materials with high initial solute contents. For example, at 200 bar and 40°C, it is possible to extract 95% of the clove oil in 45 min requiring 50% of CO_2 used for the complete extraction, whereas the complete extraction required 120 min. The external mass transfer coefficient is found to be an order magnitude higher than the internal mass transfer coefficient according to the present model, in which a sequence of two types of mass transfer coefficients are employed (Mukhopadhyay and Rajeev, 1998) to characterize the resistance to SCFE from the same natural material.

3.3.3 Effects of Axial Dispersion and Convective Flows

In supercritical CO_2 extraction from a packed bed of solids, even though plug flow is mostly assumed, an axial dispersion or back mixing of fluid flow can hardly be avoided in actual operation. For flow direction against gravity and for very low flow rates, axial dispersion needs to be included in the design and analysis of SCFE systems. Non-uniform distribution of the solvent, radial distribution of porosity, unsteady hydrodynamic flow, and radial temperature and concentration gradients are some of the factors responsible for axial dispersion. In general, axial dispersion is not desirable since it leads to inefficient extraction. Ideally it is desirable to have plug flow of the SCF solvent across the bed. At high solvent to solid feed ratios, axial dispersion is negligible. Accordingly a high solvent flow rate is preferred for large-scale operations.

In order to consider the negative effect of axial dispersion in the unsteady state mass balance of solute transported from the fluid–solid interface to the flowing bulk SCF solvent phase, Equation 3.27 is modified as:

$$\varepsilon\rho(\partial y/\partial t) = -\rho u(\partial y/\partial h) + \varepsilon D_{ax}\, \rho(\partial^2 y/\partial h^2) + A_p K\,(y^* - y) \tag{3.37}$$

where D_{ax} is the axial dispersion coefficient or axial diffusivity (m^2/s) and is evaluated from the experimental concentration profiles.

Axial dispersion is usually taken into account in terms of the Peclet number, ($u_c d_p/D_{ax}$), where d_p is the particle diameter and u_c is the characteristic velocity. Catchpole et al. (1996) experimentally evaluated Peclet number for pure fluids and found it to be independent of column length and flow direction for low velocities. On the other hand, at high velocities, dispersion was found to be more important for the short columns rather than the long columns. The following correlation was proposed by Catchpole et al. (1996) for the particle size less than 1 mm and the Schmidt number of the SCF phase in the range of 8 to 20 as:

$$\frac{1}{P_{e,f}} = \frac{0.018}{\mathrm{Re}} + \frac{10}{1 + 0.7/\mathrm{Re}} \tag{3.38}$$

from which the Peclet number based on the column length can be found as:

$$P_e = P_{e,f}\ (H/dp) \tag{3.39}$$

It was pointed out that there was no significant difference in the experimentally measured dispersion coefficient, between up flow and down flow at a constant pressure and at a constant temperature. However, with increasing dissolution of solutes from the solid bed during the extraction process, the Peclet number decreases and therefore axial dispersion increases. It is observed that the mass transfer in SCFE can not be modeled without considering axial dispersion, particularly when the flow rate is low and the flow direction is against gravity.

For SCFE from natural materials from a fixed bed of solids, the density of the SCF solvent phase may significantly change with the progress of extraction, making natural convection not so insignificant as in the case of conventional solvents. This is due to the fact that kinematic viscosity and hence the Schmidt number for the SCF phase is very low. As a consequence, the relative importance of natural convection to forced convection, in terms of the ratio of inertial forces to buoyancy forces, i.e., $Re/Gr^{1/2}$ ($Gr = gd_p^3\rho\Delta\rho/\eta^2$) is two orders of magnitude higher in an SCF at a constant value of the Reynolds number than in normal fluids. This causes a significant enhancement in mass transfer rate, when the controlling resistance to mass transfer is in the SCF phase. The rate of clove oil extraction was found to be faster when SC CO_2 flow was downward rather than when it was upward, i.e., against gravity (Dave, 1997). For a simple natural convection process, the Sherwood number, i.e., dimensionless mass transfer coefficient, becomes independent of Reynolds number and is given (Brunner, 1994) as:

$$Sh_N = A_1\ (SC\ Gr)^{1/4} \text{ for laminar natural convection} \tag{3.40a}$$

$$Sh_N = A_2\ (SC\ Gr)^{1/3} \text{ for turbulent natural convection} \tag{3.40b}$$

where A_1 and A_2 are constants, given as $A_1 = 0.5$ and $A_2 = 0.1$, though they need to be evaluated from experimental data for the systems concerned.

For forced convection and for $Re > 10$,

$$Sh_F = A_3\ Re^{1/2}\ Sc^{1/3} \tag{3.41}$$

For low flow rates, i.e., in an intermediate region, both forced convection and natural convection are important considerations. In order to consider the combined effect of buoyant forces and pressure gradients, a combined Sherwood number is calculated to account for both natural and forced convection in the gravity-assisted extraction process (Churchill, 1977) as:

$$Sh^3 = Sh_N^3 + Sh_F^3 \tag{3.42}$$

and for gravity-opposed flow direction:

$$Sh^3 = Sh_N^3 - Sh_F^3 \tag{3.43}$$

where, subscript N and F refer to natural and forced convection, respectively. However, these equations are very approximate and correlations were accordingly developed for different ranges of Reynolds and Schmidt numbers. Catchpole et al. (1990) determined k_f from:

$$Sh = 0.839\ Re^{2/3}\ Sc^{1/3} \tag{3.44}$$

in the range of 2 < Re < 55 and 4 < Sc < 16. Tan et al. (1988) recommended the following correlation for k_f:

$$Sh = 0.38\ Re^{0.83}\ Sc^{1/3} \tag{3.45}$$

which is valid for low superficial velocities in the range of 4.4×10^{-3} to 3.1×10^{-2} cm/s.

3.3.4 Shrinking Core Leaching Model

Although several attempts have been made for modeling mass transfer coefficients from a fixed bed of solid particles to the flowing SCF solvent phase, there are disparities in the values of Sherwood numbers at low flow rates since Sherwood numbers are found to be much less than 2 for single particles. Several investigators, both theoretically and experimentally, showed that there should be a limiting value of Sherwood number such as a value of 3.9 for a cubic array of spherical particles (Sorensen and Stewart, 1974), a value of 13 (Pfeffer and Happel, 1964) or 16.7 for a bed void fraction $\varepsilon = 0.4$ (Miayauchi et al. 1975).

Subsequently, Wakao et al. (1976) and later Roy et al. (1996), Catchpole et al. (1996), and Goto et al. (1998), considered the rigorous fundamental equations for determining the mass transfer coefficients from the model based on irreversible desorption followed by diffusion in porous solids through the pores, which is popularly known as the "shrinking core model," as schematically described in Figure 3.16. The fundamental equations considered are

$$\frac{\partial c}{\partial t} = D_{ax}\frac{\partial^2 c}{\partial h^2} - u\frac{\partial c}{\partial h} - \frac{3D_e}{R}\cdot\frac{1-\varepsilon_b}{\varepsilon_b}\left(\frac{\partial c_i}{\partial r}\right)_R \tag{3.46}$$

$$\varepsilon_p\frac{\partial c_i}{\partial t} = D_e\left(\frac{\partial^2 c_i}{\partial r^2} + \frac{2}{r}\frac{\partial c_i}{\partial r}\right) - \rho_s\frac{\partial c_s}{dt} \tag{3.47}$$

where c is the concentration in the bulk SCF phase, c_i is the concentration in the particle pore, c_s is the concentration in the solid, D_e is the effective diffusivity in the porous solid, D_{ax} is the axial diffusivity, ε is the bed void fraction and ε_p is the particle void fraction, and r is the radial distance and R is the particle radius. The following initial conditions are considered for solving the above equations:

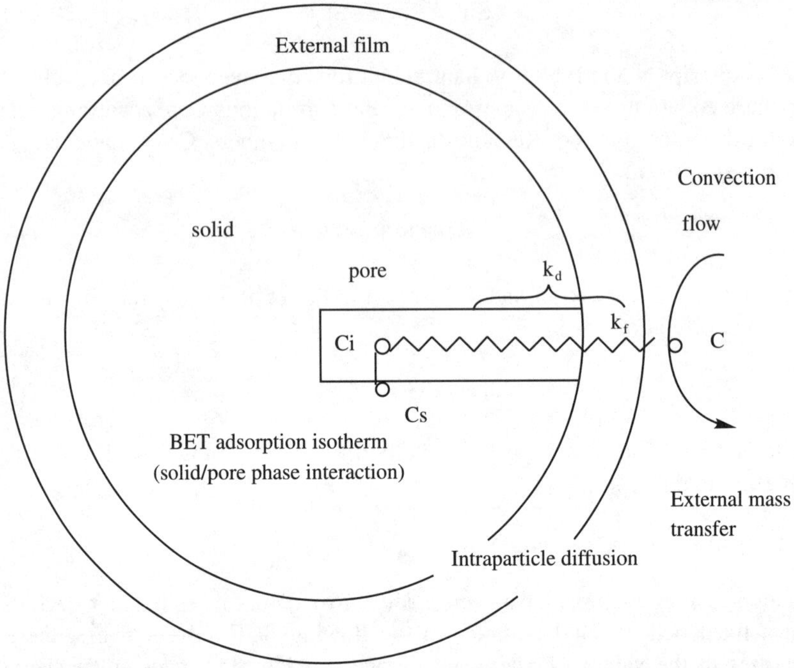

FIGURE 3.16 Schematic description of mass transfer sequences from a porous solid particle.

1. $c = 0$ at $t = 0$
2. $c_i = c_{io}$ at $t = 0$
3. $c_s = c_o$ at $t = 0$ (3.48)

and the following boundary conditions are considered:

4. $c = 0$ at $h = 0$
5. $\partial c/\partial h = 0$ at $h = H$
6. $\partial c_i/\partial r = 0$ at $r = 0$ (3.49)
7. $uc = D_{ax}\,(\partial c/\partial h)$ at $h = 0$ (3.50)

For the diffusion rate at the external surface of the solid particle:

$$D_e\left(\frac{\partial c_i}{\partial r}\right)_R = k_f(c - c_{i,R}) \tag{3.51}$$

For desorption rate at the pore surface:

$$-\frac{\partial c_s}{\partial t} = k_d(c_s/k_A - c_i) \tag{3.52}$$

where k_f is the fluid film mass transfer coefficient, k_d is the desorption rate constant, and k_A is the adsorption equilibrium constant.

Five parameters need to be evaluated according to this model, namely, D_{ax}, D_e, k_A, k_d, and k_f, out of which at least two parameters are regressed while the others are estimated from the available correlations.

There are various methods suggested and followed by different investigators for evaluating these parameters. For example, the values of D_{ax} for nonporous glass beads were related to the molecular diffusivity D_{12} (Wakao et al., 1976) as:

$$D_{ax} = 0.65\ D_{12} \tag{3.53}$$

Wakao and Kaguei (1982) provided the following correlation for D_{ax} in systems with sufficient mass transfer rates,

$$D_{ax} = (D_{12}/\varepsilon)(20 + 0.5\ ReSc) \tag{3.54a}$$

For $\varepsilon ReSc > 0.5$, Funazukuri et al. (1997) proposed the following equation:

$$D_{ax} = (D_{12}/\varepsilon)\ 1.471(\varepsilon ReSc)^{1.38} \tag{3.54b}$$

The effective diffusivity D_e depends on the nature of the micro- and macropore structures and the binary diffusivity. This can be obtained by solving Equations 3.46 and 3.47 and regressing its value from experimental extraction profiles. Alternatively, it can be approximated as Wakao and Smith (1962) suggested in the following simple correlation for estimating D_e:

$$D_e = \varepsilon_p^2\ D_{12} \tag{3.55a}$$

Since the pores do not run straight through the solid material but constitute a randomly connected network, D_e is also given as:

$$D_e = D_{12}\ \varepsilon/\psi \tag{3.55b}$$

where ψ is the tortuosity factor. If the porous structure is bidisperse, then,

$$1/D_e = 1/D_{ma} + A/D_{mi} \tag{3.56}$$

where D_{ma} is macropore diffusivity and D_{mi} is the micropore diffusivity. A is an empirical constant to be obtained experimentally and depends on the pore geometry (Wakao and Kaguei, 1982).

Based on the experimental mass transfer rates in a packed bed of pure solutes, Catchpole et al. (1992) recommended the following correlation for the fluid film mass transfer coefficient, k_f:

$$k_f = 0.82\ (D_{12}/dp)(Re^{0.66}\ Sc^{1/3}) \tag{3.57}$$

for ranges of $3 < Sc < 11$ and $1 < Re < 70$. Alternatively, k_f needs to be regressed from the experimental data along with other parameters.

Out of the five parameters listed above, the remaining two terms, namely, k_A, and k_d, can be regressed from the experimental data on the loadings of the SCF phase at different times of extraction. It is reported that k_A values are the least affected by the simulated values of D_e. However, k_f can also be regressed (instead of estimating it) assuming $k_d \cong \infty$ (as the limiting value reached at 20 cm^3/g s), as has been done by several investigators (Wakao et al., 1976). It was again confirmed that the Sherwood numbers are very low at low values of the Reynolds number. The Sherwood numbers slightly increased after consideration of D_{ax} and its effect is larger at lower Reynolds number in the range of 0.1 to 1.0.

3.4 HEAT TRANSFER IN SCF

Heat transfer is a transient phenomenon and depends not only on the thermodynamic state of the substances to, and from which, and through which medium, the change is taking place, but also on the geometrical configurations of the equipment and hydrodynamics of the flowing streams. The high pressure supercritical fluid solvent and its mixture with the extracts are made to pass through the tube side of the heat exchangers countercurrently to the heating or cooling utilities on the shell side of the heat exchangers.

Design, scale-up, and optimization of heat exchangers in supercritical fluid extraction plants pose a challenging task for two principal reasons, namely, (1) wide variations in physical and heat transfer characteristics of the SCF solvent and its mixtures over a wide range of temperature from –10 to 90°C and at high pressures ranging from 50 to 500 bar which are encountered in the process, and (2) high material and production costs of the tubes, vessels, and heat exchangers which are made from 316 or 304 Cr-Ni stainless steel. Besides, stringent design provisions are required to be made for easy cleaning, low risk of failure, compact design for minimum heat transfer area, and low pressure drop. Accordingly, knowledge and understanding of heat transfer in supercritical fluids are of utmost importance.

3.4.1 Heat Transfer Coefficients

A linear relation between the differential amount of heat transferred and the local temperature gradient across the boundaries is sufficient to represent the phenomena of heat transfer associated with supercritical fluids. For a differential element of length of the exchanger, dz, where a flowing stream of high pressure CO_2 is heated by the countercurrent heat exchange with a hot water stream, the amount of heat transferred is given as:

$$dq = U\, 2\pi r_i\, dz\, (T_o - T) = \dot{m}\, c_p\, dT \tag{3.58}$$

or

$$\frac{dT}{2\pi r_i dz} = \frac{U(T - T_o)}{\dot{m} c_p} \tag{3.59}$$

and

$$\frac{dT_o}{2\pi r_o dz} = \frac{U(T - T_o)}{\dot{m}_o c_p} \tag{3.60}$$

where T is the temperature of the fluid being heated and T_0 is the temperature of the hot utility at any point of heat exchanger; r_i and r_o are the inner and outer radii of the tube; and $\dot{m}$ and $\dot{m}_0$ are the mass flow rates of the cold and hot fluids, respectively.

The overall heat transfer coefficient, U, is given by

$$\frac{1}{U} = \frac{r_o}{h_i r_i} + \frac{1}{h_w} \ln \frac{r_o}{r_i} + \frac{1}{h_o} \tag{3.61}$$

where h_i and h_o are, respectively, the inner and outer local heat transfer coefficients, and h_w is the heat transfer coefficient across the wall.

The local heat transfer coefficient is usually calculated from the dimensionless heat transfer coefficient, namely the Nussel number (hd/λ) for forced convection which is given in terms of the Reynolds number, Re ($du\rho/\eta$), and the Prandtl number, Pr ($c_p\eta/\lambda$), as:

$$Nu = a\ Re^b\ Pr^c \tag{3.62}$$

where a, b, and c are empirical constants evaluated from experimental data in different ranges of pressure, temperature, and flow conditions. Bringer and Smith (1957) performed heat transfer experiments with near-critical and supercritical CO_2 in turbulent flow. Their experimental results were compared with those predicted from dimensionless correlations of Dittus-Boetter, Sieder-Tate, and Colburn equations. None of the correlations could give good agreement with the corresponding values measured in the near-critical region. Bringer and Smith (1957) gave the following correlation for supercritical water:

$$Nu_x = 0.0266\ (Re_x)^{0.77}\ (Pr_w)^{0.55} \tag{3.63}$$

where the subscript "x" refers to the temperature and physical state of the fluid and "w" refers to the condition at the wall.

It need be mentioned that inadequacies of the conventional forced convention correlations to predict heat transfer coefficients for supercritical fluids are due to rapid changes of properties of supercritical fluids. At the critical point, the heat transfer coefficient is five to ten times larger than that at conditions away from the critical point. A maximum in heat transfer coefficient (>1000 W/m^2K) is found at the vicinity of the critical point (Weidner, 1988). A correlation was developed for heat transfer to turbulent flow of an SCF fluid by modifying the conventional models as:

$$\frac{hd}{\lambda_w} = 0.00459\left(\frac{du\rho}{\eta_w}\right)^{0.923}\left(\frac{H_w - H_b}{T_w - T_b}\cdot\frac{\eta_w}{\lambda_w}\right)^{0.613}\left(\frac{\nu_b}{\nu_w}\right)^{0.231} \tag{3.64}$$

where H_w is the enthalpy at the wall temperature and H_b is the enthalpy at the bulk temperature, λ_w and η_w are the thermal conductivity and viscosity at the wall temperature, T_w. ν_b and ν_w are the molar volumes at the bulk temperature, T_b, and wall temperature, T_w, respectively.

Ghajar and Asadi (1986) compared the existing empirical correlations for forced convection heat transfer in the near-critical region for water and CO_2. It was pointed out that the Dittus-Boetter type heat transfer correlations could be employed for the heat transfer coefficients in the supercritical fluid region provided the property variations could be accounted for by using a property ratio method. The following correlation was suggested:

$$Nu_b = a(Re_b)^b\,(Pr_c)^c\,(\rho_w/\rho_b)^d\,(\bar{c}_P/c_{P,b})^n \tag{3.65}$$

where $\bar{c}_P$ is the integrated mean specific heat given as:

$$\bar{c}_P = (H_w - H_b)/(T_w - T_b) \tag{3.66}$$

and the constants are a = 0.0183, b = 0.82, c = 0.5, d = 0.5, and n = 0.4.

3.4.2 Effects of Free Convective Flow

Free convective fluid flow is induced by density gradients due to temperature gradient as well as concentration gradient during mass transfer. Free convective flow in SCFE process is important as the SCF solvent flow velocities are mostly kept low within the processing equipment and due to the fact that the density gradient is developed due to small changes in temperature and pressure. The thumb rule criterion (Brunner, 1994) for free convective flow to set in is

$$Gr \cdot Pr > 600 \tag{3.67}$$

At higher pressures, free convective flow is more easily established than at low pressures. For CO_2, free convective flow can set in even with a temperature difference of 3°C at 38°C (Lenoir and Comings, 1951), as the tendency for free convective flow is more pronounced near its critical temperature. The heat transfer coefficients also depend on simultaneous mass transfer. The local heat transfer coefficients for free convective flow are, in general, related to the Grashoff number and the Prandtl number as:

$$Nu = f(Gr\,Pr)^L \tag{3.68}$$

where f and L are empirical constants determined from experimental data of different mixtures at different thermodynamic state conditions. However, much fewer studies have been made and fewer correlations are available for supercritical mixtures than for the pure SCF solvents.

Adebiyi and Hall (1976) studied the effects of buoyancy on heat transfer to supercritical fluids due to significant temperature variations. It was observed that the heat transfer at the bottom of a vertical pipe is enhanced while at the top of the pipe, the heat transfer is reduced by buoyancy. The criterion for the absence of the buoyancy effects in a vertical tube is given as:

$$Gr_b\ Re_b^{-2.7} < 5 \times 10^{-6} \tag{3.69}$$

A similar criterion in the case of a horizontal tube is

$$Gr_b\ Re_b^{-2}\ (\rho_b/\rho_w)(l/d)^2 < 10 \tag{3.70}$$

where ρ_b and ρ_w are the densities of the fluid at the bulk and wall temperatures, respectively, and the Grashoff number, G_{rb}, is calculated at the condition of the bulk of the fluid as:

$$Gr = \beta_T\ \Delta_T\ gl^3/\eta^2 \tag{3.71}$$

where β_T is the thermal expansion coefficient, Δ_T is the temperature difference between the heat transferring surface and bulk of the fluid, g is the acceleration due to gravity, l is the characteristic length in the direction of flow, and η is the viscosity.

3.4.3 Heat Transfer Coefficient for Two-Phase Flow

The heat transfer coefficients for natural convection during condensation or boiling may be considered to be related to the dimensionless groups (Brunner, 1994) as:

$$(hd/\lambda) = 0.5\ (Pr\ Gr)^{1/4}\ (P_r/P_{rw})^{1/2} \tag{3.72}$$

Experimental values of heat transfer coefficients for high-pressure CO_2 compare well with the predicted values using this correlation in the range $0.5 < Gr\ Pr < 5.10^3$.

For laminar condensation on a horizontal cylinder of diameter d, the Nusselt number is given by:

$$Nu = 0.73\ (Ra_{film})^{1/4} \tag{3.73}$$

or

$$\frac{hd}{\lambda} = 0.73[\rho_L(\rho_L - \rho_v)g\Delta H d^3/\eta\lambda(T_w - T_b)]^{1/4} \tag{3.74}$$

where ΔH is heat of condensation, and ρ_L and ρ_v are the liquid and vapor phase densities.

The heat transfer coefficients in supercritical fluid CO_2 mixed with a hydrocarbon, such as decane or pentane, are found to be enhanced (Jones, 1988), when the two-phase region is approached. For example, h is found to increase from 750 $W/m^2 \cdot K$ for pure SC CO_2 at 104 bar and 51°C to 2500 $W/m^2 \cdot K$ for 12.7% decane mixed with SC CO_2 at the same condition. Similarly, the heat transfer coefficient increased from 270 $W/m^2 \cdot K$ for pure SC CO_2 at 90 bar and 55°C to 1000 $W/m^2 \cdot K$ for SC CO_2 mixed with 12.4% n-pentane. A steep rise in heat transfer coefficient was observed due to the onset of LCST condensation process for 12.7% n-decane at both pressures.

However, the experimental values of Nusselt number for two-phase systems studied by Jones (1988) were much higher than the calculated values. This enhancement was attributed to the thermocapillarity due to the temperature effect of the interfacial tension, similar to the Marangoni effect (Bejan, 1984).

3.4.4 Heat Exchanger Specifications

The specifications for heat exchangers are based on design considerations, both from the points view of ease of maintenance and of safety of operation. Fouling should be kept to a minimum during operation and the heat exchanger surfaces must be cleaned when fouling occurs. To ensure no contamination from previously extracted product, the entire network of pipelines must be cleaned before taking up any new product for extraction. These specific requirements restrict the specifications of the heat exchangers to having large tube diameters, to be vertical or slightly inclined to eliminate clogging, and to be single pass with supercritical solvent on the tube side. Construction materials and quality of seamless forgings are carefully selected and evaluated by stringent testing.

Heat exchangers are required for heating, for condensing, for pre-cooling or preheating. Two types of heat exchangers are most commonly selected for the SCFE plants, namely double-pipe heat exchangers for small plants and single-pass shell and tube heat exchangers with baffles for large plants. The single-pass shell and tube heat exchanger is compact and provides the advantage of requiring a smaller tube diameter, compared to that for double-pipe heat exchangers. For higher plant capacities, the shell and tube heat exchangers are more cost effective and the right choice. However, exact specifications are arrived at after detailed process design and economic analysis of heat exchangers.

Due to wide variations in the thermophysical properties in the near-critical to supercritical regions, calculation of heat-transfer coefficients is, in general, inherently associated with a measure of uncertainty, and in particular when an entrainer is mixed with the SCF solvent, CO_2. In view of the uncertainties in the heat transfer coefficients, even 100% excess in surface area is not considered uncommon. Sometimes the same heat exchanger may be designed for both heating and cooling, depending on the specific requirement. For example, for separating the extract from the SCF solvent, its temperature needs to be lowered at higher pressures. At lower pressures, however, temperature needs to be increased to separate out the extract from the gaseous CO_2. Sometimes heating is required after expansion of high pressure CO_2 or prior to its expansion, to avoid solidification and clogging. A heat integration network seems to be too complicated to be implemented in view of the high sensitivity of pressure and temperature on property variations. Reliable estimation of heat transfer coefficients at different lengths of the heat exchangers from the knowledge of the varying thermophysical properties over the entire regime of heat transfer, are of prime importance for the selection of heat exchanger specifications.

NOMENCLATURE

A	Cross sectional area of the column, m^2
A_p	Surface area of fluid–solid interface per unit volume of the solid bed, m^2
A_1, A_2	Constants in correlations for Sh
C	A constant
c	Concentration in the bulk SCF phase, kg/m^3
c^*	Ratio of the loading in the SCF phase to the equilibrium loading at the fluid–solid interface
c_0	Initial solute concentration, kg/m^3
c_i	Concentration in the particle pore, kg/m^3
c_s	Concentration in the solid, kg/m^3
c_1, c^2, c^3	Constants in y^* vs. x correlation
d	Diameter, m
dp	Particle diameter, m
D_{12}	Binary diffusivity, m^2/s
D_{ax}	Axial diffusivity, m^2/s
D_e	Effective diffusivity in a porous solid, m^2/s
F_H	Objective function at the exit of the bed, H
G	Molar flow rate of SCF solvent, mol/h
G_{12}	Interaction constant in the Grunberg equation
H	Total length, m
h	Height of the bed, m
h_i	Inner film heat transfer coefficient, $W/m^2{\cdot}K$
h_o	Film heat transfer coefficient, $KW/m^2{\cdot}K$
J	Molecular diffusion flux, $kg/m^2{\cdot}s$
k	Boltzman constant
k_A	Adsorption equilibrium constant
K_i	y_i/x_i = Distribution coefficient
k_d	Desorption rate constant, m/s
k_f	Fluid phase film mass transfer coefficient, m/s
K	Overall mass transfer coefficient, $kg/m^2{\cdot}s$
K_e	Overall external mass transfer coefficient, $kg/m^2{\cdot}s$
K_i	Overall internal mass transfer coefficient, $kg/m^2{\cdot}s$
K_i^*	Equivalent internal mass transfer coefficient, $A_pK_i/(1 - \varepsilon)\rho_s$, s^{-1}
K^1	Partition coefficient
L	Liquid molar flow rate, mol/h
N_A	Avogadro's number
P	Pressure, bar
q	Heat flux, W/m^2

R	Radius of the particle, m
r	Radial distance, m
r_i	Inside radius, m
r_o	Outside radius, m
T	Temperature, K
t	Time, h
U	Overall heat transfer coefficient, $KW/m^2 \cdot K$
u_c	Characteristic velocity, m/s
u	Superficial velocity, m/s
v_b	Molar volume at the normal boiling point, m^3/mol
v_1	Molar volume of component 1, m^3/mol
x	Fraction of residual solute content in the bed
x^*	Highest mass fraction of solute content in the solid bed in isothermal adsorption equation
x_0	Initial solute mass fraction in the bed
x_r	Residual solute mass fraction at the onset of unsteady state mass transfer
Y	SCF phase mole fraction (solvent-free basis)
y^*	Mass fraction of the solute loading in SCF phase in equilibrium with the residual solute in the solid bed
y	Mass fraction of solute loading in SCF phase

Greek Letters

ε	Void fraction of the bed
ε_p	Void fraction of the particle
ρ	Density of the fluid, g/l
ρ_r	Reduced density of the fluid = ρ/ρ_c
ρ_s	Density of the solid, g/l
τ	Shear stress per unit area, Pa
η	Viscosity, Pa·s
λ	Thermal conductivity, W/m·K
μ	Dipole moment, D
Ψ	Tortuosity factor
σ	Surface tension, dyn/cm
σ_1	Hard sphere diameter of the solvent molecule, m

Dimensionless Numbers

Re	Reynolds number, $Du\rho/\eta$
Sh	Sherwood number, $k_f d_p/D_{12}$
Pr	Prandtl number, $C_p\, \eta/\lambda$
Sc	Schmidt number, $\eta/\rho D$
Pe	Peclet number, $U_c d_p/D_{ax}$
Gr	Grashoff number, $gd_p^3\, \rho\Delta\rho/\eta^2$
Nu	Nusselt number, hd/λ

REFERENCES

Adebiyi, G. A. and Hall, W. B., Experimental investigation of heat transfer to supercritical pressure carbon dioxide in a horizontal pipe, *Intl. J. Heat Mass Transfer*, 19, 715–720, 1976.

Alkio, M., Harvala, T., and Komppa, V., Preparative scale supercritical fluid chromatography, Proc. 2nd Intl. Symp. SCFs, France, 1988, 389–396.

Assael, M. J., Dix, M., Lucas, A., and Wakeham, W. A., *J. Chem. Soc. Faraday Trans.* I, 77, 439, 1980.

Bartolomeo, G., Bertucco, A., and Guarise, G. B., Mass transfer in a semibatch supercritical fluid extraction contactor for liquid mixtures, *Chem. Eng. Commun.*, 95, 57, 1990.

Bejan, A., *Convective Heat Transfer*, John Wiley & Sons, New York, 1984.

Berger, D. and Perrut, M., Preparative supercritical fluid chromatography, *J. Chromatogr.*, 505, 37, 1990.

Bhaskar, A. R., Rizvi, S. S. S. H., and Harriot, P., Performance of a packed column for continuous SC CO_2, processing of AMF, *Biotechnol. Prog.*, 9, 70–74, 1993.

Bringer, R. P. and Smith, J. M., Heat transfer in the critical region, *A. I. Ch. E. J.*, 3(1), 49–55, 1957.

Brunner, G., Malchow, T. H., Struken, K., and Gotschau, Th., Separation of tocopherol from deodoriser condensates by countercurrent extraction with CO_2, *J. Supcrit. Fluids*, 4, 72–80, 1991.

Brunner, G., *Gas Extraction*, Steinkopff Darmastadt, Springer, New York, 1994, chap. 8.

Brunner, G., Industrial process development of countercurrent multistage gas extraction process, *J. Supercrit. Fluids*, 13, 283–301, 1998.

Bruno, T. J., Thermophysical property data for supercritical fluid extraction design, in *Supercritical Fluid Technology*, Bruno, T. J. and Ely, J. F., Eds., CRC Press, Ann Arbor, MI, 1991, chap. 7.

Bruns, A., Berg, D., and Werner-Busse, A., Isolation of tocopherol homologues by preparative HPLC, *J. Chromatogr.*, 450, 111–113, 1988.

Cabral, J. M. S., Novais, J. M., and Palavra, A. F., Supercritical CO_2 extraction of lipids from micro algae, Proc. 4th Intl. Symp. SCFs, 2, Nice, France, 1994, 477.

Catchpole, O. J., Sinoes, P., King, M. B., and Bott, T. R., Film mass transfer coefficients for separation processes using near critical CO_2, in Dechema, G. V. C., Eds., 2nd Intl. Symp. High Pressure Chem. Eng. Erlangen, Abstr. Handbk, 1990, 153–158.

Catchpole, O. J., King, M. B., and Bott, T. R., Separations with supercritical fluids, *A. I. Ch. E. Annu. Meet.*, Miami, 1992.

Catchpole, O. J, Grey, J. B., and Smallfield, B. M., Near critical extraction of sage, celery and coriander seed, *J. Supercritical Fluids*, 9, 273–279, 1996.

Catchpole, O. J., Berning, R., and King, M. B., Measurement and correlation of packed bed axial dispersion coefficients in SC CO_2, *Ind. Eng. Chem. Res.*, 35, 824–828, 1996.

Chrastil, J., Solubility of solids and liquids in supercritical gases, *J. Phys. Chem.*, 86, 3016, 1982.

Churchill, S. W., *A. I. Ch. E. J.*, 23, 10, 1977.

Cochran, H. D., Lee, L. L., and Pfund, D. M., Application of Kirkwood-Buff theory of solutions to dilute supercritical mixtures, *Fluid Phase Equilibria*, 34, 219, 1987.

Cygnarowicz-Provost, M., O'Brien, D. J., Maxwell, R. J., Hampson, J. W., *J. Supercritical Fluids*, 5, 24, 1992.

Dave, K. H., Effect of Flow Rate and Flow Mode of CO_2 on Mass Transfer Coefficients of Supercritical Extraction of Natural Products, M.Tech. dissertation, Indian Institute of Technology, Bombay, India, 1997.

Dawson, R. F., Khoury, F., and Kobayashi, R., Self diffusion measurements in methane by pulsed NMR, *A. I. Ch. E. J.*, 16, 725–729, 1970.

De Fillipi, R. P., Krukonis, V. J., Robey, R. J., and Modell, M., Supercritical fluid regeneration of activated carbon for adsorption of pesticides, Report EPA-600/2-80054, March, as reported in *Supercritical Fluid Extraction*, McHugh, M. A. and Krukonis, V.A., Eds., 2nd ed., Butterworth-Heinemann, Stoneham, MA, chap. 1, 1994, 16.

De Haan, A. B., de Graauw, J., Schaap, J. E., and Badings, H. T., Extraction of flavors from milk fat with SC CO_2, *J. Supercritical Fluids*, 3, 15, 1990.

Ely, J. F. and Hanly, H. J. M., Prediction of transport properties. I. Viscosity of fluids and mixtures, *Ind. Eng. Chem. Fundam.*, 20, 323, 1981.

Favati, F., King, J. W., and Mazzanati, M., *J. Am. Oil. Chem. Soc.*, 68, 422, 1991.

Feist, R. and Schneider, G. M., Determination of binary diffusion coefficients of benzene, phenol, naphthalene and caffeine in supercritical CO_2 with SFC, *Sep. Sci. Technol.*, 17, 261, 1980.

Flament, I., Keller, U., and Wunsche, L., Use of semi-preparative SFC for the separation and isolation of flavor and food constituents, in *Supercritical Fluid Processing of Food and Biomaterials*, Rizvi, S. S. H., Ed., Blackie Academic & Professional, an imprint of Chapman & Hall, Glasgow, chap. 5, 1994, 62.

Funazukuri, T., Ishiwata, Y., and Wakao, N., Predictive correlation for binary diffusion coefficient in dense CO_2, *A. I. Ch. E. J.*, 38, 1761–1768, 1992.

Funazukuri, T., Kong, C. Y., and Kagei, S., Effective axial dispersion coefficients in paced beds under supercritical conditions, Proc. 4th Intl. Symp. SCFs, Sendai, Japan, Vol. A, 1997, 5.

Gangadhara Rao, V. S. and Mukhopadhyay, M., Mass Transfer Studies for Supercritical Fluid Extraction of Spices, Proc. of the First International Symp. on SCFs, Nice (France), Societe Francaise de Chimie, Tome 2, October 1988, 643.

Gangadhara Rao, V. S., Studies on Supercritical Fluid Extraction, Ph.D. dissertation, Indian Institute of Technology Bombay, India, 1990.

Ghajar, A. J. and Asadi, A., Improved forced convective heat transfer correlations in the critical region, *A. I. A. A. J.*, 24(12), 2030–2037, 1986.

Goto, M., Roy, B. C., Kodama, A., and Hirose, T., Modelling SCFE process involving solute-solid interaction, *J. Chem. Eng. Japan*, 31(2), 171–77, 1998.

Guarise, G. B., Bertucco, A., and Pallado, P., Carbon dioxide as an SCF solvent in fatty acid refining: theory and practice, in *Supercritical Fluid Processing of Food and Biomaterials*, Rizvi, S. S. H., Ed., Blackie Academic & Professional, an imprint of Chapman & Hall, Glasgow, 1994, chap. 2.

Gubbins, K. E., *Gas Extraction*, Brunner, G., Ed., Springer, New York, 1973, 48.

Guildner, L. A., *Proc. Natl. Acad. Sci.*, 44, 1149, 1958.

Hong, I. K., Rho, S. W., Lee, K. S., Lee, W. H., and Yoo, K., Modelling of soybean oil bed extraction with SC CO_2, *Korean J. Chem. Eng.*, 7, 40–46, 1990.

Jones, M. C., Two phase heat transfer in the vicinity of a lower consolute point, in *Supercritical Fluid Science and Technology*, Johnston, K. P. and Penninger, J. M. L., Eds., ACS Symp. Ser., #406, Washington, D.C., 1989, 396.

Jossi, J. A., Stiel, L. I., and Thodos, G., The viscosity of pure substances in the gaseous and liquid phases, *A. I. Ch. E. J.*, 8, 59, 1962.

Lahiere, R. J. and Fair, J. R., Mass transfer efficiencies of column contactors in supercritical extraction service, *Ind. Eng. Chem. Res.*, 26, 2086, 1987.

Lee, B. C., Kim, J. D., Hwang, K. Y., and Lee, Y. Y., Extraction of oil from evening primrose seed with SC CO_2, in *Supercritical Fluid Processing of Food and Biomaterials*, Rizvi, S. S. H., Ed., Blackie Academic & Professional, an imprint of Chapman & Hall, Glasgow, 1994, 168.

Lembke, P., Process scale SFC – a feasibility study, Proc. 7th Intl. Symp. Supercritical Fluid Chromatography and Extraction, Indianapolis, Indiana, March–April, L-38, 1996.

Lenoir, J. M. and Comings, E. W., *Chem. Eng. Prog.*, 47, 223, 1951.

Lenoir, J. M. and Comings, E. W., *Chem. Eng. Progress*, 49, 539, 1953.

Lim, G. B., Shin, H. Y., Noh, M. J., Yoo, K. P., and Lee, H., Subcritical to supercritical mass transfer in gas-solid system, Proc. 3rd Intl. Symp. SCFs, France, 2, 141, 1994.

Lucas, K., *The Properties of Gases and Liquids*, Reid, R. C., Pransnitz, J. M., and Poling, B. E., Eds., 4th ed., McGraw Hill, New York, 1987.

Magee, J. W., Thermophysical properties of CO_2 and CO_2-rich mixtures, in *Supercritical Fluid Technology*, Bruno, T. J. and Ely, J. F., Ed., CRC Press, Boca Raton, FL, chap. 8, 1991, 329.

Marrero, T. R. and Mason, E. A., Gaseous diffusion coefficient, *J. Phys. Chem. Ref. Data*, 1, 1, 1972.

McCabe, W. L., Smith, J. C., and Harriot, P., *Unit Operations of Chemical Engineering*, McGraw Hill, New York, 1989.

McHugh, M. A. and Krukonis, V., *Supercritical Fluid Extraction*, 2nd ed., Butterworth-Heinemann Series, MA, 1994.

Mendes, R. L., Fernandes, H. L., Cygnarowicz-Provost, M., Carbal, J. M. S., Novais, J. M., and Palavra, A. F., Supercritical CO_2 extraction of lipids from micro algae, Proc. 3rd Intl. Symp. SCFs, France, 2, 477, 1994.

Miyauchi, T., Matsumoto, K., and Yoshida, T., *J. Chem. Eng. Japan*, 8, 228, 1975.

Mollerup, J., Staby, A., and Sloth, H. C., Thermodynamics of SFC, 9th Intl. Symp. Preparative and Industrial Chromatography, "PREP-92", Nancy, France, 1992.

Mukhopadhyay, M. and Rajeev, K., Parametric study and mass transfer: modelling of supercritical CO_2 extraction of clove, *Ind. Chem. Eng. Trans.*, 1, 742, 1998.

Peker, H., Srinivasan, M. P., Smith, J. M., and McCoy, B. J., *A. I. Ch. E. J.*, 38, 761, 1992.

Pfeffer, R. and Happel, J., *A. I. Ch. E. J.*, 10, 605, 1964.

Pirkle, W. H., Terfloth, G., Brice, L. J., and Gan, K., A chiral stationary phase having broad generality, in Proc. 7th Intl. Symp. Supercritical Fluid Chromatography Extraction, Indianapolis, IN, L-06, March–April, 1996.

Reid, R. C., Pransnitz, J. M., and Poling, B. E., *The Properties of Gases and Liquids*, 4th ed., McGraw Hill, New York, 1989.

Reverchon, E., Donsi, G., and Osseo, L. S., *Ind. Eng. Chem. Res.*, 32, 2721, 1994.

Roy, B. D., Goto, M., and Hirose, T., Extraction of ginger oil with supercritical carbon dioxide: experiments and modelling, *Ind. Eng. Chem. Res.*, 35, 607–612, 1996.

Sastry, S. V. G. K., Supercritical Fluid Extraction of Fragrances from Jasmine Flowers, Ph.D. dissertation, Indian Institute of Technology, Bombay, India, 1994.

Sastry, S. V. G. K. and Mukhopadhyay, M., Substrate hindrance in supercritical extraction of fragrance from jasmine flowers, Proc. 3rd Intl. Symp. SCFs, France, 2, 341, 1994.

Siebert, A. F. and Fair, J. R., Hydrodynamics and mass transfer in spray and packed liquid–liquid extraction column, *Ind. Eng. Chem. Res.*, 27, 470, 1988.

Siebert, A. F. and Moosberg, D. G., Performance of spray, sieve-tray and packed contactors for high pressure extraction, *Sep. Sci. Technol.*, 23, 2049, 1988.

Sorensen, J. P. and Stewart, W. E., *Chem. Eng. Sci.*, 29, 818, 1974.

Stiel, L. I. and Thodos, G., The thermal conductivity of nonpolar substances in dense gasesous and liquid regions, *A. I. Ch. E. J.*, 10, 26–30, 1964.

Sovova, H., Kucera, J., and Jez, J., Rate of vegetable oil extraction with supercritical CO_2 and extraction of grape oil, *Chem. Eng. Sci.*, 49, No. 3, 415–420, 1994.

Takahashi, S., Preparation of generalised chart for the diffusion coefficients of gases at high pressures, *J. Chem. Eng. Japan*, 7, 417, 1974.

Tan, C. S., Liang, S. K., and Liou, D. C., Mass transfer in supercritical fluid extraction, *Chem. Eng. J.*, 38, 17–22, 1988.

Vesovic, V. and Wakeham, W. A., Transport properties of supercritical fluids and fluid mixtures in *Supercritical Fluid Technology*, Bruno, T. J. and Ely, J. F., Ed., CRC Press, Boca Raton, FL, chap. 6, 1991, 245.

Vesovic, V., Wakeham, W. A., Olchowy, G. A., Sengers, J. V., Watson, J. T. R., and Millat, J., The transport properties of carbon dioxide, *J. Phys. Chem. Ref. Data*, 19, 763, 1990.

Wakao, N., Tanaka, K., and Nagai, H., Measurements of particle to gas mass transfer coefficients from chromatographic adsorption experiments, *Chem. Eng. Sci.*, 31, 1109–1113, 1976.

Wakao, N. and Kaguei, S., *Heat and Mass Transfer in Packed Beds*, Gordon and Breach, New York, 1982, 153 and 139.

Wakao, N. and Smith, J. M., *Chem. Eng. Sci.*, 17, 825, 1962.

Weidner, E., Enhancement of heat transfer coefficients in near-critical mixtures of propane and carbon dioxide, Proc. Intl. Symp. SCFs, France, 1, 295, 1988.

4 Flavor and Fragrance Extracts

4.1 MARKET DEMAND

For thousands of years flavors and fragrances have been recognized and used for their healing, cleansing, preservative, and mood-elevating attributes. Today their demand has greatly enhanced due to escalating interest in aromatherapy and increasing usage in improving the quality of cosmetics and perfumery products. The flavors and fragrance (F&F) industry prefers to use mostly natural, rather than synthetic substitutes, for applications in pharmaceutical and edible products. But the shortage, high price, and price fluctuations of the natural flavors and fragrances are often the compelling reasons for partially, if not fully, switching over to synthetic equivalents. Consequently natural flavors and fragrances and synthetic aroma chemicals are suitably blended as dictated by the market forces and their end uses. Flavor formulations also need to meet consumer demands for acceptable flavor profile, stability, and shelf life. The International Organisation of the Flavor Industry (IOFI) has developed a code of practice for the characterization of various flavors and fragrances, based on the following classification (Mahindru, 1992):

1. *Natural:* those flavors which are derived from natural or vegetable sources by physical or microbiological methods. The sources need not be used as food.
2. *Nature Identical:* those flavors which are chemically identical to those present in foods, herbs, and spices.
3. *Artificial:* those not yet shown to be chemically identical to those present in foods, herbs, and spices.

Human senses, along with physical instruments, are often utilized in the evaluation of fragrances and recognition of the characteristic odor profile. However, in the perfumery industry the classification of flavors and fragrances is mostly decided by the perfumer's art of sensory assessment. Consequently sensory assessment is extensively used in the various stages of product development, such as quality control and improvement and cost reduction.

The production of flavor and fragrance is an international industry which was initially confined to highly aromatic natural botanicals, such as rose, sandalwood, and jasmine, but has now rapidly expanded as more essential oils are put into practice every year. Essential oils produced from individual aromatic plants are never used directly. They are further formulated to make flavors and fragrances for a wide range of end uses, such as soaps, cosmetics, confectioneries, alcoholic and nonalcoholic beverages, perfumes, ice creams, aerosols, sprays, syrups, and pharmaceutical products. Today

TABLE 4.1
Natural Aroma Chemicals: Demand and Growth Rate in the U.S.

End Use Industry	Demand ($ Million) 1998	Demand ($ Million) 2000	% Annual Change 1998–2000
Cosmetics and toiletries	30,000	41,500	3.3
Plant-derived chemicals	394	795	7.2
Skin care	167	367	8.2
Perfume	143	240	5.3
Shampoo, cosmetics, toothpaste	84	188	8.3
Essential oils	150	258	5.5
Botanical extracts	77	147	6.6
Gums, polymers, colors	167	390	9.9

Lashkari, 1999.

many aroma chemicals are isolated in pure form from natural essential oils or synthetic nature-identical molecules. Some of these are menthol, camphor, terpineol, citronellal, geraniols, ionones, cinammic aldehyde, citral, etc. Estimated demand for aroma chemicals in the U.S., as given in Table 4.1, indicates a significant growth rate of plant-derived aroma chemicals. There will be a large gap between demand and supply by 2001 to 2002. Thus there exists significant room for new investment in this industry, which calls for proper identification of the end product, selection of the raw material, and adoption of an innovative cost-effective technology for commercial exploitation.

4.2 NATURAL ESSENTIAL OILS

Natural flavors and fragrances are the odoriferous principles found in various parts of the plant including the seeds, roots, wood, bark, leaves, flowers, fruits, balsam, and resin. They are called essential oils because they represent the characteristic essence of their origin. The chemicals responsible for the flavor or aroma are organoleptic compounds, i.e., the compounds that affect the sense organs. They are present in their sources at various concentration levels ranging from parts per billion to parts per hundred. These compounds have molecular weights normally below 300 and are relatively volatile.

Essential oils or aroma chemicals may differ greatly in their chemical constitution, but have some common characteristic physical properties, such as high refractive index, optical activity, immiscibility with water, and yet sufficient solubility to impart aroma to water. However, they are soluble in ether, alcohol, most organic solvents, and in liquid, as well as supercritical carbon dioxide (SC CO_2).

Essential oils are volatile oils different from nonvolatile fixed oils, i.e., glycerides of fatty acids. These essential oils, also called terpenoids, can be classified into two main groups:

1. The hydrocarbons consisting of terpenes, such as monoterpenes, sesquiterpenes, and diterpenes
2. The oxygenated compounds, such as esters, aldehydes, ketones, alcohols, phenols, oxides, acids, and lactones; occassionally nitrogen and sulfur compounds are also present

The term terpene is used for naturally occurring hydrocarbons derived from isoprene (2-methyl butadiene) units having molecular formula of $(C_5H_8)_n$. The terpenoids may be monocylic or bicyclic terpenoids. Terpene hydrocarbons are classified according to the value of n in the general formula of $(C_5H_8)_n$. Some of the most common naturally occurring essential oil constituents are grouped according to their molecular formula:

1. *Monoterpene hydrocarbons* ($C_{10}H_{16}$), i.e., n = 2
   Limonene (bp 176°C)
   Pinene (bp 156°C)
   Sabinene (bp 163°C)
   Myrcene (bp 167°C)
   Terpinene (bp 182°C)
   Cymene (bp 176.5°C)

2. *Sesquiterpene hydrocarbons* ($C_{15}H_{24}$), i.e., n = 3
   β-Caryophyllene (bp 288°C)
   Selinene
   Santalene (bp 252°C)
   Zingeberene (bp 128–130°C)
   Curcumene (bp 137°C)

3. *Derivatives of diterpene hydrocarbons* ($C_{20}H_{32}$), i.e., n = 4
   Phytol, ($C_{20}H_{40}O$) (bp 202–204°C)
   Abietic acid, ($C_{20}H_{30}O_2$) (mp 172–175°C)

4. *Oxygenated derivatives*

   Alcohols
   Benzyl alcohol, C_6H_5CHOH (bp 206°C)
   Santanol, $C_{15}H_{24}OH$ (bp 178°C)
   Geraniol, $C_{10}H_{17}OH$ (bp 229°C)
   Linalool, $C_{10}H_{17}OH$ (bp 209°C)
   Citronellol, $C_{10}H_{17}OH$ (bp 115°C)
   Nerol, $C_{10}H_{17}OH$ (bp 224°C)
   Terpineol, C_7H_7OH (bp 220°C at 10 mm)
   Farnesol

Aldehydes
Anisaldehyde, C_8H_8CHO (bp 248°C)
Benzaldehyde, C_6H_5CHCHO (bp 178°C)
Cinnamic aldehyde, C_6H_5CHCHO (bp 252°C)
Citral, C_9H_5CHO (bp 118°C)
Cuminic aldehyde, $C_{10}H_{12}O$ (bp 235°C)

Ketones
Cyclopentenones, $C_{11}H_{30}O$ (bp 128°C at 12 mm)
(Jasmone and Isojasmone)
Fenchone,$C_{10}H_{16}O$ (bp 193°C)
Muscone, $C_{16}H_{30}O$ (bp 128°C)
Camphor, $C_{10}H_{16}O$ (M.P. 174-179°C)
Carvone, $C_{10}H_{14}O$ (bp 230°C)

Phenol
Thymol, $C_{10}H_{14}O$ (bp 233°C)
Eugenol, $C_{10}H_{12}O_2$ (bp 253°C)
Carvacrol, $C_{10}H_{14}O$ (bp 237°C)

Acetates
Terpinyl acetate
Geranyl acetate
Citronellyl acetate
Neryl acetate
Linalyl acetate

Oxides
Cineol
Linalool oxide
Bisabolol oxide
Bisabolone oxide

The characteristic fragrances of flowers are due to the presence of volatile essential oils in their petals. These oils may occur in a free form as in the rose, or in a combined form (as glycosides) as in the jasmine. The glycosides are natural organic compounds containing sugars in combination with hydroxy compounds. These must be degraded by enzymes present in the flower in order to release the essential oils before they can be recovered. Recovery of flavors and fragrances from their sources is crucial, because of their short life span, especially for flowers that are very much short lived. Besides, due to natural enzyme reactions, there is a continuous change in the odor profile. Such flowers plucked in the morning are known to yield better odor profile than those in the evening.

The geographical, climatic, and soil characteristics, and even the part of the plant, influence the quality and quantity of the essential oil present in the raw

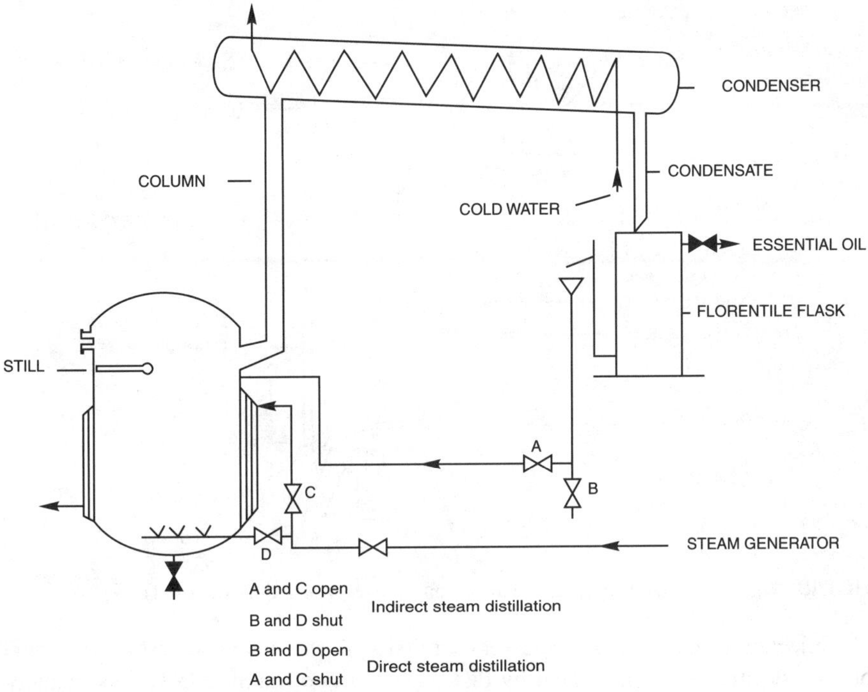

FIGURE 4.1 Steam distillation unit (Meyer-Warnod et al., 1984).

material. For example, the essential oil present in the sandalwood tree is highest in the roots, and there is an average fall from the root to the tip of the stem of 45% in oil content, with about 3% fall in santalol content (Mahindru, 1992c). Harvesting time, age of the tree, or plucking time may affect yield and quality of essential oil and active ingredients. All of these factors are taken into consideration in the selection of raw material for a particular natural extract.

4.3 NATURAL ESSENTIAL OIL RECOVERY METHODS

Recovery of essential oils from their plant sources can be carried out by a variety of both old and new processes, such as steam distillation, hydrodiffusion, enfluerage, maceration, mechanical (cold) expression, solvent extraction, and most efficiently, supercritical fluid extraction. Raw essential oils produced by any one of these methods may be required to be refined or processed further, either by a suitable combination of the above processes or by redistillation, fractionation, chromatography, crystallization, chemical treatment, etc., depending on the nature of the raw material and the desired product quality and specifications.

4.3.1 Steam Distillation

Essential oils are predominantly produced by steam distillation, using direct or indirect steam, as shown in Figure 4.1. The steam is generated in a separate boiler

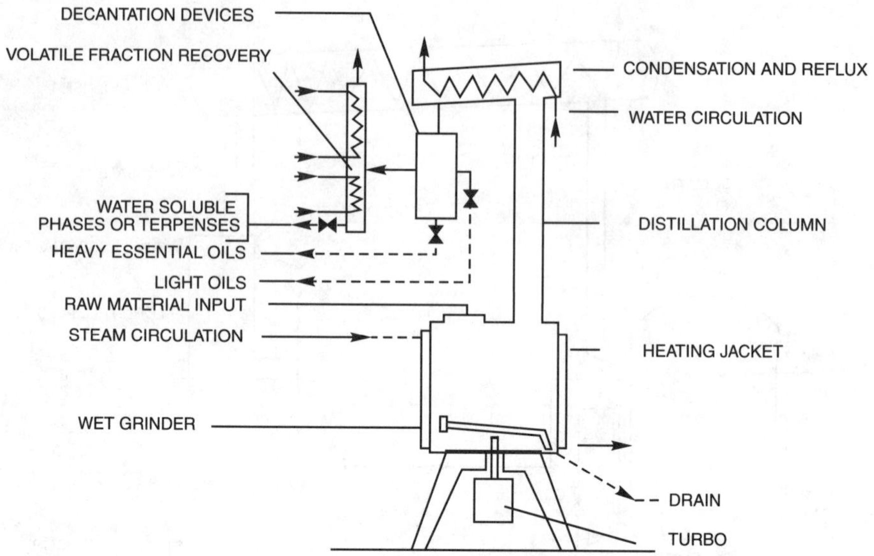

FIGURE 4.2 Accelerated hydrodistillation unit (Meyer-Warnod et al., 1984).

and injected at the bottom of the vessel filled with the botanical matter. Recovery of the essential oil is facilitated by distillation of two immiscible liquids, namely, water and essential oil, based on the basic principle that, at the boiling temperature, the combined vapor pressures equal the ambient pressure. Thus the essential oil ingredients whose boiling points normally range from 200 to 300°C, are made to boil at a temperature close to that of water. Then the steam and essential oil vapors are condensed and separated. The oil, being lighter than water, floats at the top and water goes to the bottom. Essential oils which are not volatile enough in steam, for example, 2-phenyl ethanol present in rose oil, are mostly left behind in the still with the botanical matter. Some very volatile oils may be lost in the distillation process. Further, there may be induced chemical changes, such as oxidation or hydrolysis. Hence, essential oils recovered by steam distillation are often different from those present in the original source.

For accelerating the process of steam distillation, the botanical matter is ground while it is in contact with the boiling water as shown in Figure 4.2. The botanical matter is separated from boiling water in the lower part of the still by a grid. The steam saturated with volatile aroma is condensed and the oil is separated. If the still is heated carefully, the problem of "still odor" can be avoided (Meyer-Warnod, 1984). Accelerated hydrodistillation is employed for fine powders as well as flowers.

Hydrodiffusion is a variation in the conventional direct steam distillation process where the steam enters from the top of the vessel and the oil and water mixture is condensed from the bottom as shown in Figure 4.3. This method requires less time for distillation and reduced energy or steam consumption, and is used especially for processing seeds. There is no problem of hydrolysis, since vegetable matter is never brought in contact with water, only with steam (Meyer-Warnod, 1984).

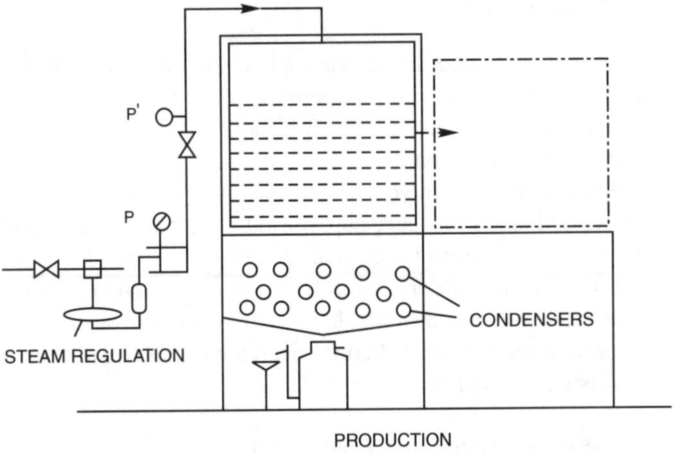

FIGURE 4.3 Hydrodiffusion unit (Meyer-Warnod et al., 1984).

4.3.2 Maceration

Maceration was one of the primitive methods used for the recovery of "absolute" or essential oil from flowers. "Absolute" describes an aromatic oil in the state of liquid or semiliquid. The oil sacs of the fragrant flowers or botanical matter were ruptured by immersing them in a molten fat or oil at 60 to 70°C, which in turn absorbed the aroma of the flowers or the substrate. Spent flowers were separated from the hot fat or oil, and the residual fat or oil was removed by hydraulic press. The treated fat was again reused for the next batch of fresh flowers or botanicals until it was saturated with aroma. This perfumed pommade was either marketed as such or subsequently extracted with alcohol to produce "absolute." Fat/oil and waxes are not soluble in alcohol. "Absolute" was produced after separation of the oil or fat and waxes in a rotary filter and the filtrate was vacuum concentrated after the elimination of alcohol. Maceration was a very tedious, time-consuming, and inefficient method, used in the past for jasmine flowers.

4.3.3 Enfleurage

Enfleurage was another primitive method of extraction of perfume from flowers and other aromatic botanicals by contacting them with cold fats. This process was predominantly used for fragrant flowers like jasmine and tube rose, which continue to emit fragrance even after plucking. The fat used for the process needed to be odorless at the beginning. The fat was spread on either side of a glass plate supported on a rectangular chassis (wooden frame) which held fresh fragrant flowers. Several such glass plates were arranged in the chassis, one above the other, sandwiching the flower in between two such layers of fat or glass plates. Spent flowers (defleurage) were replaced by a fresh charge of flowers until the fat was saturated with fragrance. This process was also a very tedious, labor-extensive, and time-consuming process, mostly replaced by the solvent-extraction method.

4.3.4 Cold Expression

This method is specifically used for recovering essential oils from lemon and orange peels or when essential oils are highly thermolabile. In this process, oil cells are broken by rolling the lemon or orange peels in hollow vessels fixed with spikes on the inside surface for the abrasion of the peel, allowing the oil to ooze out from the outside surface in the form of an aqueous emulsion, which is subsequently centrifuged. Cold expressed citrus oils have superior odor characteristics when compared to steam distilled oils, because of the nonthermal processing. They are also stable because of the natural antioxidants present, such as tocopherols, which are not soluble in steam. However, the unavoidable rise of temperature due to the mechanical friction in the process causes some thermal degradation, the result of which is that cold pressed oil is dark in color.

4.3.5 Extraction with Volatile Organic Solvents

This process is commonly employed for the production of "concrete" or total extract, and later "absolute" or essential oil from flowers, leaves, seeds, roots, and other plant tissues, using organic solvents, such as hexane, petroleum ether, benzene, toluene, ethanol, isopropanol, ethyl acetate, acetone, water, etc. The diffusive transfer of fragrance from the solid to the surrounding solvent in the solid–liquid extraction (also known as leaching) process is affected by two types of contacting devices, either a stationary or a rotatory system. In order to handle large-scale extraction, some perfumery industries have developed continuous countercurrent extraction systems requiring mechanization and automation. The operating temperature and time of extraction are specific to the nature of the botanical substance and contact device. As described in Figure 4.4a, after dissolving the fragrance, the saturated solvent is transferred to a concentrator where the volume of the solution is drastically reduced and the liberated solvent is recycled. The concentrated solution is then vacuum distilled and the residue, known as "concrete," looks like a dark-colored, waxy substance.

For transformation of "concrete" to "absolute," the first is dissolved in alcohol at 30 to 40°C and then cooled to 5 to 10°C, when waxes precipitate out. The filtrate is concentrated under vacuum to eliminate alcohol as shown in Figure 4.4b. The residue is generally liquid, forming the "absolute."

From the above process description, it is apparent that there are a number of steps involved in recovering good quality and quantity of fragrance from a given botanical substance. There may be unavoidable losses of top notes. Further some associated undesirable components may be co-extracted, depending on the polarity of components present in the vegetable matter and the polarity of the solvent. Besides, there may be thermal degradation, hydrolysis, alcoholysis, etc. which are likely to occur, affecting the quality and stability of the fragrance.

4.3.6 Choice of Solvents

Recent recommendations of the European Community's (EC) Solvents Directive indicate that only a limited number of solvents can be used for extraction of food-grade

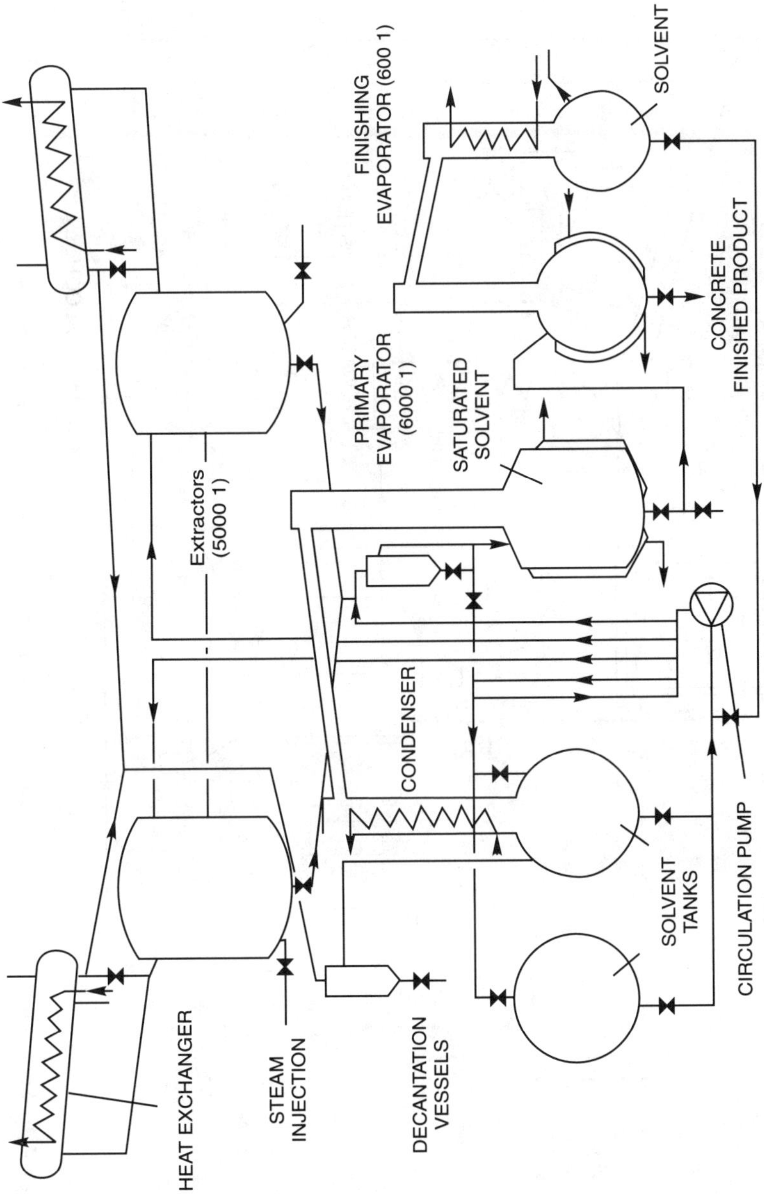

FIGURE 4.4a Solvent extraction unit for concrete (Meyer-Warnod et al., 1984).

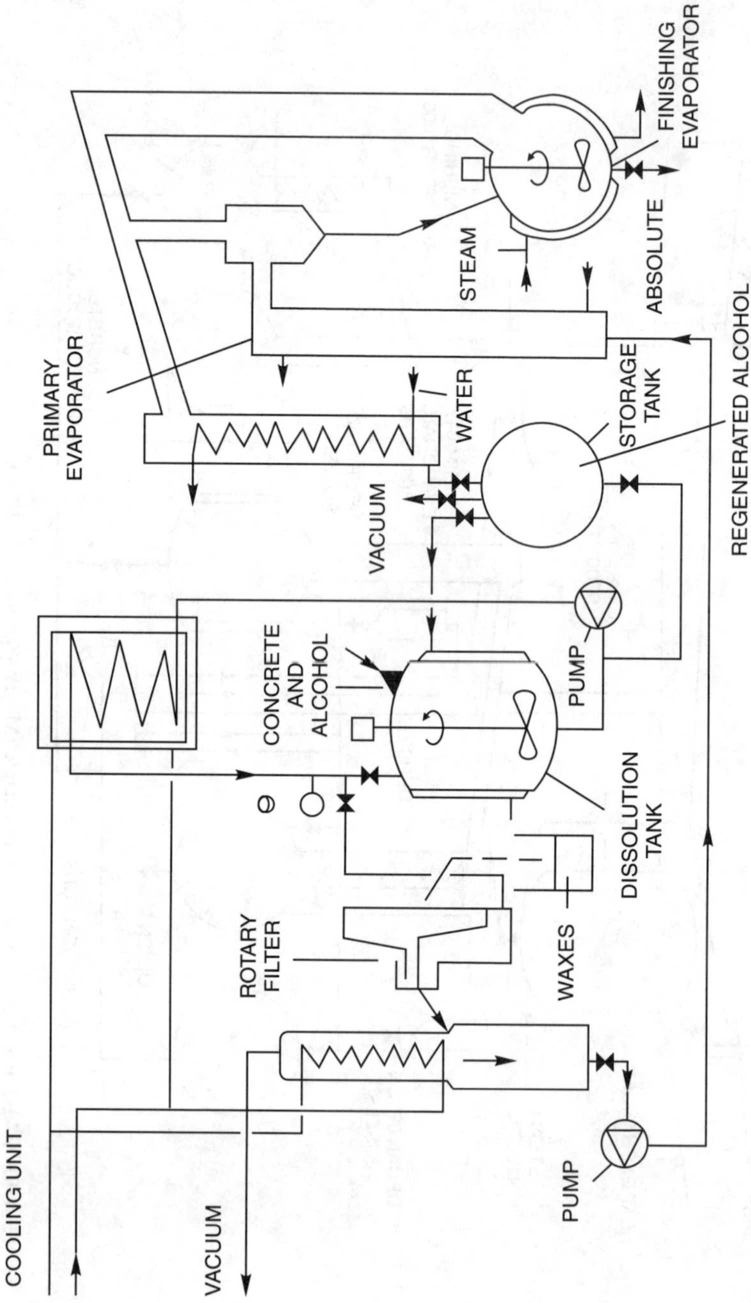

FIGURE 4.4b Solvent extraction unit for absolute from concrete (Meyer-Warnod et al., 1984).

TABLE 4.2
Volatile Solvents and ECC Allowable Limits

Solvent	B. Pt (°C)	Polarity (E°C)	Residues (ppm)	Latent heat (cal/gm)
Ethanol	78.3	0.68	TUQ	204.3
Acetone	56.2	0.47	TUQ 30[a]	125.3
Ethyl Acetate	77.1	0.38	TUQ	94.0
Hexane	68.7	0	1 ppm 5 ppm (oil/cocoa)	82
Pentane	36.2	0	1–2 ppm	84
Dichloromethane	40.8	0.32	0.1 ppm 5 ppm (decaffeinated tea) 10 ppm (decaffeinated coffee)	78.7
IPA (isopropanol)	82.3	0.63	to be reviewed	167
Water	100.0	>0.73	TUQ	540
PG (propylene glycol)	187.4	>0.73		170
Diethyl ether	34.6		2 ppm	
Carbon dioxide	–56.6	0	TUQ	42.4

Note: TUQ: Technically unavoidable quantities.

[a] 21 CFR FDA.

Moyler, 1994.

flavor and fragrances with very restricted, low (1 to 5 ppm) levels of residual solvents as tolerable limits, as listed in Table 4.2. Some useful criteria for the selection of solvents are the boiling points, polarity, latent heat of vaporization, and the allowable limit of residual solvents. The higher the boiling point at reflux condition, the higher the molecular weight profile the extract will have. From knowledge of the profile of the components present, it is possible to predict the components which are likely to be extracted, and to decide the solvents in order of their polarity. The cost of energy required for extraction is related to the latent heat of evaporation, namely, solvents with low latent heat have the advantage of being recovered using less energy than solvents with higher latent heats.

4.4 PURIFICATION OF CRUDE EXTRACT

Crude essential oils produced by any one of the methods discussed above, may need to be purified further by a number of processes, such as simple vacuum distillation, molecular distillation, liquid–liquid fractionation, etc. Chromatographic separation offers a relatively cold process with right selectivity, at the cost of high price. Supercritical CO_2 extraction, discussed later, also provides a cold process for concentrating oils with the suitable combination of desirable active ingredients.

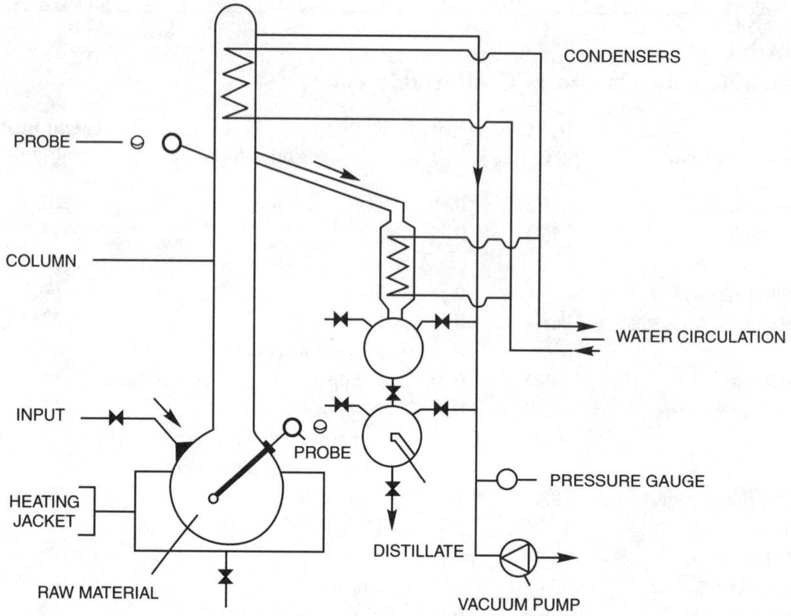

FIGURE 4.5 Vacuum distillation unit (Meyer-Warnod et al., 1984).

4.4.1 Vacuum Distillation

Steam-distilled oils often do not conform to the specifications of perfumers and/or flavorists. It sometimes becomes imperative to either reduce or reinforce certain constituents by using a vacuum-distillation column where various components can be separated according to boiling points. Temperature is gradually increased at a fixed subatmospheric pressure or vacuum is increased to lower the temperature range of operation in order to separate the more volatile ingredients first. Lowering the temperature saves the olfactive components from thermal degradation by distillation. A typical setup for the vacuum distillation process is described in Figure 4.5.

4.4.2 Molecular Distillation

The process of molecular or short-path distillation is carried out under moderate to ultra vacuum (from 10^{-4} to 10^{-8} torr) as a result of which both the operating temperature and time required for separation of the components are highly reduced. This is a secondary purification step for obtaining a lighter colored and more stable product by the process of elimination of high molecular weight undesirable constituents present in the liquid extract which was earlier recovered by the primary process of steam distillation or solvent extraction.

In this process, a heavy and a light solvent are combined with the extract to be purified and then the mixture is passed through a short-path evaporator. The most volatile constituents are collected with a light solvent on the "finger" in the middle of the evaporator and then recovered as the first distillate fraction. The

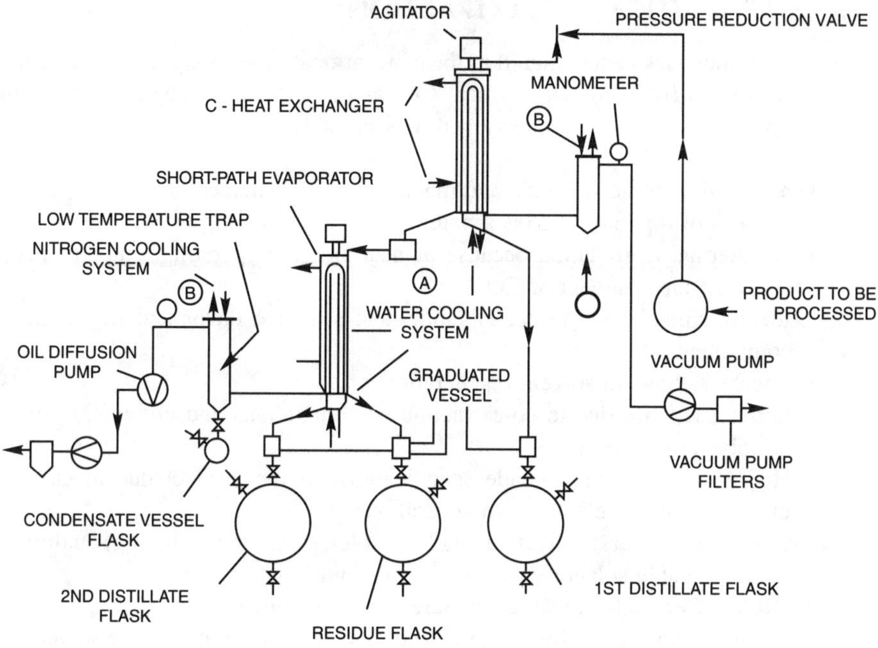

FIGURE 4.6 Molecular distillation unit (Meyer-Warnod et al., 1984).

remaining constituents are condensed on the walls of this first evaporator and then pumped into a second evaporator, as described in Figure 4.6. As before, second distillate fraction is collected in the middle of this evaporator. The residues and heavy solvent are condensed and collected in the residue tank (Meyer-Warnod, 1984). The molecular distillation products are highly expensive and are blended with other products in order to enable the perfumer to have good, long-lasting top notes in the formulation.

4.4.3 Liquid–Liquid Fractionation

The major odoriferous constituents of most essential oils are oxygenated compounds like alcohols, aldehydes, esters, ethers, ketones, lactones, phenols, and phenol ethers. Terpenes and sesquiterpenes are also present along with these compounds and their separation is essential. This is carried out either by vacuum distillation or fractional solvent extraction. Terpeneless oil is recovered with dilute alcohol or similar solvents to remove waxes and sesquiterpenes. Sometimes two partially miscible solvents, such as pentane and dilute methanol in a specific proportion are passed countercurrently through the crude extract in a liquid–liquid contactor. Pentane dissolves the terpenes whereas the oxygenated compounds get dissolved in the methanol mixture. The process is tedious and susceptible to thermal and hydraulic damage. As it depends a lot on the perfumer's skill and experience, the recent trends are towards chromatographic separation, using gas, liquid, or supercritical solvents as the mobile phase.

4.5 SUPERCRITICAL CO_2 EXTRACTION

Most of the drawbacks encountered in the conventional processes mentioned earlier are circumvented by using supercritical CO_2 as the extractant. The advantages of this relatively new method, as mentioned in Section 1.1.4, include:

- No residual toxic solvents, and much-reduced pesticides
- No loss of topnotes and back notes
- No thermal degradation because of near-ambient operating temperature and inert environment of CO_2
- Ideally suitable for thermally labile-heat sensitive flavor and fragrance components
- Energy savings in solvent regeneration
- Better shelf life due to co-extraction of antioxidants and elimination of dissolved oxygen
- High purity and tailor-made specifications of the product due to easy manipulation of selectivity of separation
- Completely natural extract having light color, transparent (shining) quality
- Nonflammable solvent with no environmental hazard
- GRAS (Generally Regarded as Safe) and non-combustible
- A wide spectrum of physical properties can be obtained in a single solvent by small variations of process parameters, such as pressure and temperature, or an entrainer, making it a flexible, versatile, and multiproduct solvent
- Faster extraction and high recovery of extracts
- Excellent blending characteristics of the extracts

It has been established that of all available and allowable volatile solvents for extraction, supercritical carbon dioxide (SC CO_2) brings out the most natural smell and taste in the extracts, bearing the closest resemblance to the original material. Some flavorists prefer to use a hydrocarbon or chlorinated solvent to extract the raw material first and subsequently purify it with SC CO_2. However this practice may suffer from loss of top notes and thermal degradation, although the cost of extraction equipment may be lowered to some extent. CO_2-extracted oils are generally more concentrated than those obtained by steam distillation or by conventional solvent extraction from the same starting material. This is due to the occurrence of lower levels of monoterpene hydrocarbons since no additional monoterpenes are formed as in steam distillation. These terpenes tend to dilute the active aromatic components and do not significantly contribute to the odor profile. A comparative study of jasmine fragrance extraction from concrete by SC CO_2 extraction and molecular distillation indicates that there is not much economic advantage in the former method over the latter. Entrainers have an advantage of increasing the polarity of SC CO_2 and can be added to the flowing CO_2 before entering the extraction equipment. Ethanol and possibly water are the most acceptable "natural" entrainers for food-grade extracts, although other organic solvents listed in Table 4.2 may also be mixed with CO_2 in small (1 to 5%) concentrations. It has been observed that the extracts contain some traces of saturated and unsaturated lipids which do not contribute to the flavor, however they can improve the solubility and hence the blending properties of some products. In formulating fragrances, these triglycerides may even act as the natural fixative.

4.5.1 Commercial Advantage

SC CO_2 extracted flavor and fragrance are significantly different from their conventional equivalents. They should be viewed as new products rather than direct replacement of conventional extracts. They are already well established as commercial products and are being produced in the U.S., Europe, Japan, China, and Australia.

There is, in general, apprehension about the high capital costs for high-pressure extraction equipment, added to the high cost of the technology. Nevertheless, energy costs in this process are lower than those incurred in steam distillation and solvent extraction, which more than offset the high capital costs involved in the SC CO_2 extraction process, even if no premium is attributed to the superiority of the extracts. SC CO_2 extraction equipment may cost 50% more than the subcritical or liquid CO_2 extraction equipment (Moyler, 1993). Again the higher costs of SC CO_2 extraction equipment are often offset by more complete extraction and the possibility of fractionation of the extract to a number of products. For achieving the desired properties of flavor, taste, color, and shelf life, it is imperative to carefully select the raw material sources, extraction and fractionation parameters, and entrainer to be added. It has been published that out of about 350,000 different species that have been identified, about 5% (i.e., 17,500) are aromatic plant, and that about 300 different plant species are used for production of essential oils for the food, flavor, and fragrance industry (Boelens, 1996a). The annual world production of volatile oils is estimated to be to the tune of 100,000 metric tonnes with a value of more than $1 billion U.S.

Out of several methods of production of essential oils, in general, the yields by CO_2 and ethanol extractions are higher than those by steam distillation. These differences are mainly due to the fact that the extracts contain nonvolatile residues. The other methods give still lower yields. Subcritical CO_2 extraction gives yields close to that by steam distillation, while SC CO_2 extraction yields match more or less like selective organic solvents. Subcritical CO_2 extracts are superior to the steam distilled extracts, as the former are closest to the natural headspace odor of the botanical. Supercritical CO_2 co-extracts certain relatively higher molecular weight lipid antioxidants which improve the shelf-life of the extracts.

The wide range of plant materials which can be extracted on commercial scale with SC CO_2 is listed in Table 4.3, along with information on yields by various methods of extraction.

It was reported (Mahindru, 1992) that of annual production and utilization of perfumery and flavoring materials, about 80% represented perfumes and the rest flavors for food, dental, and pharmaceutical products. But only 20% of the perfumes and flavors are natural. India is now a major exporter of these flavors and fragrances and is now exporting to the tune of 1500 metric tonnes per annum (MTPA) of natural flavor and fragrances, valued at about $150 million. With continuous development in supercritical extraction technology and in view of the huge reserve of botanicals

TABLE 4.3
Comparison of Percent Yields of Flavors and Fragrances from Various Natural Products

Name	Steam Distillation	CO_2 Liquid	CO_2 SCF	Solvent Extraction
Angelica	0.3–0.8	3	—	—
Aniseed	2.1–2.8	—	7	15 (ethanol)
Asafoetida	2.3–14			
Basil	0.3–0.8	1.3	—	1.5
Calamus	0.4–3.8	5.1		
Caraway	3–6	3.7	—	20
Calendula		2.3		
Carrot	0.2–0.5	1.8	3.3	3.3
Cardamom	4–6	4	5.8(7)[a]	10
Cassia	0.2–0.4	0.6	—	5 (dichloromethane)
Celery	2.5–3.0	3	—	13
Cinnamon	0.5–0.8	1.4	—	4
Clovebud	15–17	16	22	20
Chili	<0.1	4.9		10 (acetone)
Camomile	0.3–1.0	2.9		—
Coriander	0.5–1.0	1.5		
Cumin	2.3–3.6	4.5	14.0	12
Dill	2.3–3.5	3.6	—	—
Fennel	2.5–3.5	5.8		15
Fenugreek	<0.01	2	—	8
Ginger	1.5–3.0	3	4.6	7
Garlic	0.06–0.4	—	0.1–0.3	
Hop	0.3–0.5	12		20
Mace	4–15	13		40 (expression)
Nutmeg	7–16	13	—	45 (expression)
Oregano	3–4	—	5	—
Pepper	1.0–2.6	4	18(8)[a]	18
Patchouli	1.6–3.6			
Parsley	2.0–3.5	3.6	—	20
Poppy	—	2.5	—	50
Rosemary	0.5–1.1	—	7.5	5
Sage	0.5–1.1		4.3	8
Sandalwood	3–6	4.8		
Thyme	1–2	0.8	—	—
Tea	<0.01	0.2	—	35 (ethanol + water)
Turmeric	5–6	3.4		
Vanilla	<0.01	4.5		25–45
Vetiver	0.5–1.0	1.0	1.0[a]	

[a] Using 10 l pilot plant at Indian Institute of Technology, Bombay.

Moyler, 1993.

TABLE 4.4
Yield of Flower Concrete by Solvent Extraction and Absolute by Ethanol and CO_2 Extraction

Flower	Yield of Concrete from Flower (%)	Yield of Absolute from Concrete (%)	Yield of SC CO_2 Extract from Flower (%)
Helichrysum	0.90–1.15	60–70	4.4–6.6
Hyacinth	0.17–0.20	10–14	
Jasmine	0.28–0.34	45–53	0.44–0.66[a]
Lilac	0.6–0.95	35–45	
Orange flower	0.24–0.27	36–55	0.28
Rose	0.22–0.25	50–60	
Rose-de-mai	0.24–0.27	55–65	
Tube rose	0.08–0.11	18–23	
Violet	0.07–0.13	35–40	
Ylang Ylang	0.80–0.95	75–80	

[a] Sastry and Mukhopadhyay (1994).

Lawrence, 1995.

and biodiversity, exploiting this technology in India and other Asia Pacific countries may even have an edge over that in other countries, as value addition can be carried out on the diverse botanical resources available there. Thus, high value-added natural perfumes, flavors, and fragrances can be exported instead of just the basic raw materials. For dairy and confectionery products, CO_2 extracts are preferred in the formulation of high-quality natural flavors. For use in soft drinks, CO_2 extracts like ginger offer both pungency and flavor in the most stable form and can be used for bottled syrup. It has also been established that CO_2 extracted orange peel oils have fewer terpenes (limonene) and more aldehydes (citrals) as compared to the cold pressed oils.

4.6 SC CO_2 EXTRACTED FLORAL FRAGRANCE

A large number of flowers have been extracted with subcritical (liquid) or supercritical (fluid) CO_2 to concentrate and isolate components of interest. Yields of these floral extracts are compared in Table 4.4 with those by solvent extraction.

4.6.1 JASMINE FRAGRANCE

Jasmine fragrance is considered indispensible in high-grade perfumes. There is hardly any good floral or oriental perfume which does not contain a little amount of jasmine flower oil. Jasmine oil blends with practically any floral scent, lending smoothness and elegance to perfume compositions. About 10,000 flowers weigh 1 kg and a good worker spends 2 h to pluck them. Hexane extraction of 300 to 360 kg flowers yields 1 kg of jasmine concrete, which in turn produces 0.5 kg of jasmine absolute (Lawrence, 1995), whereas supercritical CO_2 extraction of 250 kg flowers yield 1 kg of jasmine oil extract in a single-step process.

Through the years, different techniques such as enfluerage, maceration, and solvent extraction have been practiced for making "pommades" and "concrete" from jasmine flowers and subsequently "absolute" after separating the waxes from the concrete (Muller, 1965). The jasmine absolute can also be obtained by supercritical CO_2 extraction of the concrete. About 40 to 50% yield of absolute can be obtained by SC CO_2 extraction of concrete, depending on the pressure and temperature conditions of the extraction and the quality of the concrete.

Recently, SC CO_2 extraction of jasmine fragrance directly from flowers has been evaluated (Sastry and Mukhopadhyay, 1994) with and without the addition of entrainer to flowing CO_2. Entrainers may also be added directly to the flowers, for reducing the affinity of the substrate or for modification of the morphology of the flowers for better release of fragrance. With proper choice of operating conditions of extraction and entrainer addition, good quality fragrance absolute can be produced directly from flowers with a very high yield (0.4 to 0.7%) on the basis of the weight of petals of flowers plucked in the morning. Comparison of extraction performance of three organic solvents, e.g., petroleum ether, benzene, and ethanol with liquid and SC CO_2 (Sastry, 1994) shows that SC CO_2 gives higher yields and imparts better quality to the extracts by selectively enriching them with the key components than any other solvent. Hindrance of the flower substrate against recovery of fragrance has been demonstrated (Sastry and Mukhopadhyay, 1994) to be dependent on the extraction process parameters. The hindrance is slightly higher at a higher temperature of 60°C than at 40°C, and there is a minimum pressure at any temperature where hindrance is minimum, as can be seen from Figure 4.7a. The polarity of the cosolvent is an important factor in the reduction of the hindrance factor, along with pressure and temperature. The hindrance factor is defined as the ratio of the neat solubility to the loading of the particular component during the static extraction from flowers at the same supercritical condition of CO_2. The effect of pressure on the hindrance factor with surface modification is seen to be different from that without the addition of an entrainer onto the flowers, as can be seen from Figure 4.7b. The addition of methanol directly onto the flowers is observed to induce as high as a four-fold decrease in substrate hindrance (Figure 4.7b). It also facilitates preferential extraction of important constituents, improvement of the overall yield, and substantial reduction of extraction time. Ethyl acetate has been found to give the best performance as an entrainer, followed by isoproponal, n-butanol, methanol, and acetone, respectively.

SC CO_2 at 120 bar and 40°C, even without an entrainer, gives extracts of floral jasmine fragrance which is superior to that obtained by any other solvents. A comparison of compositions of the extracts with liquid CO_2 and SC CO_2 with and without addition of methanol onto the flowers, is given in Table 4.5. The addition of a small amount (3 to 5 wt%) of water-soluble entrainer directly onto the flowers, prior to extraction, results in the enhancement of dissolution of fragrance components, due to morphological changes of the flower substrate and reduction of the affinity of the components. Table 4.6 shows the influence of the entrainer on the dissolution ratio enhancement, DE_i, for

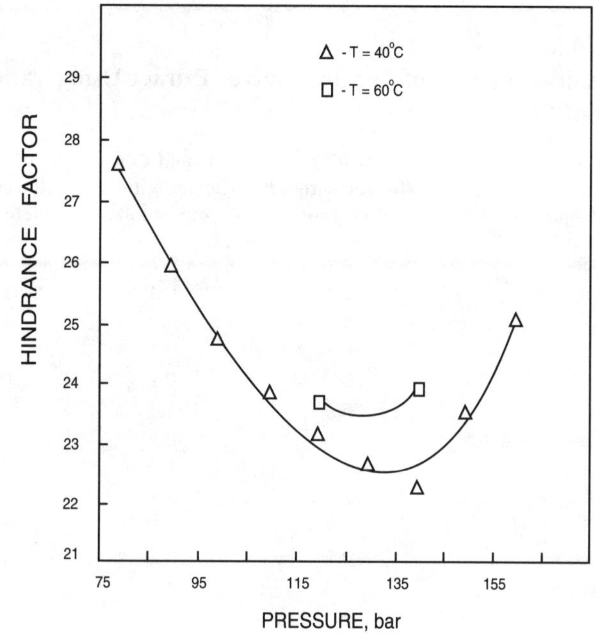

FIGURE 4.7a Effect of temperature and pressure on hindrance factor.

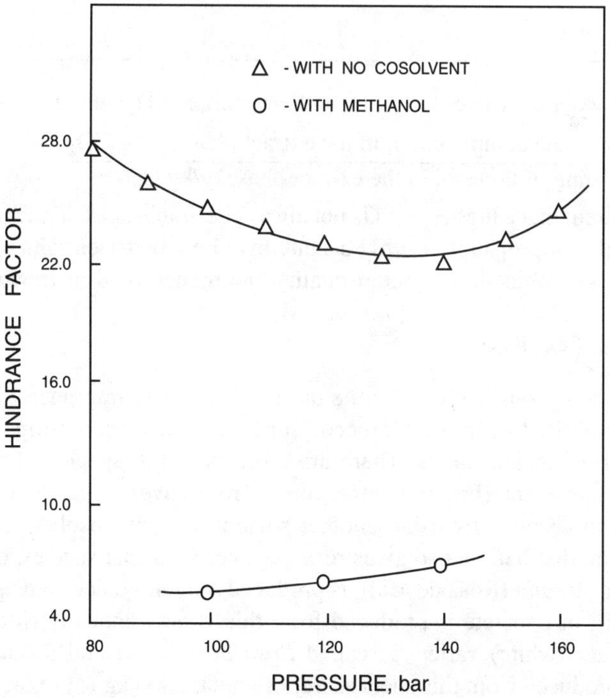

FIGURE 4.7b Effect of pressure on hindrance factor with and without cosolvent methanol.

TABLE 4.5
Composition (Wt%) of Jasmine Flower Extract Using Various Forms of CO_2

Components	SC CO_2 (flower without methanol)	Liquid CO_2 (flower without methanol)	SC CO_2 (flower with methanol)
α-Terpineol	0.9	0.7	1.2
Farnesol	3.4	0.2	12.9
Linalool	1.2	4.9	1.5
Geraniol	0.7	0.1	1.0
Nerol	0.2	0.1	0.2
Indole	26.2	11.4	39.5
Methyl anthranilate	1.5	0.9	2.9
Benzoic acid	3.0	1.5	3.5
Benzyl alcohol	1.7	3.4	2.1
p-Cresol	1.8	1.9	1.0
Linalyl acetate	0.7	3.2	1.1
Benzaldehyde	0.9	4.6	1.5
Benzyl acetate	4.8	16.7	4.2
cis-Jasmone	17.0	3.5	9.0
Eugenol	10.8	1.3	13.0
Total yield g/kg of flower	4.4	2.8	6.6

Sastry, 1994.

jasmine components, where, $DE_i = D_i$ with an entrainer/D_i without an entrainer, and

$$D_i = \frac{\text{Loading of the component, i, in the extract phase, g/kg } CO_2}{\text{Loading of p-cresol in the extract phase, g/kg } CO_2}$$

It is apparent that a higher yield is not always desirable since it affects the quality, so accordingly, isopropanol or ethyl acetate may be a better entrainer than acetone in imparting a combination of better quality and higher yield of fragrance extract.

4.6.2 Rose Fragrance

Rose fragrance is considered to be the queen of all floral fragrances. Rose flowers have been cultivated in India, Morocco, Tunisia, Italy, France, former Yogoslavia, and China from ancient times. There are more than 120 species of roses known. Like jasmine, solvent (hexane)-extraction of rose gives concrete which is then transformed to absolute by using another solvent, namely absolute alcohol. Alternatively, steam distillation also gives rose essence. A special species, cabbage rose, called *Rosa centifolia* (Rosa de Mai), is produced only in France, and approximately 550 to 800 kg of concrete is produced from this variety annually (Meyer-Warnod, 1984). Another (white) variety is called *Rosa damascena* and about 3000 kg of concrete is produced from this species, out of which 2000 kg is produced in Turkey. Solvent extraction of rose yields about 0.22 to 0.25% concrete and 0.1% of absolute,

TABLE 4.6
Variation of Dissolution Ratio Enhancement, DE_i at 120 bar and 40°C with Addition of Different Entrainers onto Flowers

Component (i)	Methanol	Acetone	Isopropanol	Ethyl Acetate	n-Butanol
α-terpineol	2.40	3.20	6.14	3.21	1.80
Farnesol	6.83	3.92	8.08	2.00	1.15
Linalool	2.25	1.95	3.20	1.86	1.55
Geraniol	2.57	5.91	4.98	4.86	1.75
Nerol	1.80	12.6	2.63	1.36	1.35
Indole	2.71	2.58	2.61	2.93	1.75
Methyl anthranilate	3.48	1.68	4.07	5.57	3.45
Benzoic acid	2.10	3.06	2.68	4.43	2.85
Benzyl alcohol	2.22	1.69	3.03	3.79	3.30
p-Cresol (Ref.)	1.00	1.00	1.00	1.00	1.00
Linalyl acetate	2.83	3.09	3.37	5.07	2.10
Benzaldehyde	2.99	6.20	3.55	1.57	1.70
Benzyl acetate	1.57	2.29	1.64	3.43	1.35
cis-Jasmone	0.95	0.85	2.18	3.86	1.55
Eugenol	2.14	2.03	2.34	2.36	3.00

Sastry, 1994.

whereas steam distillation of rose yields 0.025% of essence (Meyer-Warnod, 1984). About 50% of steam-distilled rose essence comes from Turkey, whereas the rest comes from Bulgaria and Morocco. Total world production of rose oil is at least 15 tonnes and the demand is still on the rise. *Rosa gallica* is conventionally used for commercial production of attar. While attar is enriched in beta phenyl ethanol, nonadecane, geraniol, and nerol, the damask rose is high in geraniol acetate, citronellol, geraniol, etc. The Bulgarian *R. damascena* rose oil contains total alcohols (calculated as geraniol) ranging from 66 to 80%, whereas Indian *R. damascena* contains 58 to 69% geranoils and 16 to 32% citronellol (Mahindru, 1992). For obtaining good-quality rose oil, flowers are picked early in the morning before sunrise, because volatiles decrease by 5 to 10% during the day.

It is to be noted that 2-phenyl ethanol content in CO_2 extract is 67.5%, much higher than that in the hydrodistilled oil (Boelens, 1996b). The odor value (quality and intensity) is higher for hydrodistilled oil and thus preferred by perfumers. The solvent extracted absolute contains up to 60% of 2-phenyl ethanol, but steam-distilled oil contains much less quantity of 2-phenyl ethanol since it is not steam volatile (Mahindru, 1992b).

Absolutely natural rose oil can be extracted with SC CO_2 which has all of the attributes of the solvent (hexane/ethanol)-extracted absolute in addition to the special attributes of SC CO_2 extracts, namely superior quality and no residual solvent. SC CO_2 extraction of flower concrete was carried out (Reverchon et al., 1994; Moyler et al., 1992; and Moates and Reynolds, 1991) in an extractor with an internal volume

TABLE 4.7
Chemical Composition of Rose Oils

	France[a] Rose Oil	Bulgaria Rose Hydrodistilled	Turkey Rose Hydrodistilled	Bulgarian Rose CO_2 Extract
α-Pinene	—	0.6–0.8	0.6	0.73
Ethanol	—	1.5–3.0	0.5	1.42
Linalool	0.5	2.1–2.3	0.81	0.11
2-Phenyl ethanol	1	1.7–2.0	1.85	67.53
β-Citronellol	50	27.5–28.0	45.00	7.77
Nerol	9	7.8–8.6	10.10	2.15
Geraniol	18	16.0–17.0	20.50	4.15
Geranial	—	0.5–1.0	1.34	0.33
Geranyl acetate	2	0.7–0.8	2.04	0.43
Eugenol	1	1.1–1.2	0.99	1.19
Eugenol methyl ether	2	1.6–1.7	2.85	0.71
Farnesol	—	1.4–1.5	1.38	0.11
Nonadecanes	10	14.0–15.0	3.05	3.85
Heptadecanes	—	1.5–1.7	—	0.84
Hencosanes	—	4.0–5.0	0.1	0.78
Tricosanes	—	1.0–1.2		0.23
Hexacosane	—	0.5		0.72
Rose oxide	0.6	0.5		0.15

[a] Wright, 1994.

Boelens, 1996b.

of 20 l and 1500 g of concrete was charge-mixed with 2/3-mm-diameter glass beads for maximizing the contacting surface. The CO_2 stream from the exit of the extractor was fractionated in three separators in series. The chemical compositions of various rose oils are summarized in Table 4.7.

Rose oil is used in confectionery, perfume, soft drinks, etc. Rose oil is frequently adulterated with nature-identical components, e.g., geranoil and citronellol. A typical formulation of rose flavor used for confectionery (Lawrence, 1994) contains only 2% natural rose oil, 1.5% nature-identical (citronellol, geraniol, and phenyl ethyl alcohol) constituents and the rest, the solvent base (e.g., isopropanol and propylene glycol). There are no legal restrictions on the use of rose oil in flavorings.

4.6.3 Bitter Orange Flower Fragrance

The perfume of bitter orange flower oil, also known as neroli oil, is popularly considered as the princess of floral fragrances. It is used in perfumes, toilet waters, after shaves, and eau de colognes.

Steam distillation of orange flowers yields as low as 0.08 to 0.1% neroli oil, whereas solvent (hexane)-extraction of orange blossoms yields 0.25 to 0.20% concrete; and from bitter orange concrete, in turn, is produced 36 to 55% absolute

TABLE 4.8
Comparison of Bitter Orange Flower Oil by Various Processes

Compound	Hydrodistilled Oil (%)	SC CO_2 Extract (%)
Monoterpenes	38	28
Linalyl acetate	3–5	24
Linalool	38	35
Sesquiterpene alcohols	4	<2
Nitrogen derivatives	<0.5	2

Boelens, 1996b.

TABLE 4.9
Chemical Composition (%) of Bitter Orange Flower Oils

	Hydrodistilled Oil		CO_2 Extracted Oil
Compound	(Spain) Flowers	(Tunisia) Flowers	(Morocco) Flowers
β-Pinene	10.5–13.0	14.6	8.8
Myrcene	1.4–3.1	2.1	1.1
Limonene	12.9–17.9	12.2	11.5
trans β-Ocimine	5.6–7.0	6.2	4.5
Linalool	31.4–47.1	37.8	34.6
Linalyl acetate	0.6–10.6	3.3	23.6
α-Terpineol	1.1–3.5	4.11	1.1
Nerolidol	2.1–3.4	3.4	1.2
Geraniol	0.8–2.3	2.0	0.4
Farnesol	0.7–1.6	1.6	0.4

Boelens, 1996b.

(Meyer-Warnod, 1984). From the yearly production of 1 tonne of absolute, 90% is produced in Tunisia and Morocco, and the rest in France.

The hydrodistilled oil- and the supercritical CO_2-extracted bitter orange flower oils are significantly different, as can be seen from Tables 4.8 and 4.9. The difference in the compositions of monoterpenes and linalyl acetate is attributed to the decomposition of linalyl acetate into monoterpenes during hydrodistillation. Similarly, the presence of lesser amounts of nitrogen derivatives in steam-distilled product is due to the solubility of these compounds in water, and consequently their loss to the water during hydrodistillation. The odor value, both by quality and intensity, is much higher (about twice) for CO_2-extracted bitter orange oil as compared to hydrodistilled oil. That is why the former is preferred by perfumers.

TABLE 4.10
Composition of Lavender Extract by Supercritical Extraction and Hydrodistillation

Component	SC CO_2 Extract(%)	Hydrodistillation Product (%)
1-8 Cineole	5.83	6.75
Linalool	25.29	35.31
Camphor	7.90	7.81
Borneol	2.30	2.98
4-Terpineol	3.79	3.34
Linalyl acetate	34.69	12.09
3-7 dimethyl acetate	3.08	4.38
β-Farnescene	2.23	1.00
α-Bisabolol	2.09	3.76

Reverchon et al., 1994.

4.6.4 Lavender Inflorescence Fragrance

Supercritical CO_2 extraction from Italian unground lavender flowers was performed at 90 bar and 48°C followed by a two-stage separation procedure conducted at 80 bar, 0°C, and 25 bar, –10°C, in the first and second separators, respectively. This resulted in a maximum oil yield of 1.9% by weight (Reverchon et al., 1994). However, the recovery of extract obtained by supercritical CO_2 at 110 bar and 40°C from ground Turkish lavender flowers (Adasoglu and Dincar, 1994) was 90.4% and the yield was slightly lower than merely 1.37% yield obtained by steam distillation of ground Turkish lavender (*Lavandula stoechas*) flowers having particle size –2000 to 1000 μm. This is attributed to the degradation of oil by communition and drying.

The organoleptic characteristics of the hydrodistilled and SC CO_2-extracted oils are found to be different. The linalyl acetate content was found to be 34.7% of the oil in the SC CO_2 extract, as compared to 12.1% of the hydrodistilled oil. The difference is attributed to the hydrolysis of a part of the components as seen in Table 4.10.

4.6.5 Marigold Fragrance

Like rose, marigold flowers are hydrodistilled to obtain a yellow-amber colored essential oil with a powerfully fruity (apple) sweet smell. Even leaf and stem are also hydrodistilled since all of these oils have a good market in the perfumery industry. France obtains marigold essential oil from an "English-French" variety (*Compositae patula*), whereas the variety originally grown in Mexico is striped marigold (*Compositae tenuifolia*) with a sweet smell. The native South American variety (*Compositae minuta*) is now grown all over the world and gives the highest yield. The compositions of the hydrodistilled oils from these varieties are given in Table 4.11.

The South American variety produces oil which has germicidal and microbicidal applications. It is one of the major sources of essential oil in Kenya, Australia,

TABLE 4.11
Percent (by wt) Composition of Marigold Oil by Hydrodistillation

	French Marigold			South American Marigold (Flowers)	Mexican Marigold (Flower and Plant)
Constituents	Flower	Leaf	Stem		
Tagetone	40.4	40.5	26.3	15.4	13.5
Linalool	22.1	16.6	15.2	Traces	—
Limonene	14.0	21.8	24.6	Traces	2.7
Linalyl acetate	13.8	11.0	9.1	1.8	—
Ocimene	8.5	6.0	23.7	15.0	11.0
Phenyl ethyl alcohol		—		15.4	9.1
α-Terpenol					8.8
Endesmol	—	—		2.2	3.6
β-Phenyl ethyl methyl ether					2.1
Salicyl aldehyde				7.9	2.1
Phenyl acetaldehyde				3.5	
2-hexen-1-al				3.5	
Aromadendrene				17.3	
Yield	0.027	0.137	0.037	0.5–1.0	0.027

Adapted from Mahindru, 1992a.

France, and India. The marigold oil readily polymerizes due to the presence of ocimene and tagetone, and needs to be stored with special care. The African (Aztech) marigold absolute is produced from a variety of flowers grown in Egypt and South Africa, which had its origin in Mexico. Marigold flower extracts are used both for oil and color. Yield of concrete by hexane extraction is about 0.20 to 0.22% of the flower, and the concrete, in turn, yields 30 to 32% absolute (Meyer-Warnod, 1984). It is used in "romantic" perfumes to give a soothing floral note. It also forms the bases for other floral perfumes, like hyacinth, lilac, tube rose, narcissus, etc.

4.7 SANDALWOOD FRAGRANCE

Sandalwood oil has very good fixative properties and finds applications in a classic blender fixative for use with rose, violet, tube rose, clove, and lavender orientals. It is very light in color and can be used without fear of ultimate coloration of the product where it is applied. On the other hand, it has such a delicate aroma that it can be blended in small quantities without altering the dominant fragrance. It is used in soaps, cosmetics, incense, perfumes, and confectioneries. The East Indian sandalwood roots of dead trees, or of 50 to 60 year-old trees are reported to give the highest yield of sandal wood oil, to the tune of 5.3 to 6.2%, but the santanol content of the root oil is on the lower side of 90%. On the other hand, sandalwood chips of the same type of trees yield less oil, but with a higher percentage (90 to 97%) of free santanol (Mahindru, 1992). The West Australian sandalwood, yields 2 to 4%

oil (King and Bott, 1993), having much lower free santanol (69 to 80%) content. A minimum of 90% santanol content is supposed to be present in the sandalwood oil to make it saleable as premium quality at market. Thus a proper selection of wood is required to produce sandalwood oil of santanol content more than 90%. There are two isomers of santanol: α-santanol and β-santanol; however the α-form predominates. Besides santanol, α and β santalene, along with santalyl acetate, also contribute to the overall fragrance.

Conventionally, steam distillation is employed for recovering sandalwood oil which yields 3.8% oil after 24 h of distillation, whereas liquid CO_2 extraction yields 4.9% oil in 2 h. The CO_2 process yields oil with 83.6% santanol, whereas the steam-distilled oil has a lesser santanol content of 81.8% (Mahindru, 1992c). Santanol content increases with the progress of extraction with CO_2, as the terpene content decreases. Lawrence (1995) reported that benzene extraction of sandalwood yielded 6.0 to 8.5% concrete, which in turn produces 3 to 4% oil. Liquid CO_2 and supercritical CO_2 extraction methods are superior to other methods, due to higher santanol and lower terpene contents, and also produces higher yields even though the conventional steam-distilled oil still seems to be the fragrance preferred by users.

4.8 VETIVER FRAGRANCE

Vetiver (*Vetiveria zizanioides*), native to India and Srilanka, popularly known as "khus," has been grown since ancient times for its excellent fragrance. It is a perennial nonflowering grass with spongy aromatic roots. The fibers of the aromatic grass are woven into mats, fans, and baskets that emit fragrance. The roots are also used for perfuming clothes in wardrobes and also as insect repellant. The roots contain more essential oil (1 to 2%) compared to the grass which yields merely 0.1 to 0.2% (Mahindru, 1992). The essential oil from vetiver roots, known as "the oil of tranquillity," is a high-value product which is extensively used in perfumes and cosmetics. Vetiver oil finds use as a fixative in perfumery and as a fragrance ingredient in soaps. It blends well with patchouli, rose, and sandalwood oils. The main constituents of the oil are vetiverol, vetivone, tricyclovetiverol, khusilal, khusinol, khusimol, khusol, and benzoic acid. The total alcohol, as % by wt, vetiverol is more than 70%, while the total ketone content, as % by wt, vetiverone ($C_{15}H_{22}O$) is more than 24%. The ketone-rich oils are preferred to the alcohol-rich oil by perfumers. Liquid CO_2 extraction from vetiver roots yielded 1% by wt of essential oil and the vetiver root was found to require only 2 h of extraction time compared to 8 h required for extracting volatiles by steam distillation, even though the yield was slightly lower (0.9%) in the latter case from the same roots (Mahindru, 1992d). The GC analysis of the extracts confirmed that the volatile constituents are extracted to a greater extent by liquid CO_2. Further, SC CO_2-extracted oil contains fewer terpenes but more fragrance (ketone) components compared to the steam-distilled oil. Since the quality depends on the main constituents, a better quality of oil can be produced with SC CO_2 as the extractant.

REFERENCES

Adasoglu, N. and Dincar, S., Optimization of supercritical fluid extraction of essential oil from Turkish lavender flowers by response surface methodology, Proc. 3rd Intl. Symp. SCFs, France, 329, October, 1994.

Boelens, M. H., *Proc. 10th Intl. Congr. Essential Oils Fragrances and Flavors*, Washington, D.C., Elsevier Science Publ. BV, Amsterdam, 551–565, 1988.

Boelens, M. H. and Oporto, A., Natural isolates from Seville bitter orange tree, *Perf. Flav.*, 16(6), 1–7, 1991.

Boelens, M. H., Production of essential oils, Proc. 27th Intl. Symp. Essential Oils, Vienna, Sept. 8–11, 1996a.

Boelens, M. H., The chemical and sensory properties of orange flower oil and rose oil revisited, Proc. 27th Intl. Symp. Essential Oils, Vienna, Austria, Sept. 8–11, 1996b.

Geunther, E., *The Essential Oils*, Chap. 2, Van Nostrand, New York, 17–25, 1955.

Lawrence, D. V., The flavoring of confectionery, in *Food Flavorings*, Ashurst, P. R., Ed., Blackie Academic & Professional, Glasgow, 185–209, 1994.

Lawrence, B. M., Isolation of aromatic materials from natural plant products, in *A Manual on the Essential Oil Industry*, DeSilva, K. T., Ed., UNIDO, Vienna, Austria, 57–154, 1995.

Lashkari, Z., Ethnic markets beckon producers, in *Finechem. Natl. Prod.*, 2, 2, July, 1999.

Mahindru, S. N., *Indian Plant Perfumes*, Metropolitan, New Delhi, 41, 1992.

Mahindru, S. N., *Indian Plant Perfumes*, Metropolitan, New Delhi, 234–236, 1992a.

Mahindru, S. N., *Indian Plant Perfumes*, Metropolitan, New Delhi, 221–224, 1992b.

Mahindru, S. N., *Indian Plant Perfumes*, Metropolitan, New Delhi, 226–228, 1992c.

Mahindru, S. N., *Indian Plant Perfumes*, Metropolitan, New Delhi, 239–241, 1992d.

Meyer-Warnod, B., Natural essential oils: extraction processes and applications to some major oils, *Perf. Flav.*, 9, 93–103, April/May, 1984.

Moates, G. K. and Reynolds, J., Comparison of rose extracts produced by different extraction techniques, *J. Essential Oil Res.*, 3, 289–294, Sept./Oct., 1991.

Moyler, D. A., Browning, R. M., and Stephens, M. A., Ten years of CO_2 extracted oils, Proc. 12th Intl. Congr. Essential Oils, Fragrances and Flavors, Vienna, Austria, Oct. 4–8, 1992.

Moyler, D. A., Extraction of flavours and fragrances with compressed CO_2, in *Extraction of Natural Products Using Near-Critical Solvents*, King, M. B. and Bott, T. R., Eds., Blackie Academic & Professional, an imprint of Chapman Hall, Glasgow, 140–183, 1993.

Moyler, D. A., *Food Flavourings*, Ashurst, P. R., Ed., chap. 3, 65, Blackie Academic & Professional, an imprint of Chapman & Hall, Glasgow, U.K., 1994.

Mukhopadhyay, M. and Sastry, S. V. G. K., Process for Cyclic Supercritical Fluid (SCF) CO_2 Extraction of Fragrances (absolute or essential oils) from Jasmine Flowers, Indian Patent 183454 (72/BOM/96), 1995.

Muller, P. A., The jasmine and jasmine oil, *Perfumer Essential Oil Record*, 36, 658–663, 1965.

Reverchon, E., Isolation of peppermint oil using supercritical CO_2, *J. Flavor Fragr.*, 9, 19–23, 1994.

Reverchon, E., Porta, G. D., Gorgoglione, D., and Senatore, F., Supercritical extraction and fractionation of lavender in florescences, Proc. 3rd Intl. Symp. SCFs, 2, 389, Oct. 1994.

Sastry, S. V. G. K. and Mukhopadhyay, M., Substrate hindrance in supercritical extraction of fragrances from jasmine flowers, 3rd Intl. Symp. Supercritical Fluids, France, 2, 341, 1994.

Sastry, S. V. G. K., Supercritical Fluid Extraction of Fragrances from Jasmine Flowers, Ph.D. dissertation, Indian Institute of Technology, Bombay, India, 1994.

Wright, J., Essential oils, in *Food Flavorings*, Ashurst, P. R., Ed., Blackie Academic & Professional, Glasgow, chap. 2, 24, 1994.

REFERENCES

[illegible], K. and Burfield, Organic content of supercritical fluid extraction of essential oils from Turnera diffusa [illegible], [illegible] Perf. Flav. [illegible] C... [illegible], [illegible] 1998.

Bosley, M. [illegible] and [illegible], Ed., [illegible] Bioreactors, [illegible] Washington, DC, Elsevier Science Publ., BV, Amsterdam, [illegible] 1996.

Brenner, M. W. and Owades, J. L., Manual of [illegible] Seville [illegible] Perfumery [illegible] 1996, [illegible].

Brenner, M. H., [illegible] [illegible] Perfumery and [illegible] Sept. [illegible] 1996.

Brenner, M. H., The chemical and sensory properties of [illegible] [illegible] Proc. 4th Int. Symp. [illegible] 1988.

Gaunt, [illegible] The Essential Oils, Vol. 2, Van Nostrand, [illegible] 1949.

Lawrence, B. M., The flavoring of [illegible], [illegible], [illegible] [illegible] Academic Press, [illegible] 1993.

Lawrence, B. M., Isolation of [illegible] from natural [illegible] [illegible] Perfumer and Flavorist, [illegible] 1984.

Lawrence, B. M., [illegible] products, [illegible] Perf. [illegible] [illegible].

[illegible] Plant Extracts, [illegible] 1996.

Mehrotra, S. N., Indian Plant Perfumes, Metropolitan, New Delhi, [illegible] 1970.

Mehrotra, S. N., Indian Plant Perfumes, Metropolitan, New Delhi, [illegible] 1970.

Mehrotra, S. N., Indian Plant Perfumes, Metropolitan, New Delhi, 220–221, 1972.

Mehrotra, S. N., Indian Plant Perfumes, Metropolitan, New Delhi, 220, 221, 1982.

Meyer, W. and [illegible], [illegible] processes and [illegible] [illegible] [illegible] April [illegible].

Moyler, D. A. and [illegible], [illegible] products [illegible] [illegible] techniques, [illegible] Sept/Oct 1996.

Moyler, D. A., [illegible] and Stephens, M. A., Use of CO_2 extracted [illegible] [illegible] Int. Congress Essential Oils, Fragrances and Flavors, [illegible] 1992.

Moyler, D. A., [illegible] of flavours and fragrances [illegible] in Fragrance of [illegible] Products [illegible] [illegible] Royal [illegible] [illegible] Chapman & Hall, Glasgow, [illegible] 1997.

Moyler, D. A., and [illegible] [illegible] R., [illegible] [illegible] supercritical [illegible] [illegible] Chapman & Hall, Glasgow, [illegible] 1994.

[illegible], M. and [illegible], S. N. G., [illegible] Supercritical Fluid (SCF) CO_2 extraction of [illegible] [illegible] [illegible] Indian Patent [illegible] 1993.

Muller, P. M., [illegible] Perfumer & Flavorist, [illegible].

Reverchon, E., [illegible] of [illegible] oil [illegible] supercritical CO_2 [illegible] 1992.

Reverchon, E., [illegible] [illegible] Supercritical extraction [illegible] Advances in [illegible] [illegible] [illegible] 1994.

Stahl, W. [illegible] and [illegible] [illegible] [illegible] supercritical [illegible] [illegible] natural [illegible] in [illegible] Supercritical Fluids, [illegible] 1988.

[illegible], [illegible] Supercritical Fluid Extraction of [illegible] from [illegible] Indian [illegible] [illegible], Bombay, India, 1994.

Wright, [illegible] Essential Oils and [illegible] [illegible] [illegible] Blackie Academic & Professional, Glasgow, [illegible] 1994.

5 Fruit Extracts

5.1 IMPORTANCE OF RECOVERY

Volatile fractions of fruit juices have the distinctive aroma and flavor of fruits and are popularly known as fruit extracts. Fruit extracts find uses in enhancing fruit flavors in fruit juices or as substitutes for fruit juice in fruit drinks and beverages. Fruit juice concentrates and volatile fruit extracts are being used increasingly as a base to which other ingredients are added. There has been a phenomenal growth in the consumer demand for natural fruit flavors, and their addition to end products, such as soft drinks, confectioneries, soaps, and toiletries. An important aspect of the commercial production of fruit juices is the method of processing fruit juice for separation of volatile components. For applications of flavor from citrus fruits, the volatile components present in the fruit juice are separated into oil-phase volatiles and water-phase volatiles. For most other fruits, volatile components are recovered only from concentrated juice and thus belong only to the water-phase volatiles. During the heat treatment and concentration processes, the fruit juices often undergo change in flavor profiles. In the manufacture of grape juice, it was indicated by Ohta et al. (1994) that 99.3% of the flavor of grape juice changed during its processing. For example, low-boiling volatile components, such as 2-methyl-3-buten-2-ol and ethyl crotonate were lost upon processing, as indicated by the head space analysis of fresh grape flavor. Even the high-boiling component, methyl anthranilate, was reduced to half after the concentration process. It is thus apparent that efforts should be made to recover the fruity flavor from juices prior to or during heating and concentration. A typical schematic representation of the fruit essence (extract) recovery process is shown in Figure 5.1. In addition to recovering fruity flavor from fruit juices, other related processes of significant importance include deterpenation, dealcoholization, enzyme deactivation, and stabilization of fruit juices. Different processes will be discussed for both citrus and noncitrus fruit extracts in the following sections.

5.2 CITRUS OIL RECOVERY DURING JUICE PRODUCTION

All over the world, citrus oil is one of the important products from citrus processing plants, such as from orange, lemon, lime, grapefruit, etc., not only considering the volume of operation, but also from the point of view of economics. For example, cold-pressed lemon oil is considered to be the main product in some lemon processing plants for its significant market value, whereas lemon juice is considered the byproduct. Brazil is one of the leading orange juice producers in the world and currently produces 2 million tonnes of orange juice annually. More than 16,000 tonnes of orange oil are produced worldwide, most of which are cold pressed from the peel of the fruit. Only a small quantity is produced by steam distillation. A different type of oil may be produced from concentrated orange juice. The yield of cold-pressed peel oil is 0.28% and that of juice oil is 0.008% of the fruit (Wright, 1994). It may become cloudy when

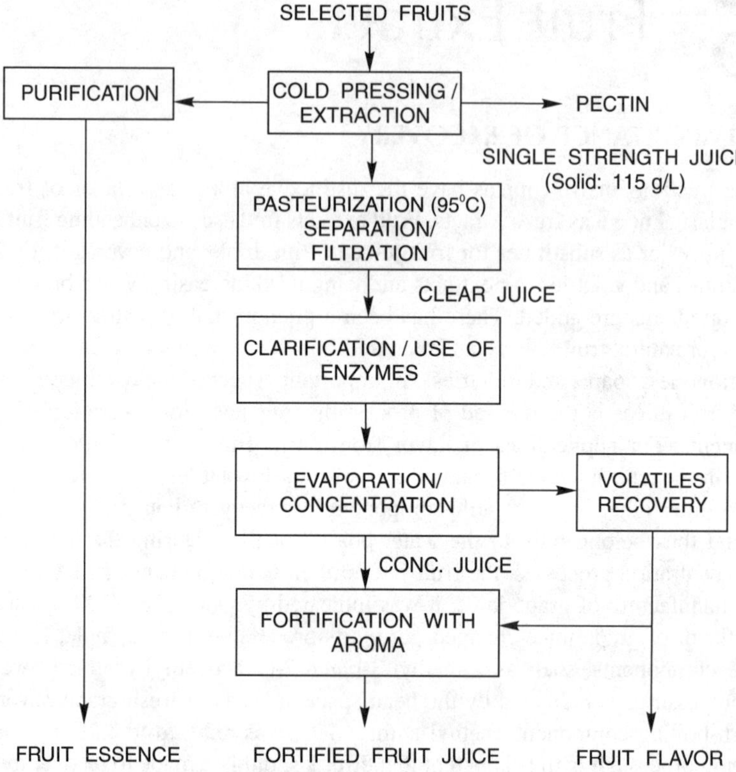

FIGURE 5.1 Schematic sequence of fruit juice processing.

chilled. Sweet orange oil is produced simultaneously with concentrated orange juice. The production of citrus oil amounts to 0.3% by wt of the total orange fruit (Mendes et al., 1998).

Citrus oil is present in oil sacs or oil glands located at different depths in the peel of the fruit. Recovery of the oil is performed mostly by cold pressing these peels. Steam or hydrodistillation and solvent extraction of dried or preprocessed peels are also quite common. However, simultaneous extraction of juice and oil is considered by many fruit processors to be economically attractive. Accordingly, the unique design of the citrus "in line extractor" by Food Machinery and Chemical Corporation (FMC) needs a special mention. It has provisions for upper and lower cups, upper and lower cutters, a strainer tube, and an orifice tube, and allows simultaneous extraction of essential oil and juice without direct contact between them (Flores and Segredo, 1996). This facilitates a sequence of operations, namely the positioning of the fruit for peeling, the cold pressing of peels after cutting, the separation of juice and other fruit parts, and the addition of water to capture the peel pieces and oil in the form of oil-in-water emulsion. The oil emulsion is separated into three phases, namely an aqueous phase, a solid phase called sludge, and an oil-rich emulsion (60 to 80% oil) phase, called cream. The oil-rich emulsion is centrifuged to obtain concentrated citrus oil.

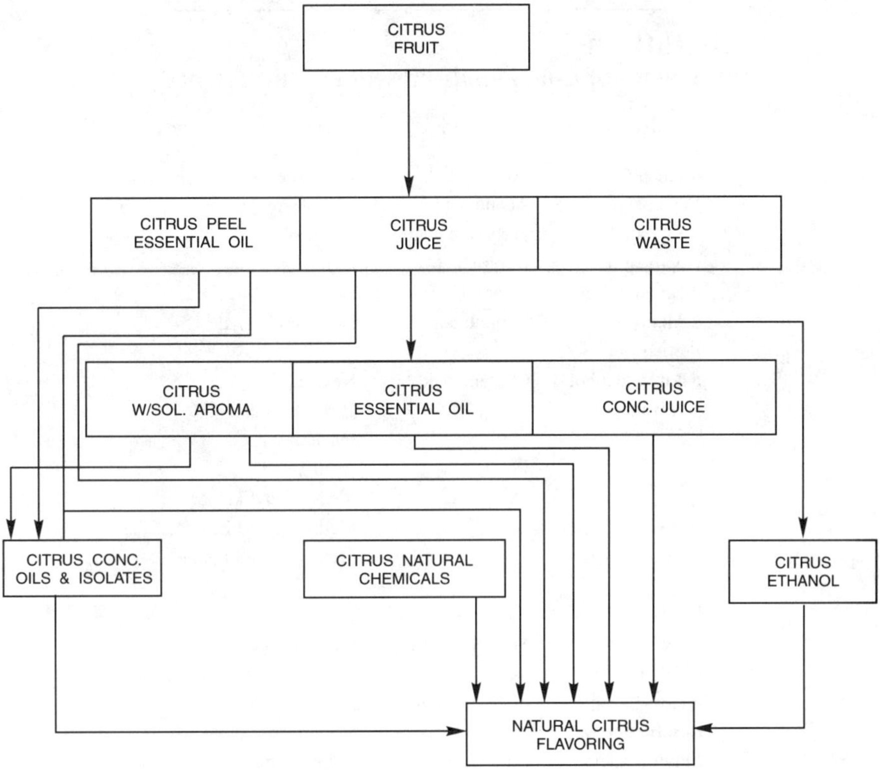

FIGURE 5.2 Products from citrus fruits (Knights, 1993).

Figure 5.2 illustrates how the three primary products, namely juice, peel oil, and orange waste, are produced, each of which may be used for fruit flavor. Peel yields 0.4 to 0.5% cold-pressed oil.

Juice after extraction may, as such, be used as a source of natural flavor. If it is concentrated, the emanated volatiles from the process of concentration may also be used as a natural flavoring substance. These volatiles are separated into two phases, water-solubilized aroma and essential oil. The water-solubilized aroma contains mainly water, ethanol, and low-boiling aromatic compounds, like acetaldehyde which may be used directly or isolated by various techniques such as evaporation, distillation, solvent extraction, chromatography, etc. The essential oil may be used either directly or further concentrated to produce a specific natural flavoring substance.

5.3 FLAVORING COMPONENTS IN FRUITS

Fruit extracts are used not only for the manufacture of natural flavors, but also for re-addition to concentrated fruit juice in order to produce more aroma and taste in the reconstituted juice, than in the original juice itself. Typical classes of compounds present in fruit essence are listed in Table 5.1.

TABLE 5.1
Classes of Compounds Present in Fruit Essence

Alcohols	Acids	Carbonyl Compounds
Methanol	Formic acid	Formaldehyde
Ethanol	Acetic acid	Acetaldehyde
n-Propanol	Propionic acid	Propanal
Isopropanol	Bulyric acid	n-Butanal
n-Butanol	Valeric acid	Isobutanal
2-Methyl Propan-l-ol	Caproic acid	Iso-valeraldehyde
n-Amyl alcohol	Caprylic acid	Hexanal
n-Hexanol		Hex-2-en-2-al
		Furfural
		Acetone
		Methyl ethyl ketone
		Methyl propyl ketone
		Methyl phenyl ketone
Esters		
Amyl formate	Methyl acetate	Ethyl acetate
n-Butyl acetate	n-Amyl acetate	Iso-amyl acetate
n-Hexyl acetate	Ethyl propionate	Ethyl butyrate
n-Butyl butyrate	n-Amyl butyrate	Methyl isovalerate
Ethyl-n-valerate	Ethyl caprolate	n-Butyl caprolate
Amyl caprolate	Amyl caprylate	

Ashurst, 1994.

Many individual components have low solubility in aqueous systems and when the concentration reaches a typical 100-fold volatile, phase separation may occur. However, the presence of ethanol in the concentrated volatiles extracts enhances their solubilities. Ethanol level may be 1 to 30% v/v. Presence of 5-6% ethanol in the recovered volatiles may be considered normal, but above this level, it is considered that fermentation has taken place. There may be commercial advantage to retain the alcohol level below the permitted level to reduce excise duty. This is often achieved by reblending the volatiles with the concentrated juice. Excluding ethanol, all the aromatic components constitute not more than 0.5% of the concentrated juice. Acetaldehyde is the second-most significant component which may be present in the water-solubilized, low-volatile aroma and may be isolated to be used directly as natural flavoring. The profile of the volatiles present in the extract is characterized by a number of factors, such as origin, genus, time of plucking and duration of storage of fruit, and the processing technology. Typical compositions of the flavors from a few popular fruits are given in Table 5.2.

For example, sweet orange oil, as obtained by cold pressing of orange peels, is a yellow to reddish yellow liquid with a characteristic orange odor, as determined by the aldehydes. The flavoring components are octanal, decanal, citrals, and esters,

TABLE 5.2
Compositions (GC Area %) of Different Fruit Flavors

Fruit Oil Constituents	Living Fruit		Picked Fruit
Peach oil (Mookherji et al., 1988)			
Ethyl acetate	6.2		—
Dimethyl disulfide	0.6		—
cis-3-hexenyl acetate	9.7		—
Methyl octanoate	34.2		7.1
Ethyl octanoate	7.4		11.0
6-Pentyl-α-pyrone	t		10.6
γ-Decalactone	2.5		39.2
Strawberry oil (Mookerji et al., 1988)			
Ethyl butyrate	0.5		9.9
Ethyl-2-methyl butyrate	—		4.2
Isoamyl acetate	0.5		4.5
Methyl hexanoate	4.6		11.3
Butyl butyrate/ethyl hexanoate	26.2		26.4
Hexyl acetate	12.7		15.7
Maltol	—		0.3
Octyl acetate	5.1		2.2
Octyl butyrate	7.1		0.3
γ-Decalactone	9.5		0.3
Grape Fruit oil ([a]Ashurst, 1994; [b]Umano and Shibamoto, 1988)			
	Fruit[b]	**Peel[a]**	**Juice[b]**
α-Pinene	2.32		0.54
Sabinene	0.76		—
Myrcene	1.93	2.0	2.10
Limonene	71.96	90.0	96.66
β-Ocimene	1.42		0.17
2-Methyl cyclopentanone	0.62		
Perillene	0.44		
cis-Limonene oxide	0.73		
trans-Limonene oxide	0.73		
Caryophylene	0.47		0.06
Sesquitepene	0.62		
α-Cadinene	3.78		
Decanal	—	0.4	0.02
Octanal		0.5	0.21
Linalool		0.3	—
Mango flavor (Wilson et al., 1988)			
	Indian Alphonso		**Srilankan Jaffna**
Monoterpenes	69.8		49.6
Sesquiterpenes	5.4		14.3

TABLE 5.2 (continued)
Compositions (GC Area %) of Different Fruit Flavors

Nonterpenoids	2.6	3.3
Car-3-ene	—	—
cis-β-ocimene	18.1	37.9
Myrcene	45.9	4.3
Limonene	0.7	2.6
α-Pinene	2.2	0.7
β-Pinene	0.5	2.9
Furanone	2 ppm	
Lactones	400 ppm	
Bitter orange peel oil (Boelens and Sindreu, 1988)		
Aliphatics	0.7	
Alcohols	0.1	
Esters	0.15	
Aldehydes	0.40	
Monoterpene hydrocarbons	95.4	
Other monoterpenes	1.8	
Sesquiterpenoids	0.8	
Aromatic & misc. compounds	0.8	
Sweet orange peel oil (Wright, 1994)		
Limonene	94	
Myrcene	2.0	
Linalool	0.5	
Octanal	0.5	
Decanal	0.4	
Citronellal	0.1	
Geranial	0.1	
Other aldehydes	0.1	
Bergamot peel oil (Wright, 1994)		
Linalyl acetate	35	
Limonene	30	
Lanalool	15	
β-Pinene	7	
γ-Terpinene	6	
Geranial	2	
Neral	2	
Geraniol	1	
Neryl acetate	0.4	
Geranyl acetate	0.2	
Bergaptene	0.2	
Lemon peel oil (Wright, 1994)		
Limonene	63	
β-Pinene	12	
γ-Terpinene	9	

TABLE 5.2 (continued)
Compositions (GC Area %) of Different Fruit Flavors

Geranial	1.5
Neral	1.0
Neryl acetate	0.5
Geranyl acetate	0.4
Citronellal	0.2
Linalool	0.2
Nonalal	0.1

mainly octyl and neryl acetate. Minor components like sesquiterpene aldehydes, alpha-, and beta-sinensal are also important contributors to the specific sweet aroma. The total aldehydes, containing mainly octanal, decanal, and citrals, and esters like octyl and neryl acetate, in orange oil are traditionally taken as the indicator of its quality. Orange juice oil has a different odor characteristic. It contains much more valencene (up to 2%). The major component in either type of oil, namely, d-limonene, makes no contribution to the odor value. Instead, it makes orange oil susceptible to oxidation, giving rise to spurious odor. It is also insoluble in water. Consequently some orange oils are further processed to remove it. Supercritical CO_2 extraction of orange peels, followed by chromatographic separation, is known to give better product yield and quality. Orange terpenes are the byproduct of these processes and find use as a solvent for flavor and fragrances. Orange oil gives an acceptable orange flavor in many applications such as confectioneries. There are no legal constraints on the use of citrus oil in flavors. Likewise, the natural flavoring components (extracts) from other fruits are also widely used as food additives and are in great demand.

5.4 STABILITY AND QUALITY

Stability of various volatile fractions from different fruits may vary significantly and should be considered in the choice of solvents, process parameters, and equipment design. Partial hydrolysis can cause changes in aroma profile and hence care is taken right from the step of fruit juice collection and concentration. Under certain circumstances, it is possible to cause damage to the nonvolatile residue which leads to undesirable changes in color (by browning) and development of unwanted cooked taste. In the past, recovery of essence from fruit juices involved flash evaporation, at atmospheric pressure, upon heating to reduce the volume of the juice by 5 to 50%. The vapor released from the flash drum was distilled to produce essence having a volume of 1/150th of the original volume of the fruit juice. Recent processes involve flash distillation under vacuum before concentration of nonvolatiles in fruit juices, also under vacuum. Large variations in the quality of fruit juice essence may occur due to difference in the processing technology. For example, ethanol is formed, if a juice is allowed to stand for some time between processing and pasteurization, and to ferment by naturally occurring yeasts that are present in the fruit. This can result in the formation of more fermentation products, such as acetic acid, esters, etc. and subsequently can cause trans-esterification of esters already present. During the process of concentration of fruit juices, further changes may take place. As a result, it is almost impossible to obtain consistency in the quality of fruit extracts to be used in a formulation. The quality therefore depends on the art and skill of the individual manufacturer. The quality of the fruit essence is monitored by a combination of gas chromatographic and sensory evaluations. Aroma and flavor, as with other essential oil products, are typically evaluated by using a "smelling strip," a firm adsorbent paper which is dipped in the aromatic product and smelled at regular intervals. Efficacy of the extract is evaluated by making a solution of 12.5% sucrose, 0.25% citric acid anhydrous, and only 0.5% of fruit extract.

5.5 CO_2 EXTRACTION PROCESSES

The use of liquid CO_2 for extraction of fruit juice concentrates was reported as early as in 1939 by Horvarth. Subsequently a large number of fruit juices and vegetables juices have been extracted with liquid CO_2. The alcohols, esters, aldehydes, and ketones that constitute the essence of fruit juices are very soluble in liquid CO_2, while water has a limited solubility (approx. 0.1% by wt) and the salts, sugars, and proteins are insoluble. Extraction with liquid CO_2 recovers the important volatile flavor components that are primarily present in the fresh fruits. Extraction of fruit juices, such as apples, oranges, peaches, pineapples, berries, and pears was studied by several researchers (Randall et al., 1971 and Schultz and Randall, 1970) who observed that most of the flavor-contributing components could be effectively extracted by liquid CO_2. Schultz et al. (1967) compared the performance of some volatile organic solvents such as isopentane, diethyl ether, and a fluorocarbon, with that of liquid CO_2 in the extraction of delicious apple essence. The important compounds responsible for apple aroma, namely, ethyl-2-methyl butyrate, trans-2-hexenel, and hexenel, were all recovered in the liquid carbon dioxide extract which was evaluated to be more pleasant and fragrant than the original head space vapor. Gas chromatographic analysis showed that the extracted organic volatiles contained acetaldehyde, hexenel, esters, and C_1–C_6 alcohols. The liquid CO_2 extract appeared to be more similar to the ether extract as can be seen from Table 5.3, whereas the fluorocarbon extract resembled the isopentane extract. The liquid CO_2 extract contained 150-fold essence because of the high selectivity of CO_2 for compounds responsible for the aroma (aldehydes, esters, ketones, and alcohols). The yields were 0.38, 0.41, 1.21, and 1.58% (g extract/g aqueous essence) for extraction with isopentane, fluorocarbon, ether, and liquid CO_2, respectively. Compared to isopentane, ether, and dichloro-tetrafluoroethane, a higher yield of volatile constituents of apple essence, such as acetaldehyde and ethyl acetate, could be recovered with liquid CO_2 at 20 to 25°C and 64 bar. Under these conditions, the solubility of water in liquid CO_2 is very low (approximately 0.1%). Liquid CO_2 processing of the juices of apple, pear, and orange by this method exhibits excellent efficiency for recovering essence which may be concentrated up to 100,000-fold, as in the case of apple fruit extract.

The extraction of essential oil from lemon peel with SC CO_2 was studied by Calame and Steiner (1980) at 300 bar and 40°C. The yield of 0.9% was reported by them, which is similar to that of cold pressing. The compositions of the oil by SC CO_2 extraction and cold pressing methods were also similar.

TABLE 5.3
Comparison of Compositions (% by wt) of Extracts of Apple Essence by Different Solvents

	Isopentane	Ether	LCO_2	Fluorocarbon
Alcohols, C_1–C_5	3	28	25	2
Alcohols, C_6	12	21	18	9
Acetaldehyde	—	trace	3	4
Aldehydes, C_6	50	30	24	45
Esters	33	19	28	38

Schultz et al., 1967.

CO_2 extraction has also been studied for concentrating flavors from vegetables, such as tomatoes, carrots, celery, and watercress, and it has been observed that CO_2 extracts are closest to the natural essence because of the recovery of all the essence-bearing components, and are more concentrated because the flavoring components are selectively extracted. However, complete extraction of the flavoring components depends on the total volume of CO_2 or extraction time (Schultz et al., 1974). Liquid CO_2 extraction of various components of fruit juices has been carried out by Schultz et al. in a countercurrent Scheibel Column (Schultz et al., 1974). These components have different distribution coefficients and are accordingly partitioned between the liquid CO_2 (solvent) phase and the bulk aqueous phase. It has been reported that under suitable conditions, liquid CO_2 gave substantial overall recovery of most of the components except methanol which remained in the aqueous phase. Terpene derivatives, such as carvone, terpinen-4-ol, and linalool can be recovered to a great extent, followed by ethyl acetate and isoamyl alcohol. The higher molecular weight, water-soluble components, such as tannins and acids, are not transferred to the liquid CO_2 phase.

5.6 DETERPENATION OF CITRUS OIL BY SC CO_2

Citrus oils are used in a variety of applications, such as flavor, beverages, food, cosmetics, soap, pharmaceuticals, etc. Citrus oils consist of mixtures of monoterpenes, sesquiterpenes, oxygenated compounds, and nonvolatiles, such as dyes, waxes, coumarin derivatives, etc.

Essential oils from citrus fruits, such as lemon, orange, grapefruit, tangerine, etc., may contain terpene hydrocarbons as high as 95 to 98% from which 95% is limonene. These terpenes do not significantly contribute to the overall flavor profile. Instead, these terpene compounds eventually degrade, resulting in an undesirable taste and odor profile. Besides, the presence of the high content of terpenes renders the product sparingly soluble in aqueous or alcohol systems, like fruit drinks and beverages. It is thus essential to separate the terpenes from the citrus oils. Deterpenation of citrus oil amounts to bringing down the terpene content to 25 to 50% by weight. The major component, d-limonene, is used for materials like spearmint oil flavor, 1-carvone, terpene resins, and adhesives.

On the other hand, the oxygenated compounds are highly odoriferous. This flavor fraction consists of aldehydes, alcohols, ketones, esters, ethers, and phenols. The total aldehyde content (about 1.5%) is used as an indicator of quality and is measured as decanal, since this is the major aldehyde. Processes such as molecular or flash distillation and solvent extraction, are conventionally used for deterpenation of citrus oils. However, a few disadvantages like thermal degradation and presence of residual solvent often make these processes unsuitable for good quality fruit drinks and beverages.

Both liquid CO_2 and supercritical CO_2 can be used for deterpenation of citrus oils on semibatch mode with and without an adsorption/desorption column and also on a continuous countercurrent mode. Kimbal (1987) extracted d-limonene from citrus juices with SC CO_2 in the pressure range of 210 to 410 bar and temperature range of 30 to 60°C, and was able to reduce the limonene content by 25% in 1 h without significant reduction of vitamin C; citric acid and amino acids as terpenes were more solubilized in SC CO_2 than the oxygenated materials. Temelli et al., (1988) used SC CO_2 to concentrate oxygenated constituents in orange oil by deterpenation in a semibatch mode and suggested 83 bar and 70°C as the SC CO_2 condition for minimum loss of odoriferous constituents in spite of low extraction yields. Even a 20-fold increase in aroma concentration of orange peel oil retains a significant quantity of terpenes.

A combination of supercritical CO_2 extraction and desorption renders deterpenation process more efficient. Yamauchi and Saito (1990) used a silica gel column with SC CO_2 at 200 bar mixed with ethanol as co-solvent to obtain a terpeneless (10% limonene) fraction containing 95% of the oxygenated (aroma) compounds. Chouchi et al. (1994) investigated deterpenation of cold-pressed citrus oils with different amounts of oxygenated compounds on a pilot plant scale using a silica column at 40°C and with a pressure gradient of 78 to 100 bar, getting excellent enrichment of aroma compounds with good yields. Dugo et al. (1995) used SC CO_2 to elute hydrocarbon terpenes from a silica gel column coated with orange peel oil. Later oxygenated compounds could be eluted by increasing the temperature and pressure of the SC CO_2 stream. However liquid CO_2 is not as selective, so extraction of orange oil from a precoated silica-gel column with liquid CO_2 at 67.5 bar and 15°C resulted in reduction of limonene from 95% to 40% (Ferrer and Mathews, 1987) as can be seen from Table 5.4.

Alternatively, continuous supercritical CO_2 deterpenation may be carried out at a relatively lower temperature of 40 to 60°C using a counter-current column. The citrus oil to be processed is introduced from the top of the column and terpenes are extracted with CO_2 which is drawn off from the top. These terpenes are separated by decreasing the pressure, and the CO_2 thus regenerated is recycled to the column from the bottom. For deterpenation of citrus oil using supercritical CO_2, it is necessary to operate the fractionating column below 110 bar at 60°C because of the formation of a homogenous phase with SC CO_2 at higher pressures. The temperature at the top of the fractionating column is usually held at a higher temperature than at the bottom of the extractor. This allows the less-volatile components to condense as a result of a decrease in their solubilities with an increase in temperature. This provides the reflux required for selective separation in the rectification column. Data from a typical pilot plant (Schultz et al., 1974) are given in Table 5.5.

TABLE 5.4
Composition of Orange Oil Percolated through Deactivated Silica Gel Bed before and after Liquid CO_2 Extraction

Constituents	Before Extraction	After Extraction
Ethyl butyrate	0.028	0.003
α-Pinene	0.479	1.745
Sabinene	0.233	0.028
Octanal/myrcene	2.014	31.47
Delta-3 carene	0.100	0.169
d-Limonene	94.69	40.22
Octanol	0.024	0.403
Nonanal/Linalool	0.550	1.867
Cironelal	0.046	3.751
α-Terpineol	0.055	0.816
Decanal	0.259	6.749
Neral	0.075	0.243
Geranial	0.078	0.367
Perilaldehyde	0.031	0.046
Lauric aldehyde	0.064	0.567
Valencene	0.485	1.058
β-Sinensal	0.018	0.158
α-Sinensal	0.008	0.053

Ferrer and Mathews, 1987.

TABLE 5.5
Pilot Plant Data on Deterpenation of Orange Oil

Aromatic content in feed oil	4.1%
Aromatic content in concentrated oil	18.9%
Recovery of aromatic fraction	90%
Solvent-to-oil flow ratio	100

Schultz et al., 1974.

The crucial parameters of the process are the extraction ratio and selectivity which increase with the increase in solvent-to-feed ratio and the pressure, both in semibatch and continuous modes of extraction.

Bergamot is an unusual citrus fruit from Italy, with a floral fruity fragrance. The bergamot peel oil is a refreshing powerful antidepressant, different from orange peel oil in the percentage of monoterpene hydrocarbons and oxygenated aroma components. Bergamot peel oil contains large amounts of volatile flavor components like linalool and linalyl acetate, influencing the selectivity of deterpenation, as can be seen from Figure 5.3 where vapor-liquid equilibrium (VLE) data for bergamot oil and orange peel oil have been compared at 60°C (Heilig et al., 1998). Bergamot

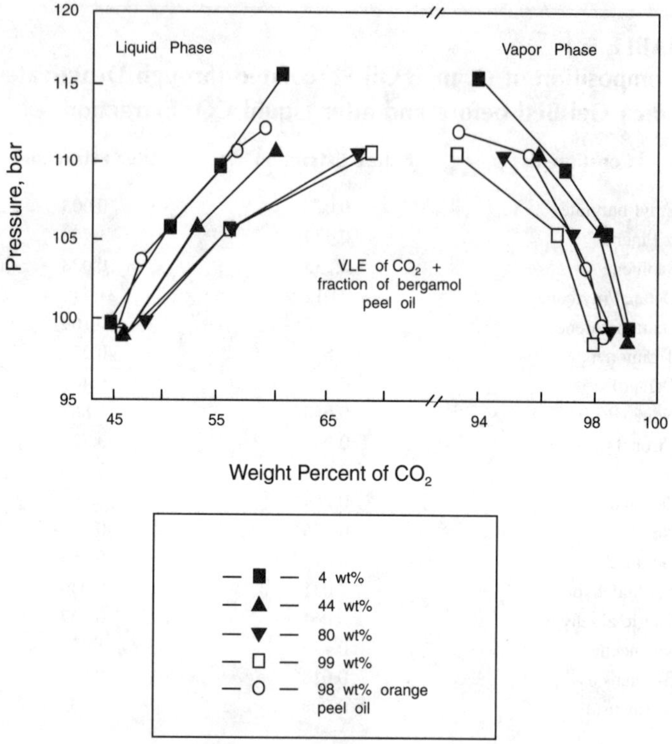

FIGURE 5.3 Vapor-liquid equilibrium data for mixtures of bergamot oil and orange peel oil at 60°C (Heilig et al., 1998).

peel oil contains only 45% terpenes and linalyl acetate is present as the main oxygenated aroma compound, which is not present in orange peel oil. Countercurrent supercritical CO_2 extraction of bergamot peel oil is of great commercial interest, since it can result in an almost terpeneless aroma fraction (Sato et al., 1994).

5.7 DEALCOHOLIZATION OF FRUIT JUICE BY SC CO_2

In the recent past, fruit juice and beverages with low or negligible levels of alcohol have been finding acceptance by the public in an attempt to prevent alcohol abuse in society. Several processes have been attempted for lowering the alcohol level or eliminating it from the traditional beverage without affecting the taste. Supercritical CO_2 extraction has been found to be an attractive process for dealcoholization of fruit juices or concentrates.

Supercritical and liquid CO_2 were utilized by Medina and Martinez (1994) to extract ethanol and aroma from cider at pressures in the range of 80 to 250 bar and temperatures in the range of 20 to 40°C using a pilot plant with CO_2 flow rate of 10 kg/h and cider feed rate 1 l/h having ethanol concentration of 6.2 wt%.

A number of components have been found to be responsible for the aroma of cider. The components which have been identified are methanol, 1-propanol, ethyl-2-methyl

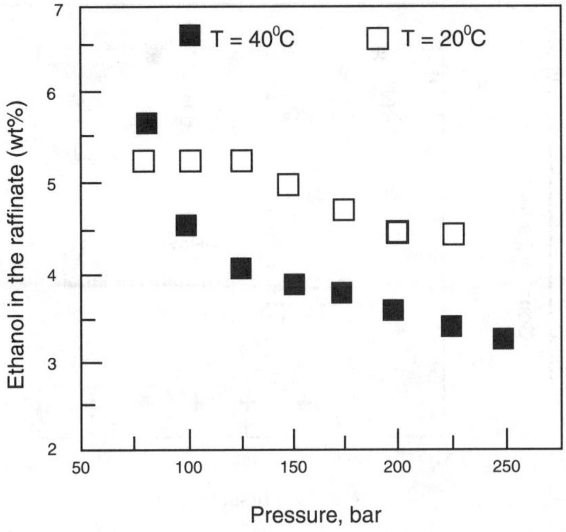

FIGURE 5.4a Effect of pressure and temperature on the extraction of ethanol (Medina and Martinez, 1994).

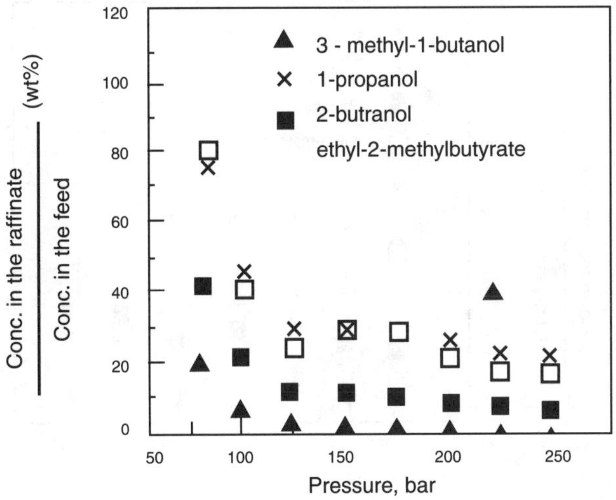

FIGURE 5.4b Percentage of compounds related to the aroma of cider in the raffinate after extraction with supercritical carbon dioxide at 40°C (Medina and Martinez, 1994).

butyrate, 2-butanol, 3-methyl-1-butanol, 1-hexanol, acetic acid, and 2-phenyl ethanol. As shown in Figures 5.4a–e, SC CO_2 is a more selective solvent compared to liquid CO_2 in lowering the concentrations of these compounds in the raffinate (fruit juice). Acetic acid and 2-phenyl ethanol are not recovered by either SC CO_2 or liquid CO_2, whereas methanol can be fully extracted with either of them. Most of the volatiles are recovered by SC CO_2 at 125 bar and 40°C. However, 250 bar and 40°C seemed to be the best condition for removal of ethanol with higher recovery efficiency with a higher flow rate of CO_2. This suggests that the volatile aroma fraction can be extracted at

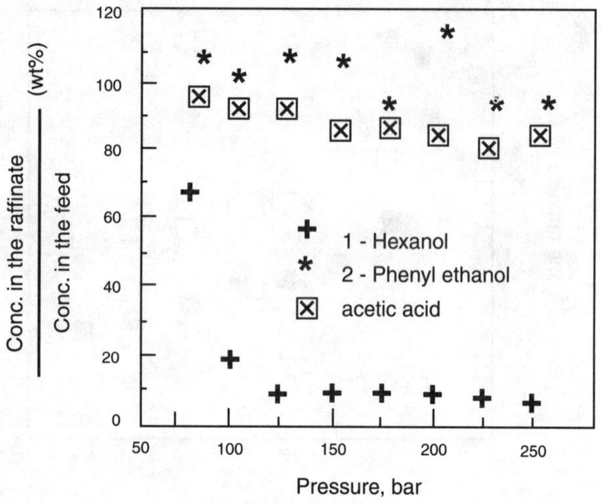

FIGURE 5.4c Percentage of compounds related to the aroma of cider in the raffinate after extraction with supercritical carbon dioxide at 40°C (Medina and Martinez, 1994).

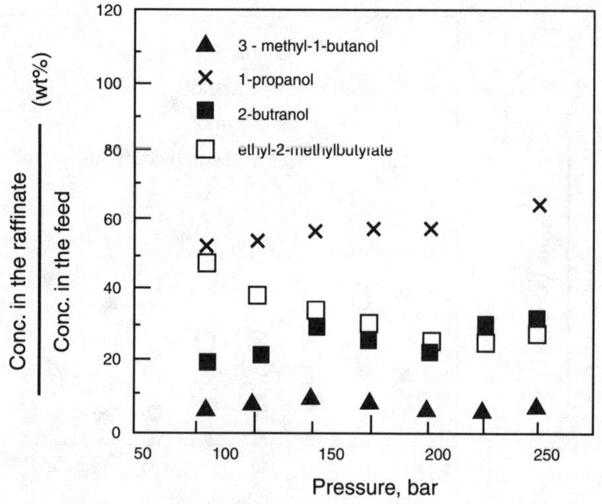

FIGURE 5.4d Percentage of compounds related to the aroma of cider in the raffinate after extraction with liquid CO_2 at 20°C (Medina and Martinez, 1994).

moderate pressure of 125 bar, followed by dealcoholization at higher pressures of 250 bar with subsequent re-addition of aroma to the raffinate to make the cider contain a very low level of ethanol.

5.8 ENZYME INACTIVATION AND STERILIZATION BY SC CO_2

It is a common practice to carry out heat treatment of fruit juice at 90°C for 1 min in order to kill microorganisms, to inactive the enzymes, and to avoid loss of fruit

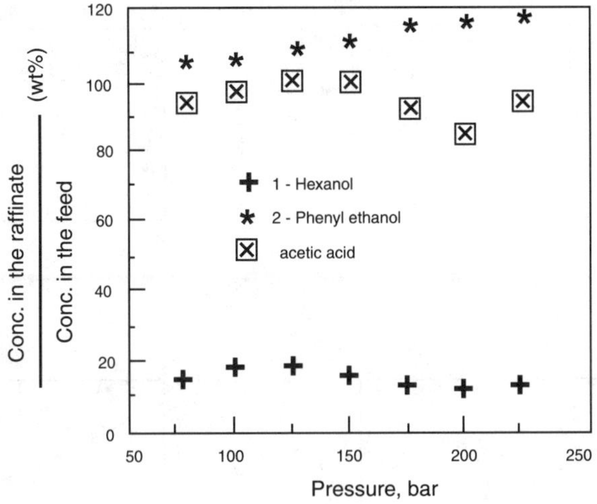

FIGURE 5.4e Percentage of compounds related to the aroma of cider in the raffinate after the extraction with liquid CO_2 at 20°C (Medina and Martinez, 1994).

juice cloud. This results in loss of aroma and darkening of color. Alternatively, supercritical CO_2 may be used for the inactivation of enzymes in fruit juices. Pectinesterase (PE) enzyme is known to cause sedimentation in orange and pineapple juice as does polyphenoloxidase (PPO) in apple juice. Treatment of fruit juices with supercritical CO_2 immediately after juice extraction inactivates enzymes. Apple juice was treated (Balban et al., 1993) with SC CO_2 at 337 bar and 45°C for 4 h with temperature control with the treatment resulting in an increase in PPO activity in the first hour. However it was followed by a drastic decrease of 8% in a time period of 4 h.

Orange juice was treated with SC CO_2 (Balban et al., 1993) at pressures from 69 to 344 bar and temperatures from 35 to 65°C for 15 to 180 min with temperature control. Extent of deactivation is dependent on pressure, temperature, and time, as can be seen in Figure 5.5. The temperature control (TC) samples (continuous line) show very little PE inactivation at 40°C, whereas SC CO_2 processing at the same temperature and 310 bar pressure resulted in 43.2 to 100% PE inactivation beyond temperature control levels, as can be seen from Figure 5.5. The combined effect of supercritical pressure and temperature decreased the time required to inactivate the enzyme at any temperature (Arreola et al., 1994). Further PE inactivation is more sensitive with SC CO_2 with respect to temperature than with CO_2 at atmospheric pressure. The inactivation rate constants are higher for SC CO_2-treated samples than for TC samples, implying that the rate of PE inactivation is faster for SC CO_2-treated samples. The inactivation rate constant increased from 1.77×10^{-2} min^{-1} at 1 bar pressure to 8.26×10^{-2} min^{-1} at 138 bar and to 9.45×10^{-1} min^{-1} at 310 bar (Arreola et al., 1994). The cloud and color of fresh orange juice, when simply treated with SC CO_2 at 290 bar and 45°C for 2 h, are very much improved without affecting the sensory quality, as can be seen in Figure 5.6. TC samples were found to be gelled after 45 days while SC CO_2 samples remained fluid/stabilized and retained their enhanced cloud.

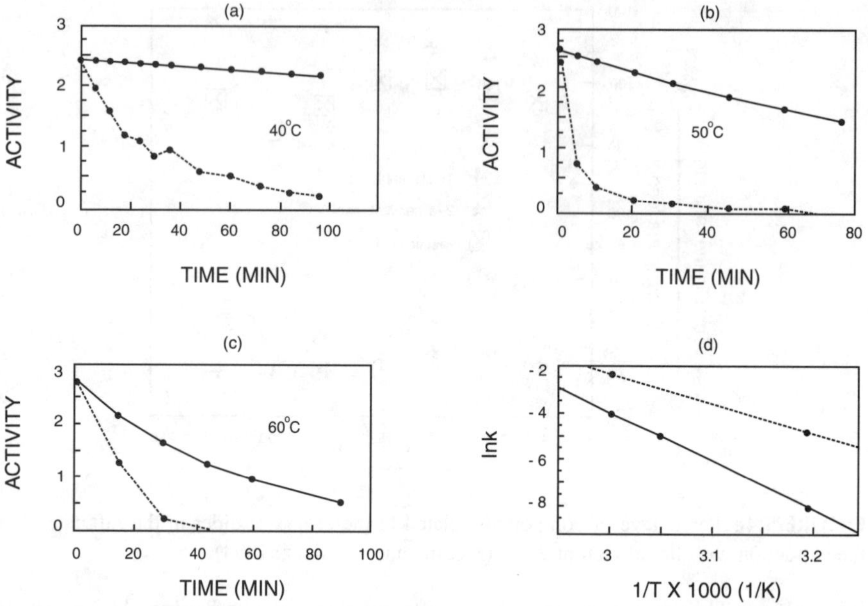

FIGURE 5.5 Inactivation of PE in SC CO_2-treated Valencia orange juice at 310 bar and different temperatures and temperature control samples (Arreola et al., 1994).

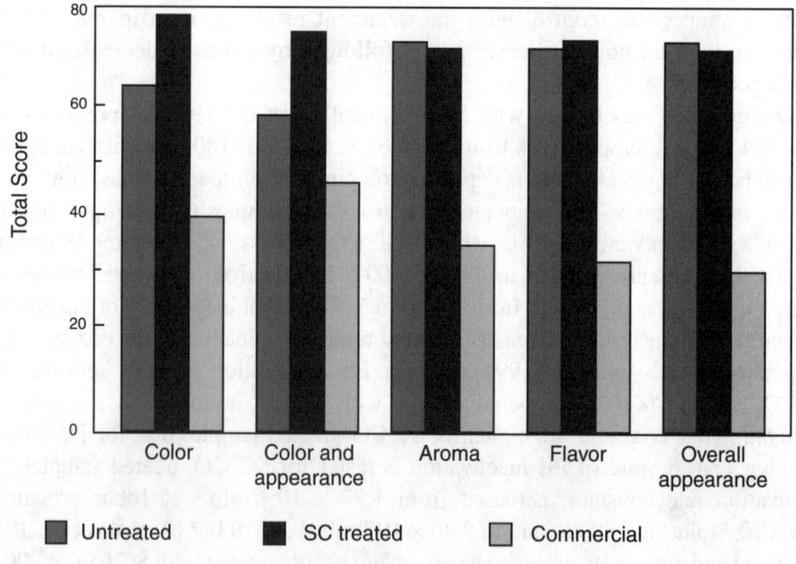

FIGURE 5.6 Results of sensory evaluation tests. Total score is the sum of 30 evaluation results (3 best, 2 second, 1 least liked). Average of two evaluations (Arreola et al., 1994).

SC CO_2 processing of fruit juices has the additional advantage of reducing the microbial numbers and the rate of microbial growth of several microorganisms (containing 70 to 90% water), including *Lactobacillus*, *Listeria*, and *Salmonella* (Molin, 1983). Dissolution of high-pressure CO_2 in water leads to formation of carbonic acid, thereby temporarily lowering the pH to 3 to 4. On depressurization of the SC CO_2-treated fruit juice, the original pH is restored and the residual amounts of CO_2 quickly diffuse out. More ascorbic acid is retained and total acidity is preserved, over and above cloud and color are enhanced and stabilized even in the presence of residual PE, without any adverse effects on sensory quality of the SC CO_2-treated fruit juice. However dry cells are not sterilized by SC CO_2 under the same conditions. The sterilizing effect of CO_2 on different types of bacteria was reported by Kamihira et al. (1987) at 203 bar and 35°C. The inhibitory effect increases with decrease in temperature and increase in pressure, possibly due to higher solubility of CO_2 in water and the lowering of pH or possible chemical interaction between the microorganism and CO_2. This low-temperature sterilization and enzyme inactivation process using SC CO_2 is especially attractive for fruit juice processing as the fruity flavor components are thermally labile and can get easily degraded at thermal inactivation temperatures.

REFERENCES

Arreola, A. G., Balban, M. O., Marshall, M. R., Wei, C. I., Peplow, A. J., and Cornell, J. A., Supercritical CO_2 processing of orange juice: effects on pectinesterase, microbiology and quality attributes, in *SCF Processing of Food and Bio Materials*, Rizvi, S. S. H., Ed., Blackie Academic & Professional, Glasgow, 1994.

Ashurst, P. R., Fruit juices, in *Food Flavourings*, Ashurst, P. R., Ed., Blackie Academic & Professional, Glasgow, 86, 1994.

Balban, M. O., Pekyardima, S., Chen, C. S., Arreola, A. C., and Marshall, M. R., Enzyme inactivation in pressurized CO_2, in Proc. 6th Intl. Congr. Eng. Food, Chiba, Japan, 855, May, 1993.

Boelens, M. H. and Sindreu, R. J., *Essential Oils from Seville Bitter Orange in Flavours and Fragrances — A World Perspective*, Lawrence, B. M., Mookerjee, B. D., and Willis, B. J., Eds., Elsevier Science Publishers, New York, 551, 1988.

Calame, J. P. and Steiner, R., *Chem. Ind.*, 12, 399–402, 1980.

Chouchi, D., Barth, D., and Nicoud, R. M., Fractionaton of citrus cold pressed oils by supercritical CO_2 desorption, Proc. 3rd Intl. Symp. SCFs, 2, 183, 1994.

Dugo, P., Mortello, L., Bartle, K. D., Clifford, A. A., Breen, D. G. P. A., and Dugo, G., Deterpenation of sweet orange and lemon essential oils with SC CO_2, *Flavour Fragr. J.*, 19, 51–58, 1995.

Ferrer, O. J. and Mathews, R. F., Terpene reduction in cold pressed orange oil by frontal analysis-displacement adsorption chromatography, *J. Food Science*, 52, 801–805, 1987.

Flores, J. H. and Segredo, G. T., Citrus oil recovery during juice extractions, *Perf. Flav.*, 21, 13–15, 1996.

Heilig, S., Budich, M., and Brunner, G., Countercurrent SCFE of bergamot peel oil, Proc. 5th Meet. SCFs, France, 2, 445, 1998.

Kamihira, M., Taniguchi, M., and Kobayashi, T., Sterilization of microorganisms with SC CO_2, *Agric. Biol. Chem.*, 2, 407, 1987.

Kimball, D. A., Debittering of citrus juices using SC CO_2, *J. Food Science*, 52, 481–482, 1987.

Knights, J., Natural Flavourings — what opportunities does the directive definiton offer?, *Proc. of Food Commodities and Ingredients Group Symp. on the Use of Natural Flavors and Colors in Food Products*, London, Society of Chemical Industry, U.K., 20 April, 1993.

Medina, I. and Martinez, J. L., Dealcoholation of cider by SC CO_2, Proc. 3rd Intl. Symp. Supercritical Fluids, 2, 401, 1994.

Mendes, M. F., Stuart, G., Oliveira, J. V., and Uller, A. M., Simulation of the SCFE fractionation of orange peel oil, Proc. 5th Meeting on SCFs, France, 2, 465, 1998.

Molin, G., The resistance to carbon dioxide of some food related bacteria, *Eur. J. Appl. Microbial. Biotechnol.*, 18, 214, 1983.

Mookerjee, B. D., Trenkle, R. W., Wilson, R. A., Zamping, M., Sands, K. P., and Mussinan, C. J., Fruits and flowers: live vs. dead — which do we want?, in *Flavours and Fragrances — A World Perspective*, Lawrence, B. M., Mookerjee, B. D., and Willis, B. J., Eds., Elsevier Science Publishing, Amsterdam, 415, 1988.

Ohta, H., Nogata, V., and Yoza, K., Flavor change of grape juice during processing, in *Developments in Food Engineering*, Part 2, Matsuno, T. Y. R. and Nakamura, K., Eds., Blackie Academic & Professional, U.K., 900, 1994.

Randall, J. M., Schultz, W. G., and Morgan, A. I., Extraction of fruit juices and concentrated essences with liquid CO_2, *Confructa*, 16, 10–19, 1971.

Sato, M., Goto, M., and Hirose, T., Fractionation of citrus oil by supercritical fluid extraction tower, Proc. 3rd Intl. Symp. SCFs, 2, 83, 1994.

Schultz, W. G. and Randall, J. M., Liquid CO_2 for selective aroma extraction, *Food Technol.*, 24, 1282–1286, 1970.

Schultz, T. H., Floth, R. A., Black, D. R., and Schultz, W. G., Volatiles from delicious apple essence — extraction methods, *J. Food Sci.*, 32, 279–283, 1967.

Schultz, W. G., Schultz, T. H., Cartson, R. A., and Hudson, J. S., Pilot plant for extraction with liquid CO_2, *Food Technol.*, 66, 32–36, 1974.

Temelli, F., Chen, C. S., and Braddock, R. J., Supercritical fluid extraction in citrus oil processing, *Food Technol.*, 145–150, June, 1988.

Umano, K. and Shibamoto, T., A new method of head space sampling: grape fruit volatiles, in *Flavours and Frangrances — A World Perspective*, Lawrence, B. M., Mookerjee, B. D., and Willis, B. J., Eds., Elsevier Science Publishers, Amsterdam, 981, 1988.

Wilson, C. W., Shaw, P. E., and Knight, Jr., J., Importance of selected volatile compounds to mango (*Mangifera indica L.*) flavor, in *Flavour and Fragrances — A World Perspective*, Lawrence, B. M., Mookerjee, B. D., and Willis, B. J., Eds., Elsevier Science Publishers, Amsterdam, 283, 1988.

Wright, J., Essential oils, in *Food Flavorings*, Ashurst, P. R., Ed., Blackie Academic & Professional, U.K., 24, 1994.

Yamauchi, I. and Sato, M., *J. Chromatogr.*, 505, 237–246, 1990.

6 Spice Extracts

6.1 IMPORTANCE OF RECOVERY

Spices may be defined as a class of strongly flavored or aromatic substances obtained from tropical plants, commonly used as condiments and utilized for their flavor and preservative qualities. Spices can be universally grown. They are generally suitable and often adaptable as a small holder backyard crop in most countries that produce and export spices, including India and Indonesia. Cultivation of spices is encouraged as a secondary or even tertiary cash crop, essentially to supplement incomes of small farmers. Some spices can even be cultivated on steep slopes of mountains that are unsuitable for other crops.

Though usage of raw spices has been known to mankind since antiquity, it is only over the last three decades that production of spice extracts and their formulations has enabled the spice-processing industry to run a profitable business. Both individual spice extracts and their formulations are used not only to camouflage undesirable odors in food, but also to add flavor to stimulate appetite and to imbibe preservative and therapeutic values in food, soft drinks, beverages, confectioneries, and health tonics. If a flavor need be added slowly to a food item, such as a traditional curry dish being cooked, then a simple ground or powdered spice is the first choice. But if a customer requires an instant spice aroma in a sauce preparation, then the spice extract is the only choice, since it quickly and consistently disperses into the sauce. In the present era of quickening life styles and fast foods, the significance of spice extracts needs no emphasis.

Spice extracts are of at least two kinds: the one responsible for aroma or flavor, called an essential oil or simply essence; and the other, a higher boiling fraction responsible for taste or pungency of the spice, called oleoresin. The spice aroma or essential oil is traditionally obtained by steam distillation of the ground spice or by steam distillation of the extracts obtained by solvent extraction of the ground spice. Table 6.1 lists a few examples of spice extracts that are produced commercially with their percentage yields of essential oils and oleoresin from ground spice as reported by Marion et al. (1994). Table 6.2 lists the spices with their harvesting time and botanical nature. There may be variations in quality and yield of spice extracts due to variations in the origin of spices and their harvesting time. Figure 6.1 outlines the various alternative steps involved in the production of spice extracts.

The quality and yield of extracts also depend upon the preprocessing operations, the technique of extraction, and the nature of the solvent, which, in turn, are decided based upon the desired specification of the end product in terms of its aroma, flavor, and solubility. Each extract plays a specific role in the formulation and its selection is the key to the product development which needs to achieve the ultimate aim of proper value addition.

TABLE 6.1
Commercial Spice Extracts

Spice	Essential Oils Min–Max (%)	Oleoresins (%)
Anise	1.0–4.0	
Caraway	3.0–6.0	
Cardamom	4.0–10.0	
Carrot	0.5–0.8	
Cassia	1.0–3.8	
Celery seed	1.5–2.5	
Cinnamon	1.6–3.5	
Clove bud	14.0–21.0	
Coriander	0.1–1.0	
Cumin	2.5–5.0	
Curcuma	2.0–7.2	7.9–10.4
Dill (seeds)	2.5–4.0	
Fennel	4.0–6.0	
Garlic	0.1–0.25	
Ginger	0.3–3.5	3.5–10.3
Marjoram	0.2–0.3	—
Mace	8.0–13	22.0–32.0
Nutmeg	2.6–12	18.0–37.0
Pepper	1.0–3.5	5.0–15.0
Pimento berry	3.3–4.5	6.0
Saffron	0.5–1.0	
Savory	0.5–1.2	14.0–16.0
Vanilla	—	29.9–47.0

Marion et al., 1994.

6.2 CLASSIFICATION OF SPICES

Selection of extraction technique and processing conditions for the production of spice extracts having various product specifications largely depends upon the type and nature of the spices. Spices may be classified based upon their botanical flavor characteristics as given in Table 6.3.

6.3 THERAPEUTIC PROPERTIES OF SPICES

Spices have been known to mankind from ancient times for their significant health promotive and protective benefits. It is fascinating to note that about 80% of the world's population use natural medicines derived from plant products, such as herbs and spices, containing a variety of compounds, many of them being biologically active (Duke, 1994) and generally recognized as safe (GRAS). There is hardly any spice which does not have at least some medicinal effect. Most commonly used spices are all proven to be medicinal, for example, black pepper, cayenne, cinnamon,

TABLE 6.2
Spices Plant Part and Harvesting Time

Spice	Part	Peak Months
Anise	Seed	September/October
Basil	Leaf	May/June
Caraway	Seed	July, August
Cardamom	Seed	April, May
Cassia	Bark	June, July
Celery	Seed	June, July, August
Chili	Fruit	January, February, March
Cinnamon	Bark	June, July
Clove	Buds	September, October
Coriander	Seed	July, August
Cumin	Seed	June, July
Dill	Seed	June, July
Fennel	Seed	June, July
Fenugreek	Seed	June, July
Ginger	Root	January, February, March
Nutmeg	Seed	February, March, April
Mace	Seed	February, March, April
Oregano	Leaf	June, July
Paprika	Fruit	September, October
Pepper (Indonesia)	Fruit	September, October, November
Pepper (India)	Fruit	November–April
Pimento	Berry	August, September
Rosemary	Leaf	May, June
Thyme	Leaf	May, June
Turmeric	Root	March, April, May

Moyler, 1994b.

garlic, ginger, licorice, onion, chives, etc. Table 6.4 lists some of the spices with their synergistic therapeutic benefits, based upon the number of bioactive components present, pertaining to particular biological actions.

To start with, garlic can be ranked at the top of all medicinal spices that can be easily afforded by most people suffering from hypertension, high levels of cholesterol, respiratory and urinary tract infections, and digestive and liver disorders. Garlic is one of the effective natural antibiotics. Pasteur verified its antiseptic qualities as early as 1858. Garlic inhibits pathogenic bacteria, ameba, fungi, and yeast, at a concentration of only 10 ppm (Duke, 1994). Aqueous extracts significantly inhibit both gram-positive and gram-negative bacteria, better than onion extracts. Japanese studies have shown that garlic and its allies, such as onion, chive, etc., have antioxidant properties. The compound called alliin present in them is known to have anti-aggregation property, besides its ability to reduce the cholesterol in the serum and liver. Ajoene is the most potent antithrombotic compound present in garlic. Garlic also has antispasmodic qualities. It also reduces gas formation in the stomach. It is used for diphtheria,

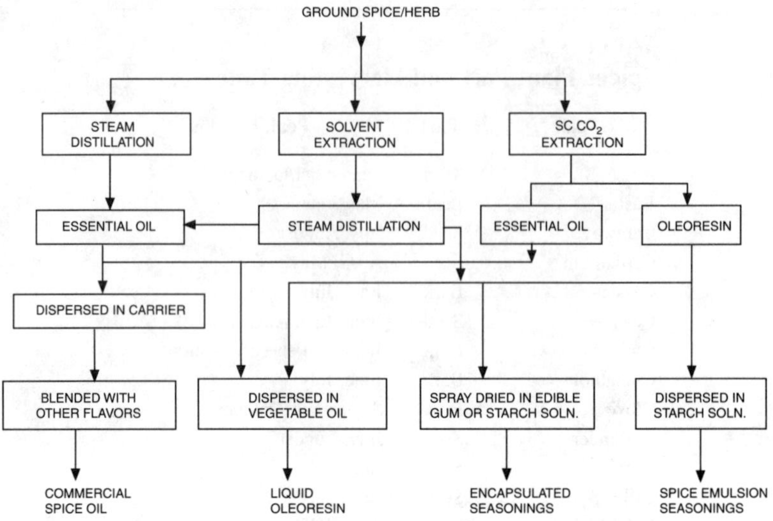

FIGURE 6.1 Various alternative steps for spice extraction.

TABLE 6.3
Classification of Spices Based on Flavor

Spices	Flavor Characteristics
Pimento, ginger, pepper, mustard	Sting and burnt flavor
Juniper berries, nutmeg, mace, cardamom, fenugreek	Smell of aromatic chemicals, such as terpenes, camphor, viandox, etc.
Caraway seed, cumin, coriander, celery, and lovage	Highly penetrating or mild fruity smell
Cinnamon and cassia	Aromatic rinds rich in cinnamon aldehyde
Cloves, cinnamon leaves	Eugenol smell
Paprika, saffron, turmeric	Intense color
Vanilla, angelica, licorice	Pleasant smell
Onion, garlic	Heavy sulfur note

hepatitis, ring worm, typhoid, and bronchitis. It is thus not unwise to say "a garlic clove a day keeps the doctor away." It is highly popular both in the East and the West.

It is well known that onion possesses significant antidiabetic activity due to the presence of diphenylamine (Duke, 1994). Onion also has 1 ppm of prostaglandin A and has high contents of vitamins A, B, and C. It has many properties of garlic and helps to prevent colds and infections. It is used as a blood cleanser and weight regulator.

Next to garlic and onion, ginger is rated as the next most potent medicinal spice. It shows heart-stimulating properties, besides bactericidal, antidepressant, antinarcotic, antihistaminic, and hypoglycemic qualities. Therefore, it is offered in every home in India and the Far East as a remedy for colds, cough, depression, asthma,

TABLE 6.4
Number of Bioactive Components Responsible for Specific Therapautic Benefits of Domestic Spices

Spice (Part)	Therapautic Benefits (Correlated to Number of Bioactive Compounds)
All spice (Fruit)	P(8), AC(5), AI(3), AB(2), AU(2), AA(2), H(3), AD(2), AAM(2), AG(2), AM(2)
Black mustard (Seed)	P(8), AC(10), AS(3), FC(3), AB(6), AO(4), AV(4), AG(3), L(2)
Black pepper (Fruit)	P(31), AC(21), AS(16), AI(8), FC(10), AB(14), AO(4), AV(6), ST(7), AA(4), AN(8), H(8), AG(5)
Cardamom (Seed)	P(2), AC(8), AI(2), HG(2), AM(2), HP(3)
Cassia (Bark)	P(10), AC(7), AS(5), ADB(3), AO(3), AU(4), AV(3), AA(3), AD(3), AM(3), I(4)
Cinnamon (Bark)	P(29), AC(14), AS(10), AI(7), FC(10), AB(11), AU(5), AV(6), ST(8), AA(4), H(8), AG(4), ADB(3)
Clove (Bud)	P(11), AC(10), AS(4), AI(6), FC(5), AO(3), AU(5), ST(3), AN(2), AA(3), AG(2)
Coriander (Fruit)	P(40), AC(27), AS(9), AI(8), FC(11), AB(20), AO(7), AV(12), ST(8), H(7), AG(6), AAM(5)
Cumin (Fruit)	P(27), AC(11), AS(5), AI(7), FC(6), AB(11), AO(5), AU(5), AV(7), ST(6), H(6), AG(3), AAM(3)
Garlic (Bulb)	P(23), AC(21), AS(6), FC(8), AB(13), AO(9), AU(6), AV(5), ST(5), AA(9), AD(5), HG(6), HP(5)
Ginger (Rhizome)	P(43), AC(25), AS(11), FC(18), AB(17), AO(6), AU(13), AV(6), ST(11), H(7), HP(8)
Licorice (Root)	P(45), AC(26), AS(23), AI(12), FC(21), AB(20), AO(10), AU(6), AV(8), ST(6), AN(9), AA(5), E(8)
Nutmeg (Seed)	P(32), AC(15), AS(11), FC(14), AB(15), AU(4), AV(4), ST(6), AN(5), AA(6), E(4), H(6), AG(3)
Poppy (Seed)	P(5), AO(3), AU(6), HPT(4), AD(5), AM(4)
Sesame (Seed)	ADB(4), P(7), AC(17), AB(5), AO(7), AD(7)
Turmeric (Rhizome)	P(15), AC(9), AS(4), AI(5), FC(7), AB(8), AO(3), AU(6), AV(3), AN(3), H(4), AG(3), I(4)
Vanilla (Fruit)	P(20), AC(7), AS(9), FC(9), AB(7), AO(7), AU(3), AV(3), AN(5), E(4)

Note:

P	Pesticidal	E	Expectorant
AC	Anticancerous	H	Herbicidal
AS	Antiseptic	AG	Analgesic
AI	Antiinflammatory	AD	Antidepressent
FC	Fungicidal	HG	Hypoglycemic
AB	Antibacterial	AM	Antimigraine
AO	Antioxidant	ADB	Antidiabetic
AU	Antiulcerous	AAM	Antiasthmatic
AV	Antiviral	L	Laxative
ST	Sedative	I	Immunostimulant
AN	Anesthetic	HP	Hepatoprotective
AA	Antiaggregant		

Beckstrom-Sternberg and Duke, 1994.

tuberculosis, high cholesterol levels, low blood pressure, motion sickness, etc., and is widely used both internally and externally.

Cinnamon and clove are also very commonly indicated for a variety of ailments. Both varieties of cinnamon, namely *Cinnamonum cassia* and *Cinnamonum verum*, are popular because of their bactericidal, fungicidal, and viricidal properties. Both clove bud and cinnamon leaves contain significant amounts of eugenol and eugenol acetate which are antioxidants. They also possess fungicidal and pain-killing properties. Clove oil has been indicated for its antibacterial activity and also to cure toothache, flatulence, indigestion, nausea, hyperacidity, etc. The oil diluted to 1/80th strength has been found to inhibit tuberculosis. Clove oil mixed with a clove of garlic and honey helps alleviate the painful spasmodic coughs in tuberculosis, asthma, and bronchitis. Cinnamaldehyde, the key constituent of cinnamon oil obtained from cinnamon bark, relieves fever, insomnia, allergy, even lowers blood pressure, in addition to its sedative properties. Cinnamon like clove, turmeric, and bay leaf shows anti-diabetic activity and can be used for sparing insulin.

Cumin is a native of Egypt and Syria, but is now grown in Southeastern Europe, North Africa, India, and China. Cumin seeds are commonly used in every household in India. Cumin oil is a rich source of thymol. The key active ingredient of this oil is cuminaldehyde, which can be readily converted to thymol by chemical reduction. The cumin oil containing about 16 to 22% cuminaldehyde shows antimicrobial and antifungal activity. It cures hoarseness of voice, piles, dyspepsia and diarrhea, jaundice, insomnia, even debility due to fever. It also improves appetite and energy. It is very useful in the treatment of colds and fevers. Thymol is used as a curative against hook worm infections and as an antiseptic agent making it a part of many formulations.

Usage of cardamom in the form of additives in food and medicine has been practiced all over the world. There are two varieties — the larger and the smaller, the chemical compositions of which are different. There are also variations in the chemical constituents depending on the geographical area of cultivation and time of harvesting. Its smooth volatile oil is used as an excellent muscle relaxant due to its carminative and antispasmodic properties. It is also recommended as a home remedy for indigestion, nausea, or bronchial infections, in ayurvedic or unani formulations.

Turmeric rhizomes are commonly used in every household in India because of the diverse therapeutic values, so much so, that it is considered to be an auspicious ingredient in many Hindu rituals. Its antiinflammatory activity is attributed to the active principle of curcumin, which is similar in action to cortisone; besides, turmerin and phenyl butazone help in the treatment of edema of joints or rheumatoid arthritis. Even sore throat, bronchial asthma, and respiratory tract infection can be controlled by using the turmeric extract in low concentrations. It is also recommended for the cure of urticaria/skin allergy, viral hepatitis, and wounds, due to its antiseptic properties.

Consumption of chili is known to be beneficial to gastrointestinal tract and sensory system. Chilis are a good source of many essential nutrients, such as vitamin C, provitamin A, E, B1 (thiamine), B2 (riboflavin), and B3 (niacin). Chilies are known to stimulate the flow of saliva and gastric juices for ingestion. It is also reported that chili raises body temperature, relieves cramp, soothens gout, and treats asthma, coughs, and sore throats when used in very, very dilute concentrations. The recovery of capsaicin and dihydrocapsaicin from different varieties of capsicum is useful for medicinal applications, such as for curing headache, colic, toothache, neuralgia, and rheumatism. Capsaicins also show antiinflammatory, antioxidant, and antitumor activity, and may even help prevent cancer. Recent studies reveal that capsaicinoids may also have cancer-prevention properties. There is evidence that chili possesses antibacterial activity (Duke, 1994).

Both black and white varieties of pepper are the most prominent spices which have been valued for both medicinal and economic importance. Unripe dried fruits form black pepper, whereas drying and peeling off the outer skin of presoaked ripe fruits form white pepper. Black pepper contains more flavoring components than white pepper. Both peppers contain many alkaloids including piperine and other pipernoids. White pepper is used to treat malaria, cholera, dysentery, diarrhea, stomachache, and other digestive problems, whereas black pepper is used in skin care, muscle and joint pains, and in improving blood circulation and respiratory systems. The use of pepper for treating diseases has been well documented. The key bioactive molecule, piperine, present in pepper has major pharmacological impacts on the nervous and neuromuscular systems. It has a sedative effect, helps in digestion, and cures frost bite.

6.4 SPICE CONSTITUENTS

It is commonly known that spices inhibit oxidation. Common spices exhibiting antioxygenetic properties are clove, turmeric, allspice, rosemery, mace, sage, oregano, thyme, nutmeg, ginger, cassia, cinnamon, savory, black and white pepper, aniseed, basil, etc.

The components present in spices have been studied individually. The spectrum of compounds present in commonly used domestic spices having immense therapeutic values are listed in Table 6.5. The volatile constituents of spices constitute the essential oil and contribute to characteristic aroma of the spices and the nonvolatile constituents form the oleoresin. The oil constituents may be classified into four major groups, monoterpenes, diterpenes, sesquiterpenes, and oxygenated compounds, as mentioned earlier. The miscellaneous compounds belonging to the last group, such as esters, ketones, alcohols, and ethers, are very specific to the species or the genus of the spice plant. Although they are present in very small quantities, they are the substances responsible for the characteristic flavor of the spice and their absence may sometimes change the aroma completely.

The nonvolatile constituents or large molecular-weight compounds of the spice extracts are fatty acids, resins, paraffins, waxes, and alkaloids, which form the bitter or pungent part of the spice and are termed as oleoresin. Oleoresin is viscous and resinous, and contains most of the ingredients that are responsible for the medicinal attributes of the spice. For example, piperine, the compound responsible for the bitter taste of black pepper, is a nonvolatile substance present in oleoresin. It does not contribute to the aroma of black pepper and is not present in the essential oil.

TABLE 6.5
Biologically Active Constituents in Common Spices

Spice	Part of Plant	Bio-Active Constituents (ppm)
Garlic (*Allium sativum*)	Bulb	Ajoene, allicin, alliin, allistatin-I, allistatin-II, arginine (6000–15000), ascorbic acid (100–800), choline, citral, diallyl disulfide, geraniol, glutamic acid (8050–19320), linalool, niacin (4–17), scorodin-A, tryptophan (660–1584)
Ginger (*Zingiber officinale*)	Root	Acetaldehyde, ascorbic acid (0–310), asparagine borneol (55–1100), bornylacetate (2–50), camphene (25–550), chavicol, 1-8 cineole (30–650) citral (0–13500), dehydrogingerdione, geraniol (2–50), gingerdiones, gingerols (18200), hexahydrocurcumin, limonene, linalool (30–650), methionine (670–735), myrcene (2–50), pinene (5–200), selinene (35–700), shogaols (1800), zingerone, zingibain, tryptophan (630–690)
Clove (*Syzygium aromaticum*)	Bud	Anethole, benzaldehyde, carvone, caryophylene (7400–8160), chavicol (465–510), cinnamaldehyde, elagic acid, eugenol (108000–120000), eugenol acetate (36000–40000), furfural, gallic acid, kaemferol, linalool (1), methyl eugenol (310–340)
Cassia and Cinnamon (*Cinnamonum cassia* & *C. verum*)	Bark	Benzaldehyde (25–100), camphene, camphor, caryophyllene (135–1315), 1,8 cineole (165–1800), cinnamaldehyde (6000–30000), cuminaldehyde (5–100), p-cymene (55–445), eugenol (220–3520), farnesol (3–10), furfural (3–10), limonene (45–180), linalool (230–950), methyl eugenol, myrcene (5–20), niacin (8), pinene (20–235), piperitone (7–25), safrole, terpineol (1–260)
Cumin (*Cuminum cyminum*)	Fruit	Anisaldehyde (835), ascorbic acid (0–75), bornyl acetate (35), delta-3-carene (270), beta-carotene (5), carveol (435), caryophyllene (140–320), 1,8 cineol (40–135), copaene (30), p-cymene (810–12600), farnesol (830), limonene (60–695), linalool (30–315), methyl chavicol (30), myrcene (35–120), niacin (45), pinene (10–6600), piperitone (170), terpinene (25–11800), terpinen-4-01 (30), terpineol (30–275)
Coriander (*Coriander sativum*)	Fruit	Anethole (1–2), ascorbic acid (180–6290), borneol (2–50), camphor (100–1300), carvone (20–25), caryophyllene (1–8), 1–8 cineole, p-cymene (70–725), ferrulic acid (460–1360), geraniol (30–440), limonene (30–1230), linalool (4060–16900), pinene (50–13750), terpineol (30–40), vanillic acid (220–960)
Cardamon (*Elettaria cardamomum*)	Fruit	Borneol (30–8000), camphene (10–30), camphor (5–20), 1,8 cineole (525–56000), citronellal, citronellol (10–40), p-cymene (130–28000), geraniol (45–140), limonene (595–9480), linalool (1285–8000), myrcene (335–3000), nerol (10–30), nerylacetate, pinene (70–3000), terpinen-4-ol (250–23200), terpinene (20–140)

TABLE 6.5 (continued)
Biologically Active Constituents in Common Spices

Spice	Part of Plant	Bio-Active Constituents (ppm)
Turmeric (*Curcuma domestica*)	Rhizome	Ascorbic acid (0–290), bisdesmethoxy curcurmin (60–27000), borneol (15–350), camphor (100–720), 1,8,cineole (30–720), cinnamic acid, curumin (10–38500), p-cymene, niacin (5–60), p-tolmethyl-carbinol (500–1750), turmerone (1800–43200)
Black pepper (*Piper nigrum*)	Fruit	Ascorbic acid (10), benzoic acid, borneol, camphor, carvacrol, carveol, caryophyllene, 1,8 cineole, cinnamic acid, citral, citronellal, p-cymene, eugenol, limonene, linalool, myrcene, myristicin, pinene, piperidine, piperine, safrole, terpinen-4-ol
Black mustard (*Brassica nigra*)	Seed leaf	Allyl isothiocyanate (6510–11760, seed), arginine (1810–26657, LF), ascorbic acid (235–4000, LF), β-carotene (30–475, LF), erucic acid (770–11340, LF), methionine (230–3390, LF), niacine (3–48, LF), tryptophan (270–3975, LF)
Saffron (*Crocus sativus*)	Flower	β-carotene, 1-8 cineole, crocetin, crocin (20000), delphinidin, hentria contane, kaemferol, lycopene, myricetin, naphthalene, pinene, quercetin
Licorice (*Glycryrrhiz glabra*) "Yashtimadhu"	Root	Acetic acid (2), anethole (1), betaine, choline, O-cresol, estragole, eugenol (1), ferulic acid, glycryrhizic acid (100000–240000), guaiacol, kaemferol, linalool, mannitol, niacin (70)
Mace and nutmeg (*Myristica fragans*)	Seed	Borneol (4200–25600), 1,8 cineole (440–3500), p-cymene (120–960), elemicin (20–3500), eugenol (40–320), furfural (15000), geraniol, limonene (720–5760), linalool, methyl eugenol (20–900), myrcene (740–5920), myristicin (800–12800), pinene (3000–4000), safrole (120–2720), terpinen-4-ol (600–4800), terpineol (120–9600)
Thyme (*Thymus vulgaris*)	Plant	Borneol (15–1460), bornyl acetate (15–540), caffeic acid, camphene (15–270), delta-3-carnene (510), beta-carotene (20–25), carvarol (15–18720), chlorogenic acid, 1,8 cineole (80–4590), p-cymene (145–20800), geraniol (0–10660), limonene (15–5200), linalool (180–17420), methionine (1370–1980), myrcene (35–675), niacin (50), pinene (15–1600), rosmarinic acid (5000–6000), terpinen-4-ol (70–8320), terpineol (35–6500), thymol (15–24000), tryptophan (1860–2000), ursolic acid (15000–18800)
Red Pepper/chili (*Capsicum corrals*)	Fruit	Arginine (400–8000), ascorbic acid (350–20000), asparagine, betaine, capsaicin (100–2200), beta-carotene (0–460), chlorogenic acid, hesperidin, methionine (100–1900), niacin (5–170), oxalic acid, tryptophan (100–2000)
Vanilla (*Vanilla plantifolia*)	Fruit	Acetaldehyde, acetic acid, anisaldehyde, benzaldehyde, benzoic acid, creosol, eugenol, furfural, guaiacol, vanillic acid, vanillin (13000–30000), vanillyl alcohol

Duke, 1994.

6.5 PRODUCTION OF SPICE EXTRACTS

Recovery of essential oils from spices by steam distillation has been going on for centuries. However the oils produced this way may contain artifacts formed during the processing, and may be low in oil recovery, since some of the components are not steam volatile. Alternatively, organic solvents are used to extract essential oils from ground spices. But the quality of the extracts by the solvent extraction process is restricted by the presence of residual solvent, artifacts formed by thermal degradation during the recovery of the solvents, or by coextraction of undesirable components due to the polarity of the solvent. A polar solvent is likely to extract most polar components from spices, some of them may even be undesirable. In food and pharmaceutical applications, carbon dioxide (CO_2) is the most favorable fluid because it is a natural solvent, ideally suitable for natural products. Supercritical (SC) or subcritical CO_2 extraction methods have now been accepted by the industry for commercial production of essential oils and oleoresins from spices. Oleoresins are extracted by SC CO_2 from spices at relatively higher pressures. On the other hand, essential oils are recovered at relatively lower pressures, which can be added directly to soft drinks to form a clear solution. The traditionally extracted oleoresin includes undesirable resin which precipitates and makes the solution cloudy, requiring an additional step of filtration. Even steam-distilled oil forms an immiscible layer due to the presence of monoterpene hydrocarbons, which to a large extent inhibit the solubility of oil in soft drinks or other beverages. The remarkable advantage of SC CO_2 over organic solvents is its selectivity which facilitates the recovery of spice extracts with desirable constituents and superior blending characteristics. Another important advantage of SC CO_2 extraction is that the raffinate ground spice left after extraction is absolutely free from any contaminants and artifacts. Thus it has a market value for its high fiber and protein content.

The capital equipment cost of a SC CO_2 extraction plant is approximately 1.5 times that of a subcritical CO_2 plant, the latter being used only for the production of natural essential oils from spices (Moyler, 1993). But simultaneous extraction of oleoresins and essential oils at pressures from 250 to 350 bar, followed by their stage-wise selective fractionation at supercritical and subcritical conditions, in addition to a higher premium due to the superior quality of the products, may overcome the extra cost. The selective solubility in SC CO_2 at the optimum process condition facilitates extraction of all useful ingredients from spices, resulting in the production of extracts close to that in nature. SC CO_2 is often mixed with small quantities of food-grade entrainers, such as ethanol, ethyl acetate, and water, to adjust the polarity of the extractant to give specific product profiles. An entrainer is also selected based on the consideration that it can be left behind in the extracted product, or used with allowance made for dilution level. Calame and Steiner (1987) compared the extraction yields obtained by steam distillation and SC CO_2 extraction with an entrainer as given in Table 6.6. They established the point that yields are much higher in the case of SC CO_2 extraction, but there are other factors responsible for the recovery of the extract which are not included in this table. For example, particle size, methods of drying and grinding, times of extraction and storage after harvesting, and even

TABLE 6.6
Comparison of Extraction Yields by Steam Distillation and $SC\ CO_2$ Extraction with Entrainer

Spice	Steam Distillation Yield (%)	$SC\ CO_2$ Extraction			
		Entrainer	Extractor P/T (bar/°C)	Separator P/T (bar/°C)	Yield (%)
Allspice	2.5	Ethanol	300/40	55/37	5.3
Basil	0.5	Ethanol	200/40	56/15	1.3
Cardamom	4.0	Methyl acetate	150/60	50/9	5.8
Coriander	0.6	Ethanol	300/40	54/13	1.3
Ginger	1.1	Ethanol	300/40	52/11	4.6
Juniper berry	1.5	Hexane	300/60	52/11	7.2
Marjoram	2.06	Ethanol	250/40	50/35	1.7
Oregano	3.0	Ethanol	150/40	55/14	5.4
Rosemary	1.44	Ethanol	250/60	53/12	7.5
	1.34	Hexane	250/60	53/12	7.5
Sage	1.1	Methyl acetate	200/40	53/12	4.3
Thyme	1.85	Hexane	150/46	50/9	2.1

Calame and Steiner, 1987.

the geographical origin of the raw spice can make a difference. According to the experience of the author, higher yields may be obtained from some of these spices with $SC\ CO_2$ even without an entrainer, as can be seen later in Table 6.8.

Not only the yields of the extracts, but also the organoleptic (sensory) characteristics are different for extracts obtained by different methods. But the criteria for selection of the best process condition and solvent is based on the requirement of quality and the consumer's preference. The positive attributes of CO_2 extracts were discussed in Chapter 4. Briefly, CO_2 extracts have more top notes, more back notes, no off notes, no degradation, more shelf life, and better aroma and blending characteristics than steam-distilled and hexane extracts, as can be seen in Table 6.7. The extract obtained by solvent (e.g., hexane) extraction is called oleoresin and contains all the ingredients that are soluble in the organic solvent, including the volatile oils and the resins. Some triglyceride lipids present in spices are co-extracted and act as Nature's own fixative which can be exploited in the making of formulations.

For example, the yield of oil by steam distillation of cumin (2.5%) is less than that by $SC\ CO_2$ extraction (3.5%) at 120 bar and 40°C (Gangadhara Rao and Mukhopadhyay, 1988). A comparison of the composition of clove extracts obtained by liquid CO_2 extraction, ethyl ether extraction, and steam distillation indicates that liquid CO_2 and ethyl ether extracts are similar, though liquid CO_2 extract has the characteristics of both essential oil and oleoresin (Meireles and Nikolov, 1994). Liquid or $SC\ CO_2$ extracts are always transparent and contain a higher number of active components, close to that found in the original natural material. Recently $SC\ CO_2$ extraction followed by fractional separation was carried out on a variety

TABLE 6.7
Comparison of Spice Constituents (Area %) Recovered by Various Methods

Constituents	Distillation (%)	L CO_2 (%)	SC CO_2 (%)	Hexane (%)
Ginger extract (analyzed by GC)				
α-Curcumene	10.0		3.7	2.3
α-Zingiberene	44.0		19.6	12.1
β-Zingiberene	8.0		3.4	2.0
β-Bisabolene	8.3		3.7	2.4
β-Sesquiphellandrene	17.8		7.9	4.9
Zingerone	0.8		0.7	0.3
Ginger extract (analyzed by HPLC)				
6-Gingerol	0.2		16.4	0.9
8-Gingerol	0.3		3.1	0.7
10-Gingerol	—		3.8	0.8
6-Shogaol	0.3		2.8	6.3
8-Shogaol	—		—	1.6
Cumin extract[2] (analyzed by GC)				
α-Pinene	—		1.1	—
β-Pinene	13.0		21.0	
ρ-Cymene	13.0		9.4	
γ-Terpinene	24.8		20.0	
Cuminaldehyde	16.0	20.3[1]	21.0	11.4
Cymol	33.4[2]	26.7[1]	15.2	13.5
Clove extract (analyzed by GC)				(ethyl ether)
Eugenol	76.4	77.1	71.8[3]	73.3
Eugenol acetate	5.6	4.9	11.1[3]	4.6
β-caryophyllene	5.8	8.5	9.3[3]	10/4

[1] Mahindru, 1992.
[2] Gangadhara Rao, 1990.
[3] Pilot plant experiment at Indian Institute of Technology, Bombay.

Meireles and Nikolov, 1994.

of spices using a 10-l capacity extractor pilot plant designed and built at the Indian Institute of Technology (IIT), Bombay. The composition of the active ingredients in the extracts separated at the two separators, were analyzed by either GC or GC MS, as indicated in Table 6.8 (except for pepper, where piperine was quantified by the UV method). The yields and compositions were compared with those of hexane-extracted products from the same spices, as shown in Table 6.8. It can be seen that the yields were better from SC CO_2 extraction for clove, cumin, and black pepper. In most cases, the concentrations of the active ingredients were higher in supercritically extracted products.

TABLE 6.8
Comparison of Percent (by wt) Yields and Active Ingredients in Extracts Obtained with SC CO_2 Hexane

	SC CO_2 Extraction (200 bar, 40°C)			Solvent (Hexane) Extraction	
Spices (Active Ingredient)	Yield (%)	Essential Oil (%)	Oleoresin (%)	Yield (%)	Extract (%)
Clove	23.8			16.8	
Eugenol		71.8	—		70.7
Eugenol acetate		11.1	—		11.3
		(Analyzed by GC, 10% FF AP)			
Cumin	21.0			12.2	
Cymol		15.2	—		13.5
Cuminaldehyde		15.3	—		11.4
		(Analyzed by GC MS, DB-5)			
Coriander	3.6			20.0	
Linalyl acetate		7.8	—		5.8
D-linalool		13.0	—		—
		(Analyzed by GC MS, DB-5)			
Ginger	4.6			4.9	
Zingiberene		26.7	1.6		31.6
Gingerol		5.65	10.1		5.4
		(Analyzed by GC MS, DB-5)			
Cinnamon	3.0			5.11	
Cinnamicaldehyde		77.5		—	45.0
		(Analyzed by GC, SPB-1)			
Pepper	4.6			5.0	
Piperine		—	53.0		46.4
		(Analyzed by UV method)			
Ajwain	4.5			5.18	
Thymol		63.6	—		24.6
		(Analyzed by GC OV-101)			

6.6 SC CO_2 EXTRACTION AND FRACTIONATION

Spice extracts usually constitute a complex mixture of volatile essential oils, waxes, triglycerides, resinous and other miscellaneous materials contributing to aroma, flavor, and pungency, appropriate for specific applications. Accordingly, for customized applications, SC CO_2 extraction of spices requires fractional separation of selected groups of constituents. This can be achieved in two ways, by either stagewise extraction followed by separation upon depressurization of the solution of the extract in SC CO_2, or single-stage extraction at a very high pressure (or density)

TABLE 6.9
Comparison of Celery Seed Essential Oil Composition (%) by Supercritical CO_2 Extraction and Hydrodistillation

Class of Compounds	SC CO_2 Extraction	Hydrodistillation
Monoterpenes	16.1	57.6
Oxygenated monoterpenes	0.2	0.6
Sesquiterpenes	19.7	23.3
Phthalides	56.8	15.2
Others	4.8	0.3

Zhang et al., 1997.

followed by fractional separation upon stage-wise depressurization. In the former method, the volatile oil is first extracted at relatively milder conditions and subsequently the nonvolatile oleoresins are extracted at relatively more severe conditions. In the latter method, the finely ground spice is more or less completely extracted at a relatively more severe condition to recover both oleoresin and volatile oil simultaneously and efficiently, so that the time of extraction is greatly reduced. The extract-laden SC CO_2 is subsequently depressurized in two or three separators at predetermined conditions in order that specific products are selectively precipitated and collected. The second method offers significant advantages since the quality of the product is improved and the batch time for extraction is reduced, resulting in higher production capacity and in turn, increased cost effectiveness for the SC CO_2 extraction technology. The simultaneous fractionation at precisely selected conditions allows production of customized fractions and the elimination of undesirable contaminants from them. The specific advantages of this SC CO_2 extraction and fractionation process with respect to a few common spices are described in the following subsections.

6.6.1 Celery Seed (*Apium graveolen*)

A comparison of chemical composition (Table 6.9) of celery seed oil by hydrodistillation and SC CO_2 extraction (Zhang et al., 1997) at 100 bar and 40°C indicated that the hydrodistilled (HD) oil contained mostly monoterpenes, whereas the SC CO_2-extracted oil contained mostly phthalides. SC CO_2 extract contained some additional components, such as fatty acids, which were not present in the HD oil. Monoterpenes constituted 57.6% of HD oil, whereas SC CO_2 oil contained 56.8% phthalides (Zhang et al., 1997). The low level of phthalides (15.2%) in HD oil was attributed to their high boiling points and low volatility in steam. More than 10 h of hydrodistillation was necessary for their complete recovery. Phthalides are cyclic esters or lactones with the outstanding odor characteristics of celery. The odor of the SC CO_2 extracted oil is more intense and less terpenic. Therefore SC CO_2-extracted oil is preferred to HD oil to impart celery flavor. With SC CO_2 extraction carried out at a relatively moderate pressure of 100 bar at 40°C from celery seeds, the yield of essential oil was

TABLE 6.10
Comparison of Composition of Essential Oils from Celery Seeds and Leaves by SC CO_2

Component	Celery Seeds (100 bar, 40°C)	Chinese[a] Celery Seed (100 bar, 40°C)	Celery leaves (90 bar, 40°C)
Limonene	3.7	14.9	33.4
β-Selinene	33.8	17.6	3.0
α-Selinene	5.3	1.8	0.5
Butyl phthalide	19.8	5.5	2.8
Sedanenclide	—	22.4	—
Bedanolid	—	28.8	—
Germacrone	21.0	—	45.4

[a] Zhang et al., 1997.

Della Porta et al., 1998.

TABLE 6.11
Comparison of Composition in Light and Heavy Fractions of Chili Extract

Products	Capsaicin (%)	Dihydro-Capsaicin (%)	Total Capsaicin (%)
Raw material	0.21	0.14	0.39
Light fraction	8.10	4.05	13.50
Commercial product	1.83	1.52	3.93
Heavy fraction	0.57	0.31	0.95
Ratio L/H	14.2	13.1	14.2

Nguyen et al., 1998.

merely 2.03% of the charged material (Della Porta et al., 1998). The yield of essential oil from celery leaves by SC CO_2 extraction at 90 bar and 40°C was much lower (0.04%). The compositions of the essential oils from celery seeds and celery leaves by SC CO_2 extraction were found to be significantly different, as can be seen in Table 6.10. Celery seed extracts contained more paraffins and fatty acid methyl esters than the celery leaf extract.

6.6.2 Red Chili

SC CO_2 extraction of red chili is carried out in the pressure range of 300 to 500 bar and 80 to 100°C with simultaneous fractionation of the extracts into two fractions. The light fraction includes most of the capsaicin, i.e., the compound responsible for the hotness of the spice, besides the essential oil, whereas the heavy fraction contains triglycerides and the color compounds, besides a small quantity of capsaicin (Nguyen et al., 1998), as shown in Table 6.11. The distribution of capsaicin in the two fractions can be adjusted by selecting the conditions of fractionation in the separators.

TABLE 6.12a
Comparison of Light and Heavy Fractions of SC CO_2 Extracted Ginger Oleoresin

Product	6-G (%)	8-G + 6-S (%)	10-G + 8-S (%)	Ratio $\frac{8G + 6S}{Total}$	Total Extract (%)
Raw material	0.87	0.14	0.27	0.11	1.28
Heavy fraction	13.95	2.58	4.37	0.12	20.90
Commercial product	2.81	5.83	1.19	0.52	11.12
Light fraction	1.43	0.61	0.36	0.25	2.40
Ratio H/L	9.8	4.2	12.1	0.5	8.7

Note: (G) Gingerol; (S) shogaol.

Nguyen et al., 1998.

6.6.3 Paprika

Paprika is useful in industry for its natural color. For SC CO_2 extraction of paprika, most of the color compounds are collected in the heavy fraction, while aroma is collected in the light fraction, as will be elaborated in Chapter 9. It is reported (Nguyen et al., 1998) that the color value of SC CO_2 extracted product could reach as high as 7200 on the standard scale specified by the American Spice Trade Association (ASTA), New Jersey, whereas a normal commercial product is characterized to have a color value in the range from 1000 to 2000.

6.6.4 Ginger

The ginger extract using SC CO_2 is fractionated into two fractions, namely, the essential oil-enriched light fraction and the oleoresin-enriched heavy fraction, the compositions of which are compared in Tables 6.12a and b. The gingerols (G) and shogaols (S) are the compounds responsible for the pungency of ginger and they are mostly collected in the heavy fraction of the SC CO_2 extract, as can be seen in Table 6.12a. Shogaols, being the oxidation products of gingerols, are present in a very lesser quantity in the SC CO_2 extracted fractions. A product of desired specification can be formulated by combining the two fractions in a suitable proportion (Nguyen et al., 1998).

6.6.5 Nutmeg

SC CO_2 extraction and fractionation of nutmeg can yield good quality nutmeg butter as the heavy fraction with very little volatile oil, and nutmeg oil as the light fraction in which the undesirable hallucinatory compound, myristicin, is present in negligible concentration (Nguyen et al., 1998). The presence of this compound in nutmeg oil has forced some countries to ban this product. It is possible to use SC CO_2 to produce nutmeg oil product devoid of this compound.

TABLE 6.12b
Comparison of Compositions (%) of Light and Heavy Fractions of SC CO_2 Extracted Ginger Essential Oil

Product	Raw Material	Heavy Fraction	Light Fraction	Ratio L/H
Essential oil (ml/100 g)	2.0	4.4	98.8	22.5
α-Pinene	2.5	0.5	2.6	5.2
Camphene	7.0	1.6	7.3	4.6
Cineole	8.4	2.3	8.6	3.7
Limonene	1.2	0.3	1.2	4.0
Zingiberene	21.8	17.4	22.7	1.3
Bisabolene	8.7	8.6	8.8	1.0
Sesquiphellandrene	11.9	12.7	11.9	0.9

Nguyen et al., 1998.

6.6.6 Pepper

When black pepper is supercritically extracted and fractionated using CO_2 into two fractions, the light fraction may be completely freed from piperine, while the heavy fraction may be enriched up to 60% piperine, the active ingredient of pepper (Nguyen et al., 1998).

Besides the concentration of the specific component, all SC CO_2 fractionated products are of superior quality. The rate of extraction at 500 bar is almost double the rate at 300 bar and 60°C. The production capacity of the fractions may be enhanced four times at 500 bar with cascading mode of extraction by using four extractors. Thus the operating cost of extraction can be reduced to one fourth of that obtained by the traditional supercritical extraction plant. The current commercial practice is to follow this technique to improve the efficiency and cost effectiveness of SC CO_2 extraction of major spices.

6.6.7 Vanilla

Natural vanilla fragrance is extracted from cured vanilla beans. Green vanilla beans are cured to bring about hydrolysis of the glucosides present in the beans to generate vanillin and other flavor and fragrance components. The curing process changes the green vanilla beans into dark brownish soft beans. The current commercial extraction method uses aqueous alcohol of 35 to 40 vol% in concentration at a temperature as high as 87°C in a number of steps making the extract thermally degraded. SC CO_2 extraction of cryogenically ground dried beans resulted in a 10.6% yield of oleoresin at 110 bar and 36°C, which is even higher than 8.4 to 5.3% yields by alcohol extraction (Nguyen et al., 1998). The vanilla oleoresin contained as high as 16 to 36% vanillin by SC CO_2 extraction, which amounted to 74 to 97% recovery of the total vanillin content. Other flavor and fragrance constituents in the natural vanilla extract are p-hydroxy benzaldehyde, vanillic acid, and p-hydroxybenzoic acid. The quality of the extract is, however, characterized by its vanillin content. The compositions of the natural vanilla extracts by the traditional alcohol extraction and SC CO_2 extraction methods are compared in Table 6.13. The highest purity vanillin could be obtained by SC CO_2 extraction of water-presoaked beans, though the yield was only 3%. On the other hand, cryogrinding apparently releases more compounds and accordingly the yield was also high (10.6%). The purity of the alcohol extract as well as percent recovery of vanillin (61%) are lower than those of the supercritically extracted product. Even the color of the extract is superior in the case of SC CO_2 extract, which is yellow compared to the dark brown color of the alcohol extract.

6.6.8 Cardamom

SC CO_2 extraction of cardamom requires the much higher pressure of 100 bar as compared to subcritical propane, which is as low as 20 bar, to yield the same amount of essential oil. Addition of ethanol to SC CO_2 does not have much effect on the increase in the yield, but increases the co-extraction of pigments, as can be seen in Table 6.14. Reduction in the pressure of SC CO_2 usually reduces the contents of β-carotene, chlorophyl, and pheophytin in the extract.

The amount of pigment extracted is significantly more when subcritical propane is used as the extractant than SC CO_2. However, better recovery of aroma (Table 6.15) is possible with SC CO_2 at 100 bar and 35°C, as reported by Illes et al. (1998).

6.6.9 Fennel, Caraway, and Coriander

Recovery of the active components from fennel, caraway, and coriander by different methods of extraction has been compared in Table 6.16. It is clear that SC CO_2 extracts are richer in active components owing to better selectivity of the extractant.

6.6.10 Garlic

SC CO_2 extraction of valuable ingredients from garlic is comparable to that by hexane (Nawrot and Wenclawiak, 1991). The major components of garlic oil are diallyl disulfide (30%), diallyl trisulfide (30%), and diallyl sulfide (15%). Alliin, a major garlic active ingredient, is known to degrade to allicin by an enzymatic reaction, and other garlic components are also susceptible to oxidation with temperature. A comparison of high-performance liquid chromatography (HPLC) and GC analysis of extracts obtained by solvent extraction with a variety of solvents with varying polarity with that by SC CO_2 indicated that the former contained a larger number of components. This is attributed to degradation of the components in solvent extraction. Clinical tests also indicated that the SC CO_2 garlic extract has more potent bioactivity, close to that of raw garlic (Nawrot and Wenclawiak, 1991).

6.6.11 Cinnamon

Two types of essential oils are produced from two different parts, viz., the leaf and bark of a cinnamon tree. Cinnamon leaf oil is mainly produced in Sri Lanka. It is

TABLE 6.13
Comparison of Compositions of Vanilla Extracts

Solvent (Beans)	Supercritical CO_2 (120 bar, 33°C) Dry	Cryoground	Water Soaked	Ethanol + H_2O (Water Soaked)
p-Hydroxybenzoic acid (area %)	0.2	0.1	0.1	1.1
Vanillic acid (area %)	0.1	1.3	0.1	1.1
p-Hydroxy benzaldehyde (area %)	0.6	1.9	0.9	2.7
Vanillin (mass %)	21.0	16.1	36.3	20.0[a]
Unknown	0.0	2.4	0.0	8.0

a Water-free basis.

Nguyen et al., 1991.

TABLE 6.14
Yield of Cardamom Oil and Pigment Content by SC CO_2 and Propane

Process Conditions	Yield (%)	β-Carotene (μg/g Oil)	Chlorophyll (μg/g Oil)	Pheophytin (μg/g Oil)
SC CO_2 (80 bar/25°C)	5.65	0.8	0.65	—
SC CO_2 (100 bar/35°C)	5.45	2.1	0.30	—
SC CO_2 (200 bar/35°C)	5.95	3.9	0.36	0.33
SC CO_2 (300 bar/35°C)	6.65	5.8	4.53	2.36
CO_2 + ethanol (100 bar/25°C)	5.28	1.64	9.65	2.10
Ethanol		0.80	11.95	2.60
Propane (50 bar/25°C)	7.24	18.6	10.80	4.80
Propane (20 bar/25°C)	6.85	16.2	3.40	2.10

Illes et al., 1998.

TABLE 6.15
Peak Area (× 10^3) of Aroma Constituents of Cardamom Oil by SCF

	β-Pinene	Cineole	Linalool	α-Terpinol	Borneole
CO_2 (80 bar, 25°C)	16.1	295	34.8	47.8	356
CO_2 (100 bar, 35°C)	27.6	450	73.5	91.2	579
CO_2 (300 bar, 35°C)	17.4	341	32.7	46.4	340
Propane (20 bar, 25°C)	15.5	286	25.6	36.9	304
Propane (50 bar, 25°C)	26.9	386	72.1	82.7	521
CO_2 + Ethanol (100 bar, 25°C)	6.5	198	5.8	8.9	112

Illes et al., 1998.

TABLE 6.16
Recovery of Active Components from Fennel, Caraway, and Coriander by Various Methods

Active Component	SC CO_2 80 bar, 28°C	SC CO_2 100 bar, 30°C	SC CO_2 200–300 bar, 35°C	Ultrasound Water	Hexane	Steam
Fennel						
Fenchon	10.7	13.1	9.2	21.9	16.3	0.3
Estragol	1.6	0.5	1.5	6.6	3.1	1.7
trans Anethole	68.2	50.8	72.5	70	70	77.6
Caraway						
Limonene			33.5	32.0	33.3	30.1
D-Carvone			56.9	54.0	54.3	50.2
Coriander						
Linalool	20–30	15	80–85	67	80	79

Then et al., 1998.

TABLE 6.17
Comparison of Cinnamon Leaf Oil and Cinnamon Bark Oil

Component	Leaf Oil[a] (%) Steam Distillate	Bark Oil (%) Steam Distillate	Bark Oil (%) $SCCO_2$ Extract (200 bar, 60°C)	Bark Oil (%) $SCCO_2$ + Ethanol Extract (200 bar, 60°C)
Eugenol	85–95	3.3	2.0	2.8
Caryophyllene	6	traces	2.1	1.6
Cinnamicaldehyde	3	87.8	81.9	86.8
Isoeugenol	2	1.9	1.2	0.4
Linalool	2	0.1	0.9	0.8
Cinnamyl acetate	2	3.6	5.1	1.8
o-Methoxy cinnamic aldehyde		1.3	1.6	2.6

[a] From Wright, 1994.

Calame and Steiner, 1987.

also produced in India and Seychelles. Most of the cinnamon oil is produced from leaves, the bark oil amounts to only 15% of total production. Leaves yield 1% oil. However, the root bark yields 3% oil (Mahindru, 1992). Comparison of the compositions of extracts from Srilankan cinnamon bark and leaves is given in Table 6.17. Steam distillation of the cinnamon bark yielded 1.4% oil, whereas SC CO_2 extraction at 200 bar and 60°C resulted in 1.5% yield. However, addition of ethanol as entrainer increased the yield to 2.6% (Calame and Steiner, 1987).

The leaf oil is rich in eugenol, which makes it a substitute for clove oil, and may be used for conversion to vanillin. Bark oil is more valuable than the leaf oil, though both find wide uses in flavoring and pharmaceutical industries.

6.7 MARKET TRENDS

North America and Western Europe are the two regions with a huge import demand for most spices. East European countries are significant outlets for pepper, Latin American countries are significant importers of cinnamon and cassia, while the countries of the Middle East are major markets for cardamom, accounting for over 80% of the total world consumption of this spice. In the Asian and Pacific region, the major consumer of imported spices is Japan, followed by Australia and New Zealand. Indonesia is a substantial importer of spices, being the world's leading user of clove. Spices are traded in a variety of forms. It is estimated that by far the bulk (over 90%) is traded internationally in whole form, i.e., unground. The remainder is made up of spice oils, spice oleoresins, and ground spices, the latter consisting mainly of ground paprika, spice mixtures, and curry powder. Pepper is almost the principal spice, in terms of both volume and value. Next in importance is the capsicum group, consisting of paprika, chilies, and ayenna pepper. Three spices, whose trading volume is low, but have high unit values, are vanilla, saffron, and cardamom.

India is a major spice producing and processing country in the world. In 1997 to 1998, India exported 228,821 MT of spices valued at Rs. 1,408 crores ($378.72 million U.S.) and spices like pepper, ginger, turmeric, coriander, cumin, fennel, and spice oils and oleoresins have been well exported. Spice oils and oleoresins share about 16% of the value of the total spices export. U.S. is the single largest market for black pepper with an average consumption of 40,000 tonnes. The growth in the export of value-added spice oils and oleoresins has been significant at a rate of 11%. The major exported spice oils are pepper oil, aniseed oil, celery seed oil, and kokam oil.

In search of low volume–high value products, many manufacturers and also raw material importer cum exporters are now seeking a switch over to the business of spice oils and oleoresins and their formulations. The end-users of oleoresins and essential oils are mostly the pharmaceutical, food, beverage, and confectionery industries. As oleoresins contain nonvolatile constituents and are in the form of pastes or viscous oils, it is not easy to convert them into free-flowing powders, unless a carrier is used. Commercial oleoresins are converted into the form of emulsions, spray dried, encapsuled, or simply spread or plated onto salt or starch to make them more easily available. Essential oils are often blended to reinforce or reconstitute flavor. Because of their potency, most spice oils are blended as solutions in a convenient solvent, such as isopropyl alcohol, and are dispersed onto a suitable carrier, such as lactose or maltodextrin product. Table 6.18 gives an indication of the demand of whole spices and oleoresins in 1997 and their values and value additions.

Table 6.19 shows the approximate prices of the spice oils and their countries of origin. The CO_2 extracted products are chemically stable and microbiologically safe and have an edge over the ground spices which can easily suffer microbiological contamination. Table 6.20 indicates that spice exports are consistently increasing in India, which commands 49% in the global trade and 24% in terms of total value (Dubey, 1999).

TABLE 6.18
U.S. Imports[a] of Spices Essential Oils and Oleoresins in 1997

	Amount MTPA	Value Million U.S. $
Oleoresin		
Paprika	262	8.5
Black pepper	322	6.8
Others	575	11.0
Subtotal (oleoresins)	1160	26.3
Essential Oils		
Cassia oil	330	8.05
Clove oil	441	2.47
Nutmeg oil	246	5.08
Onion/Garlic oils	56	2.30
Subtotal (essential oils)	30195	323.60
Spices		
Allspice (pimento)	1060	2.15
Anise	1211	2.71
Basil	2765	3.58
Paprika	7064	15.87
Caraway	3139	2.81
Cardamom	256	0.91
Cassia & cinnamon (Gr.)	16430	28.54
Clove	1431	1.90
Coriander	3100	2.70
Cumin	6559	10.56
Curry & curry powder	626	1.90
Dill	598	0.77
Fennel	3418	4.07
Garlic (dehydrated)	5952	2.25
Ginger	13448	14.38
Mustard (ground)	59006	24.62
Nutmeg	1924	4.18
Onions (dehydrated)	318	0.61
Black pepper	45319	155.41
Poppy	5237	6.00
Turmeric	2043	2.76
Vanilla beans	2198	42.33
Mixed spices/others	4532	12.01
Subtotal (spices)	289041	523.48

Note: Horticultural & Tropical Products Division, FAS/USDA.

U.S. Department of Commerce, March 1998.

TABLE 6.19
U.S. Import Price of Essential Oil from the Country of Origin

Name of Oil	Origin	Product No.	Size	Price ($)
Aniseed	Spain	OA105	1/3 oz.	12.50
Caraway	France	OC205	1/3 oz.	12.00
Cardamom	Indonesia	OC255	1/3 oz.	11.00
Cinnamon	Madagaskar	OC405	1/8 oz.	14.00
Clove	Indonesia	OC555	1/3 oz.	6.00
		OC55L	3.3 oz.	44.00
Cumin	France	OC804S	1/3 oz.	15.00
Fennel	Spain Wild	OF105	1/2 oz.	7.00
Ginger	Indonesia	OG405	1/3 oz.	8.50
Nutmeg	Indonesia	ON405	1/3 oz.	7.5
Pepper	Madagaskar	ON40L	3.3 oz.	52.00
		OP60S	1/3 oz.	16.00

Note: Source: Price Catalog: Our Finest Essential Oils on Internet (http://members.aol.com/Natscents/ns.htm, Dec. 29, 1998).

TABLE 6.20
Estimated Export of Spices from India (1998–1999)

Spice	Quantity, MT	Value (Million U.S. $)
Pepper	30,350	131.06
Cardamom (small)	260	3.93
Cardamom (large)	1,125	2.26
Chili	34,750	31.82
Ginger	9,300	9.41
Turmeric	21,500	20.32
Coriander	18,300	9.63
Cumin	10,950	14.22
Celery	3,325	1.96
Fennel	5,200	3.50
Fenugreek	7,700	3.45
Garlic	3,375	1.07
Mint oil	2,775	18.72
Spice oleoresins	2,250	57.81

Note: Source: Dubey, 1999.

REFERENCES

Beckstrom-Sternberg, S. M. and Duke, J. A., Potential for synergistic action of phytochemical in spices, in *Spices, Herbs and Edible Fungi*, Charalambous, G., Ed., Development in Food Science Series, Elsevier Science Publishers, Amsterdam, 201–222, 1994.

Calame, J. P. and Steiner, R., Supercritical extraction of flavours, in *Theory and Practice of Supercritical Fluid Technology*, Hirata, M. and Ishikawa, T., Eds., Tokyo Metropolitan Univ., 275, 1987.

Della Porta, G., Reverchon, E., and Ambrousi, A., Pilot plant isolation of celery and parsley essential oil by SC CO_2, Proc. 5th Meet. SCFs, France, 2, 613, 1998.

Dubey, S. K., Ed., Estimated export of spices from India, EX-IMPO, Action Bulletin, Mumbai, India, p. 5, April 27, 1999.

Duke, J. A., Biologically active compounds in important spices, in *Spices, Herbs and Edible Fungi*, Charalambous, G., Ed., Development in Food Science Series, Elsevier Science Publishers, Amsterdam, 225–250, 1994.

Gangadhara Rao, V. S. G., Studies on Supercritical Extraction of Spices, Ph.D. dissertation, Indian Institute of Technology, Bombay, 1990.

Gangadhara Rao, V. S. G. and Mukhopadhyay, M., Selective extraction of spice oil constituents by supercritical carbon dioxide, Proc. Annu. Conv. Indian Inst. Chem. Eng., Baroda, India, Dec., 1988.

Illes, V., Daood, H., Karsai, E., and Szalai, O., Oil extraction from cardamom crop by sub and supercritical carbon dioxide and propane, Proc. 5th Meet. Supercritical Fluids, France, 2, 533, 1998.

Moyler, D. A., Extraction of flavors and fragrances with compressed CO_2, in *Extraction of Natural Products Using Near-Critical Solvents*, King, M. B. and Bott, T. R., Eds., Blackie Academic & Professional, Glasgow, 154, 1993.

Moyler, D. A., Oleoresins, tinctures and extracts, in *Food Flavorings*, Ashurst, P. R., Ed., Blackie Academic & Professional, Chapman & Hall, Glasgow, 54, 1994a.

Moyer, D. A., Spices — recent advances in spices, in *Spices, Herbs and Edible Fungi*, Charalambous, G., Ed., Development in Food Science Series, Elsevier Science Publishers, Amsterdam, 34, 1994b.

Mahindru, S. N., Indian plant perfumes, *Metropolitan*, New Delhi, India, 1992.

Marion, J. P., Audrin, A., Maignial, L., and Brevard, H., Spices and their extracts: utilization, selection, quality control and new developments, in *Spices, Herbs and Edible Fungi*, Charalambous, G., Ed., Development in Food Science Series, Elsevier Science Publishers, Amsterdam, 71, 1994.

Meireles, M. A. A. and Nikolov, Z. L., Extraction and fractionation of essential oils with liquid CO_2, in *Spices, Herbs and Edible Fungi*, Charalambous, G., Ed., Development in Food Science Series, Elsevier Science Publishers, Amsterdam, 171, 1994.

Nawrot, N. and Wenclawiak, B., Supercritical fluid extraction of garlic followed by chromatography, Proc. 2nd Intl. Symp. SCFs, Boston, MA, 451, 1991.

Nguyen, U., Anstee, M., and Evans, D. A., Extraction and fractionation of spices using SCF CO_2, Proc. 5th Meet. Supercritical Fluids, 2, Nice, France, 523, 1998.

Nguyen, K., Barton, P., and Spencer, J. S., Supercritical carbon dioxide extraction of vanilla, *J. Supercritical Fluids*, 4, 40–46, 1991.

Then, M., Daood, H., Illes, V., and Bertalan, L., Investigation of biologically active compounds in plant oils extracted by different extraction methods, Proc. 5th Meet. Supercritical Fluids, Nice, France, 2, 555, 1998.

Zhang, J., Fan, P., Guo, Z., Liu, L., Huang, C., and Zeng, J., Volatile compounds of a SCF extract of Chinese celery seed, Proc. 4th Intl. Symp. Supercritical Fluids, Sendai, Japan, 235, 1997.

7 Herbal Extracts

7.1 IMPORTANCE OF RECOVERY

From time immemorial, man has relied on the fascinating world of herbal medicines for two important reasons, namely, for getting healthy and staying healthy in a natural way. An herb is a plant with no woody stem above the ground. Commonly the term herb refers to aromatic plants whose leaves, stems, flowers, or seeds are used in cookery or medicine. Herbal remedies have gained popularity all over the world. The four most commonly used herbs are mint, parsley, sage, and thyme. Table 7.1 lists some commonly used medicinal plants and their uses. The pain killing drug morphine is derived from poppy; digitalin, a heart stimulant, is extracted from fox gloves; and colchicine, a drug used in the treatment of rheumatic arthritis, is obtained from cochicum (Peterson, 1995). Apart from their culinary use, herbs may be used to make aromatic mixtures with dried petals of fragrant flowers. Potpourris can be made from various mixtures of dried flowers and herbs. Aroma therapy, the therapeutic use of blended essential oils, is practiced to uplift physical and emotional strength and to relieve stress and fatigue. A mixture of rosemary, lemon, and basil is very good for memory enhancement. Ginkgo biloba extract is the most popular herbal medicine for alleviating many symptoms associated with aging, memory loss, depression, and senility, including Alzheimer's disease.

A vast reserve of herbs or medicinal plants found in India and China had formed the basis of the traditional system of remedies, according to an ancient health science, called ayurveda. Brahmi (*Centella asiatica*) is an herb which has been consumed for more than 3000 years, as revealed by ancient scriptures, and is regarded as the most important rejuvenative herb in ayurveda. By stimulating brain tissues and revitalizing the nerves, it is claimed to expand understanding and intellectual capacity, to improve memory and concentration, and to sharpen both mental and physical reflexes (Deora, 1992).

It is not difficult to rationalize the recent trend of other countries — both developed and developing, who are looking to the therapeutic benefits of herbal products and gaining confidence in the traditional system of natural medicines. According to a regional survey made a decade ago, it was estimated that out of 2000 drugs used in curing human ailments in India, about 1500 were herbal medicines, even though there were about 15,000 medicinal herbs available then (Jain, 1996). An order of magnitude difference is due to lack of information on identification of the useful species and also absence of clinical or pharmacological tests on a majority of them. Efforts need to be intensified in this direction with a view to exploiting huge herbal plant resources for the benefit of the human race.

The current global demand for herbal medicines is estimated to be to the tune of 12 billion U.S. dollars, of which approximately half is met by China. The World Health Organisation (WHO), however, predicted a business potential of \$5 trillion in herbal health care products worldwide (Business Line, February 20, 1999). Indian herbal industry is reported to be growing at a fast rate from Rs. 23,000 crores in

TABLE 7.1
Some Common Medicinal Plants and Their Uses

Plant	Botanical Name	Product	Uses
Ashvagandha (roots)	*Withania somnifera*	Alkaloids	Rejuvenator and diuretic
Mint	*Mentha arvensis*	Menthol	Pharmaceuticals
Peppermint	*Mentha piperita*	Peppermint oil	Flavor and pharmaceuticals
Yew tree (bark)	*Taxus brevifolia*	Taxol	Anticancer agent
Bergamot mint	*Mentha citrata*	Bergamot oil	Perfumery
Spearmint	*Mentha spicata*	Spearmint oil	Food and flavor
Vetiver	*Vetiveria zizanioides*	Vetiver oil	High grade perfumes
Patchouli (leaves)	*Pogostemon patchouli*	Patchouli oil	Fixative
Hops (flower)	*Humulus tupulus*	Bitter	Beer
Periwinkle (leaves and roots)	*Catharanthus roseus*	Leaf Alkaloids (vinblastine and vincristine)	Anticancer agents
		root alkaloid (raubasine)	Hypotensive agent
Poppy	*Papver somniferum*	Opium alkaloids (codeine, morphine)	Stimulant
Pyrethrum	*Chrysanthemum cinerariaefolium*	Pyrethrins	Natural insecticide
Wild yam	*Dioscorea deltoidea*	Diosgenin	Progesterone
Liquorice (roots rhizome)	*Glycyrrhiza glabra*	Glycyrrhizinic acid	Sweetening agent
Papaya	*Carica papaya*	Papain	Curative and digestive
Orange (peel)	*Citrus aurantium*	Pectin	Treatment of diarrhea
Cochicum	*Gloriosa superba*	Colchicine	Relief from gout pain
Chamomile (flowers)	*Chamaemelum nobile*	Esters of angelic and tiglic acids	Digestive, antispasmodic and antiseptic agent
Chinchona	*Chinchona calisaya*	Quinine alkaloid	Curing malaria and stomach-ache
Worm seed	*Artemesia brevifolia*	Santonin	Purgative, curing round-worm
Senna	*Cassia angustifolia*	Anthroquinone	Laxative
Fox gloves	*Digitalis lanata*	Digoxin/digitalin	Cardiac stimulant
Fever few	*Tanacetum parathenium/ Chrysanthemum parathenium*	Thujone	Treatment of fevers, arthritis, migraine, worms

Adapted from Choudhury, 1996.

1998 to as high as Rs. 40,000 crores by 2003 A.D. (Lashkari, 1999). An important factor responsible for this rising trend is presumably due to the relaxation of the regulatory norms in several western countries, particularly in Europe. One of them is the acceptance of the total herbal extract by consumers, as it is no longer considered mandatory to isolate the pure active ingredients from herbal extracts. It is now believed that the herbal extracts with assured quality have better efficacy and total bioactivity than the pure active components. The world's attention is now focused on the development of efficient recovery methods and herbal products as "alternative natural medicines" for the cure and prevention of a variety of ailments.

7.2 HERBAL REMEDIES

For the treatment of many illnesses and complaints for both acute (short-lived) and chronic (prolonged) conditions, homeopaths and biochemics offer time-tested, safe and effective remedies based on much-acclaimed natural extracts which thus ensure Nature's sure and simple way to good health without any side effects.

Herbalists' approach to treating human ailments is to use skill or knowledge about the effects of plants on a person's health. The vast range and unique combination of constituents in a specific plant can provide the remedies suitable for the individual patient. The principal constituents that are responsible for herbal remedies include alkaloids, bitters, flavonoids, carotenoids, mucilages, resins, glycosides, volatile oils, vitamins, and minerals. Alkaloids are substances containing nitrogen and were originally defined as vegetable alkali. They tend to have fairly strong actions on a wide range of body tissues. For example, caffeine present in tea and coffee is an alkaloid responsible for their stimulating action on the nervous system. Capsaicin is one of the principal compounds present in red chili belonging to the capsaicinoids group with an alkaloid character that stimulates digestion and circulation. Other well-known medicinal alkaloids are quinine, codeine, morphine, nicotine, etc. The bitter taste of some botanicals, such as bitter gourd, contain herbal medicines that are responsible for stimulating appetite or for increasing the activity of stomach and liver. Flavonoids present in many fruits and vegetables are quite effective antioxidants and are frequently included in remedies to help cure problems related to circulation. The most effective flavonoids are quercetin, rhamnetin, kaemferol, rutin or vitamin P, and quercitrin. In rosemary, carnosic acid has been described as the most active antioxidant constituent. Due to its antiaging action, β-carotene, which is a precursor to vitamin A, is genetically enhanced in fruits and vegetables, such as sweet potato, carrot, and tomato. Efforts made by the USDA in breeding tomatoes with 11 to 18 times more beta carotene will enable Americans to have longer lives. It is known that turmeric contains two ingredients called curcumin and turmerin, that have anti-inflammatory and wound-healing properties. The Indians have long been applying turmeric in powder and paste form to wounds and ulcers as well as taking it internally with water, milk, and honey.

Mucilage is a starch-related substance that renders therapeutic action from its slippery texture. This helps give soothing effects to inflammation in the inner lining of the digestive system. Resins like balsam are the product of the sap of some trees that seeps out when their bark is cut. Most resins are used for their anti-infective properties.

The sap of sugar cane or the root of the sugar beet contains sugar. Roots of several herbs, such as licorice, contain sugar-related substances, called glycosides. A glycoside has a sugar part joined to another part that is usually responsible for therapeutic action. It is the nonsugar parts that have the remedial effect, but the sugar part helps with their initial absorption. For example, there are about six related glycosides present in lily. When given in isolation, none of them achieves the same beneficial effect as when the whole herb is given. Digitalin/digoxin is a glycoside which is a powerful cardiotonic present in foxgloves. Another medicinal glycoside is salicin found in the bark of certain willows. It was salicin that inspired the idea to mimic it and synthesize aspirin.

Tannins present in plants are valued medicinally for their astringent properties, i.e., the ability to constrict the skin pores or blood vessels. Plants characteristically produce varieties of essential oils which can have many medicinal applications.

Almost all essential vitamins can be naturally synthesized by plants which can be included in a well-balanced diet to replace the synthetically produced vitamins. Even antibiotics are produced in certain species by plant metabolism: sulfur compounds in garlic, glycosides in mustard, and alkaloids in water lily. The flower Vinca rosa contains two valuable ingredients, namely, vincristine and vinblastine, which are prescribed to fight certain types of cancer. It is not clear at the moment why plants produce these chemicals if they are not required by the plants themselves. Is it only for protection of human/animal race and indirectly for their survival?

In general, medicinally active substances tend to be concentrated in a specific part of the plant, at certain stages of its growth, at a certain season or in certain types of soil. Hence the prescription of herbal medicines requires expert knowledge on accurate identification, pharmacological tests, dosage, purity, and toxicity levels. While some herbs are used as natural medicines, others are used in foods as supplements. The distinction between herbal foods and herbal medicines is narrowly defined.

7.3 RECOVERY METHODS

Health-protective and disease-preventive attributes of herbs have given herbs a wide range of use. Accordingly, health promoting drinks and health foods are becoming very popular all over the world. Irrespective of whether herbs are used for medicines or foods, they must be safe. Recent trends indicate that herbal extracts are more effective, more convenient, and safer to use than the whole herb itself. Currently more attention is being given to recover the nonvolatile compounds in addition to the volatile oil and to examine their therapeutic benefits. Extraction with CO_2 has several advantages over traditional methods of steam distillation and solvent extraction, because CO_2 does not alter the delicate balance of components in natural products. Today SC CO_2 extraction can be utilized to recover a large number of biologically active compounds from a wide variety of common herbs as discussed below.

7.3.1 Antioxidative and Antimicrobial Constituents

Natural extracts from several medicinal herbs are known to have strong antiseptic, antioxidative, and antimicrobial activities and have been in use in both cosmetic and therapeutic formulations. A comparison of performances of SC CO_2 extraction with steam distillation and solvent extraction of different herbs like thyme (*Thymus vulgaris*), rosemary (*Rosmarinus officinalis*), chamomile (*Matricaria chamomilla*), parsley (*Apium petroselinum*), sage (*Salvia officinalis*), and oregano (*Oreganum*

TABLE 7.2
Comparison of Thyme Extract Compositions (% Area by GCMS)

Component	Distillation with Steam	Extraction with Hexane	Extraction with Alcohol	Extraction with SC CO_2
α-Pinene	0.6	—	—	0.5
Camphene	1.0	—	—	0.9
β-Pinene	1.7	—	—	1.3
p-Cymene	12.9	1.0	—	10.7
γ-Terpinene	11.2	1.2	—	7.1
Linalool	2.2	0.8	0	1.5
Thymol	57.0	56.7	53.5	62.1
Carvacrol	6.4	1.4	3.7	1.7

Simandi et al., 1998.

TABLE 7.3
Comparison of Rosemary Extract Compositions (% Peak Area by GCMS)

Component	Distillation with Steam	Extraction with Hexane	Extraction with Alcohol	Extraction with SC CO_2
α-Pinene	12.3	5.8	10.5	11.7
β-Pinene	1.4	2.7	1.4	1.4
1,8 Cineole	8.5	7.3	8.2	10.5
Camphor	14.4	9.0	15.1	11.5
Borneol	11.7	6.6	14.2	7.5
Bornyl acetate	2.7	1.7	2.8	1.4
β-Caryophyllene	2.5	traces	2.9	1.9

Simandi et al., 1998.

virens) indicates that the Soxhlet extraction with alcohol results in the maximum yield of plant extracts, as unwanted compounds like waxes, tannins, chlorophyll colors, and minerals are coexterted (Simandi et al., 1998). The overall yields obtained by SC CO_2 extraction are similar to those by hexane extraction. The major components and their compositions (area %) are presented in Tables 7.2 to 7.6. Sensory evaluation indicates that SC CO_2 extracts of rosemary and chamomile are significantly better than the solvent extracts for applications in cosmetic preparations, such as base cream and body milk. On the other hand, SC CO_2 extract of thyme has intense odor, which makes it unsuitable for cosmetics formulations. SC CO_2 (95 bar, 50°C) extraction of parsley seeds and leaves followed by fractional separation was carried out (Della Porta et al., 1998) on a pilot scale to separate cuticular waxes from essential oils in the separators. The yield of essential oil from the seeds was slightly higher

TABLE 7.4
Compositions (% Peak Area by GCMS) of Different Chamomile Extracts

Component	Distillation with Steam	Extraction with Hexane	Extraction with SC CO_2
Farnesene	8.0	4.3	4.2
Bisabolol oxide II	7.7	2.7	3.3
α-Bisabolol	10.1	4.1	4.3
Chamzulene	1.8	—	—
Bisabolol oxide I	44.8	46.7	46.6
Dicyclo ethers	13.4	16.6	13.4

Simandi et al., 1998.

TABLE 7.5a
Comparison of Compositions (GC Area%) of Parsley Extracts

Parsley	SC CO_2 300 bar/35°C	SC CO_2 200 bar/35°C	Ultrasound Water	Hexane	Steam Distillation
α-Pinene	14.2	17.1	15.2	3.2	16.18
β-Pinene	19.8	15.8	14.2	5.2	15.32
n-Pentyl benzene	5.2	—	1.5	—	—
Myrcene	—	1.16	0.5	0.54	1.15
Ocene	—	0.17	0.2	0.55	0.14
Myristicin	10.9	10.11	9.8	6.3	9.6
Apiole	52.3	48.76	52.3	80.04	49.5

Then et al., 1998.

(0.5%) than that from the leaves, which was 0.4% of the charged material. A comparison of the compositions of the essential oils extracted from seeds and leaves, as given in Table 7.5b, indicates that the two key components, estragole and anethole, are present in larger quantities in the seed extracts, whereas a large number of components could be detected in the leaf extract.

A similar comparison of compositions of essential oils extracted by SC CO_2 from dried sage leaves with fresh leaves, calyx, and petals of sage (*Salvia officinalis*), is presented in Table 7.5c. It can be seen that the essential oil composition is strongly dependent on the dry or fresh state of the leaves and also on the part of the plant species. For example, the concentration of thujone isomers was highest (37.9%) in the dried leaf oil and lowest (5.9%) in the calyx oil, where β-caryophyllene and α-humulene were found in maximum concentration in fresh leaf oil, but lowest in dried leaf oil (Lemberkovics et al., 1998). Since the essential oil composition varies with time during the SC CO_2 extraction (e.g., sesquiterpenes are present in higher

TABLE 7.5b
Comparison of Composition (GC MS Area%) of Essential Oils by SC CO_2 at 95 bar and 50°C from Parsley Seeds and Leaves

Components	Essential Oil from	
	Parsley Leaves	Parsley Seeds
α-Pinene	2.0	—
β-Pinene	1.6	—
Myrcene	3.6	—
p-Cymene	3.2	0.1
γ-Terpinene	6.6	1.9
Dimethyl styrene	16.7	0.1
1,3,8-p Menthatriene	1.6	0.2
Estragole	12.2	33.8
Carvone	1.0	4.3
p-Anisaldehyde	—	2.3
trans-Anethole	5.3	23.8
β-Chamigrene	0.4	1.7
β-Selinene	0.6	3.5
Myristicin	16.7	0.7
Apiole	5.4	5.1
Linoleic acid methyl ester	3.4	5.4
n-Tricosane	1.4	0.3

Della Porta et al., 1998.

TABLE 7.5c
Comparison of Sage Oil Composition (Area %) by SC CO_2 Extraction from Different States and Parts of the Plant

Component	Dried Leaf	Fresh Leaf	Calyx	Petal
α-Pinene	9.9	10.9	2.08	4.57
Camphene	0.4	0.22	0.64	0.55
β-Pinene	4.7	4.12	14.1	24.4
Limonene	1.1	0.20	0.38	0.53
Eucalyptol	5.3	5.29	13.9	4.61
α,β-Thujone	37.9	16.92	19.86	5.89
Borneol	18.4	1.67	3.80	1.13
β-Caryophyllene	1.8	13.0	6.05	3.75
α-Humulene	2.0	22.9	9.57	7.55
α-Caryophyllenol	3.0	22.7	24.7	19.0

Lemberkovics et al., 1998.

TABLE 7.6
Comparison of Essential Oil Composition (% by wt) of *Oreganum virens* (Oregano) by Steam Distillation and SC CO_2 Extraction

Component	HD (Intact)	HD (Crushed)	SC CO_2 (Intact)	SC CO_2 (Crushed)
α-Pinene	0.9	0.4	0.6	0.4
β-Pinene	0.8	0.3	0.6	0.3
β-Myrcene	2.1	1.1	1.5	1.1
α-Terpinene	4.5	2.9	3.4	2.1
p-Cymene	3.8	3.0	3.3	2.2
γ-Terpinene	16.8	11.8	17.2	10.3
Linalool	2.9	3.0	2.4	2.5
α-Terpineol	15.4	19.9	15.1	20.5
Thymol	16.5	22.3	16.7	24.8
Carvacrol	12.0	14.9	20.0	18.3
Caryophyllene	4.1	4.1	2.9	3.8
Others	20.3	16.4	16.1	13.7

Gasper et al., 1998.

concentrations in later fractions), it is possible to produce sage oil which is rich in either monoterpenes or sesquiterpenes by proper blending of the two fractions.

Essential oil from oregano herbs belonging to the Labiate family is characterized by the presence of a high content of carvacrol, besides α-terpeneol, γ-terpinene, and thymol. This oil is known to have a strong antimicrobial activity and a high therapeutic value due to its high cytotoxicity levels. Gasper et al. (1998) compared the performance of SC CO_2 (80 bar, 40°C) extraction with hydrodistillation from intact and crushed bracts of *Origanum virens* L. Composition of the essential oils obtained by the two processes are compared in Table 7.6. Though hydrodistillation of leaves (intact) resulted in a higher yield (1.46%) compared to the SC CO_2 extraction of crushed leaves (1.06%) for a particle size (1.0 to 1.4 mm), the active ingredients were recovered in higher concentrations in the SC CO_2 extract. However, it was concluded that SC CO_2 extraction of more finely ground herbs can yield both higher yield and better quality of essential oil, close to that present initially in the herb (Gasper et al., 1998).

In India and the Far East, basil (*Ocimum basilicum*) is widely used in ayurvedic medicines where it is called "tulsi." It is known to have wide-ranging applications, such as confectionery, condiments, dental creams, mouth washes, etc. In the West, it is used for rheumatic pain, skin irritations, and nervous disorders, where it is known as the "cooling herb." There are four different varieties of basil plants and the active components present include methyl chavicol (70 to 88%), d-linalool, camphor, methyl cinnamate, besides eugenol and thymol. SC CO_2 extraction of ground basil leaves at 200 bar and 40°C and stage-wise separations at 56 bar and 20°C and at 25 bar and –5°C resulted in a yield of 1.05% by weight of the charged material, which increased to 1.30% on addition of ethanol as entrainer (Calame and

TABLE 7.7
Comparison of Yields (wt%) by Different Methods and Solvents

Plant Material	SC CO_2 40°C, 200 bar	SC CO_2 + Ethanol	Ethanol	Hexane	Steam (vol/wt) % Distillation
Celery seeds	5.7	5.1	12.4	8.0	3.9
Paprika	2.2	13.4	24.3	13.1	—
Elder fruit	—	2.0	41.3	13.0	—
Dandelion roots	0.6	1.0	9.2	1.1	<0.1
Garden sage	1.3	2.5	16.1	3.0	0.5
Milfoil herb	1.0	2.1	8.0	2.0	0.3
Garlic	0.3	0.6	10.1	0.5	0.2

Peplonski et al., 1994.

Steiner 1987). This was from basil leaves which yielded only 0.5% by steam distillation. Extraction yields of medicinal plant materials, e.g., celery seeds, paprika, elder fruit, dog rose fruit, dandelion roots, garden sage, and garlic by SC CO_2 with and without an entrainer at 200 bar and 40°C (Peplonski et al., 1994) are compared with solvent extraction and steam distillation results in Table 7.7. These extracts have superior organoleptic and antimicrobial activities.

It is clear that SC CO_2 at a relatively high pressure with a polar entrainer is required to recover the polar constituents of the medicinal plant materials, though the yields of extraction with ethanol are the highest, but the active components in the extracts get diluted due to co-extraction of several undesirable polar components as can be seen from Table 7.8.

A beautiful, tall, evergreen eucalyptus tree, native to Australia and now cultivated in the hilly areas of Brazil, India, and China, yields 0.75 to 1.25% essential oil from leaves. This oil, 70 to 85% of which is 1,8 cineole, is used as both flavoring agent and for bioactive principles in antimicrobial and antiseptic formulations. SC CO_2 extraction of *Eucalyptus globulus* leaves gave an oil with high content of spatulenol (43.2%) with p-cymene (16.8%) with the highest yield at a pressure of 300 bar and at a temperature of 40°C. The 1,8 cineole content in the oil was reported as 4.5% (Garau and Pittau, 1998).

Tea tree (*Melaleuca alternifolia*), a native of Australia, is a small tree or shrub with needle like leaves and bunches of yellowish or purple flowers. The name is derived from usage of its leaves by the aboriginal people of Australia for preparation of a kind of herbal tea. Tea tree oil is a wonderful antibacterial and antifungal essential oil obtained from the leaves and twigs of tea tree or shrub. The principal constituents of the oil are terpinene-4-ol (up to 30%), cineol, pinene, terpinenes, cymene, sesquiterpenes, etc. It is well known as anti-infectious and antiseptic oil because it is very active against all three organisms, namely, bacteria, fungi, and viruses. Extensive scientific research has confirmed that it is a very powerful immunostimulant, that is, it is more powerful against infections when the body is faced with these organisms. Other uses include soaps, toothpaste, deodorant, after shaves, and gargles. It blends well with spice oils, such as clove, nutmeg, and also sage, rosemary,

TABLE 7.8
Comparison of Celery Seed Extract Compositions (GC MS Area %) by Various Methods

Components	CO_2 (200 bar, 40°C)	CO_2 + Ethanol (200 bar, 40°C)	Ethanol	Hexane	Steam Distillation
Myrcene	0.4	traces	—	0.4	0.9
β-Pinene	0.6	traces	—	0.7	1.5
Limonene	32.8	11.7	5.1	38.0	80.3
Cyclohexadiene	2.7	0.6	0.2	2.5	2.9
β-Selinene	2.1	2.8	1.7	2.0	2.3
3-n-Butylphthalide	2.4	4.4	3.2	3.0	0.7
Sedanolide	31.7	41.2	23.8	32.0	0.6
Dihydrosedanolide	20.1	37.8	16.5	21.4	—
Benzoic acid ester	—	—	26.6	—	2.7
C_{16} acid ester	3.1	1.5	20.6	—	2.2
Others	4.1	—	2.3	0.7	5.9

Peplonski et al., 1994.

and marjoram oils (Lawless, 1995). SC CO_2 at pressures in the range of 100 to 110 bar and at temperatures of 45 to 50°C can be used for extraction of the highly potent tree tea oil.

Mentha or mint leaves are characterized by certain essential oils or iridoids, important for their antitumorous, immunostimulating, antibiotic, and other pharmacological properties. Menthol and menthyl acetate are considered to be the key components of one species of genus mentha, whereas trans-sabinene hydrate, trans-sabinene hydrate acetate, terpinene-4-ol and α-terpineol are considered the key components of another species of genus mentha. SC CO_2 extracts have a composition very close to that in the plant (Barth et al., 1998) and the yield of extract varied between 2.2 and 3.2% in the pressure range of 90 to 100 bar at 50°C.

Saffron (*Crocus sativus*) is a native of Southern Europe, but has been grown in Northern India since ancient time. Saffron flowers give 0.6 to 1.4% essential oil, based on the recovery method employed and the origin of saffron (Mahindru, 1992). It has a beautiful light orange color due to the presence of crocin, the colorant compound. The principal constituent of the oil is safranal (2,6,6 trimethyl-1, 3-Cyclohexadien 1-al). Saffron is a highly valuable natural product which finds a wide variety of applications in the pharmaceutical industry ranging from fairness creme to rejuvenator and memory enhancer, apart from its uses as flavoring and coloring agent. It is widely used in the East as well as in the West. Since it is very expensive and has a low level of extractable content, its recovery needs to be very efficient and fast. Accordingly, SC CO_2 is recommended for the extraction of saffron oil.

Recently the extract of *Ginkgo biloba* leaves has become one of the most popular herbal medicines consumed in the world today. It is prescribed for depression, cerebrovascular insufficiency, impotency, and senility, including Alzheimer's disease.

TABLE 7.9
Comparison of Quality of Marigold Extracts

Component	% (by wt) in Extract by Alcohol	$SC\ CO_2$
Faradiol	0.06	5.0
Faradiol monoester	0.10	12.0

Ronyai et al., 1998.

Ginkgo biloba leaves extract contains 24% polar flavonoids. The major constituents of the extract are quercetin and isorhamnetin. They can be separated and isolated by supercritical fluid chromatography (SFC) using a mobile phase of SC CO_2 at 250 bar and 50°C, mixed with 10% ethanol and 0.5% phosphoric acid to improve the polarity of eluent. A phenyl-bonded silica column was found to be the best (and the ODS column as the worst) for the separation of the major flavonoids (Zhimin et al., 1997). It is also apparent that SC CO_2 mixed with 10% ethanol and 0.5% phosphoric acid can recover the extract enriched with the flavonoids to improve the recovery efficiency of the active ingredients as well as to isolate them subsequently by using the preparative SFC method.

Henna (*Lawsonia inermis*) is a household herb commonly used in India as well as in European and Middle Eastern countries. The leaves of the plant have several medicinal applications. Henna leaves are also used as a pleasant orange natural dye for coloring of hair and decorating palms, fingers, nails, feet, toes, even the beard. The medicinal properties of henna leaves are utilized as a prophylactic against skin diseases, sore throat, hair loss, headache, depression, burning sensation in the feet, etc. The essential oil from henna flowers are utilized in perfumery and embalming (Mahindru, 1992). SC CO_2 extraction can yield all three types of products, namely, essential oil, oleoresin, and natural color simultaneously.

7.3.2 Antiinflammatory Constituents

Marigold (*Calendula officinalis*) flower contains some biologically active components besides essential oil and natural color, making its extracts phytotherapeutically very valuable. The most important attribute of this extract is its antiinflammatory activity due to the presence of triterpenoids, among which the most potent ingredient being faradiol. SC CO_2 extraction of dried marigold flower at 450 bar and 60°C resulted yields varying between 5.3 to 11%, depending on variety. Extraction yield at a lower pressure of 200 bar was merely 1.2%, which increased to 2.8% with ethanol as an entrainer. The amount of faradiol monoester in the SC CO_2 extract is two orders of magnitude higher than that in the alcohol extract, as seen in Table 7.9. Since the antiinflammatory activity is proportional to the faradiol monoester, SC CO_2 extract is therapeuticlly much superior to the alcohol extracts.

Sesquiterpene-γ-lactones are valuable nonvolatile bioactive terpenoids which are used for the treatment of inflammation, migraine, and several other disorders. Sesquiterpene-γ-lactone (parathenolide) is such an active compound, present in fever-

TABLE 7.10
Comparison of Feverfew (*Chrysanthemum parathenium*) Essential Oil Compositions (wt%)

Compound	Steam Distilled Oil	Essential Oil from SC CO_2 Extract (400 bar, 60°C)
α-Pinene	1.1	—
Camphene	4.4	5.4
β-Pinene	0.4	0.1
Limonene	0.3	0.1
Eucalyptol	1.9	2.1
γ-Terpinene	0.1	—
Linalool	0.2	0.1
α-Thujone	1.2	0.3
Camphor	53.9	70.9
Borneol	0.2	—
Chrysanthenyl acetate	26.9	29.1

Kery et al., 1998.

few (*Chrysanthemum parathenium*), which could be recovered in a significant proportion (0.73% by wt) in SC CO_2 extract. The yield of extract from feverfew was 5.2% at 400 bar and 60°C. The essential oil (0.2 to 0.6 ml/100 g) obtained by steam distillation of SC CO_2 extract contained 70.9% camphor and 29.1% chrysanthanyl acetate which are higher in the SC CO_2 extract than in the steam distilled product, as seen in Table 7.10. Another important sesquiterpene-γ-lactone, known as cnicin, is available in blessed thistle (*Cnicus benedictus*) and could be extracted by SC CO_2 by adding 4% methanol. The lactone fraction could be enriched to contain 70% cnicin. It was reported that the SC CO_2 extracts (Kery et al., 1998) were cleaner, safer, and more potent than the solvent(hexane and ethanol)-extracted products.

7.3.3 Anticancerous Alkaloids

Pyrrolizidine alkaloids are well known for their anticancerous activity. Monocrotaline ($C_{16}H_{23}NO_6$, mol wt = 325.3) is one such alkaloid present in *Crotalaria spectabilis* seeds, containing 4.4% extractable material. SC CO_2 mixed with ethanol as the cosolvent was employed by Schaeffer et al. (1989) at pressures in the range of 185 to 274 bar and temperatures from 35 to 55°C to recover monocrotaline content as high as 24% in the extracts from the seeds containing 1.9 wt% monocrotaline and 2.5 wt% lipids. With further modification in the process, namely by utilizing the cross-over phenomena of solubility behavior with temperature, the purity could be enhanced up to 50%. Upon incorporating ion-exchange resins, the purity of the alkaloid was further improved to 94 to 100% monocrotaline which could be used for chemotherapy. The neat solubilities of monocrotaline are significantly low in SC CO_2 at pressures in the range of 88.6 to 274.1 bar, which ranged from 6×10^{-6} to 4.4×10^{-5} in mole fraction at temperatures in the range of 35 to 55°C. However with 5 to 10 mol%

ethanol addition to SC CO_2, the solubilities improved by twenty-five times. The solubility of monocrotaline when extracted from seed material was, however, 50 to 93% less than the corresponding neat solubility. The selectivity of monocrotaline recovery increased drastically after the depletion of lipids. It increased with ethanol concentration and with increasing temperature and decreasing pressure. SC CO_2 extracts of *Crotalaria spectabalis* contain both monocrotaline and nonpolar lipids. Cosolvent may be retained in complex substrates, possibly by its absorption in the intraparticle lipids. Accordingly, multistage separations may be necessary. A cationic exchange resin removed monocrotaline from the SC phase, which could be recovered by flushing the resin with ethanol and then with NH_4OH to yield 94% monocrotaline (Schaeffer et al., 1989).

The use of ion-exchange resins combined with SC CO_2 extraction results in an attractive isolation technique, provided care is taken not to deactivate the resin. A similar approach can be used for recovering other alkaloids from the solid substrates of crushed seeds.

Maytansine (*Maytenus senegalensis*) is another alkaloid present in several bioactive plant materials (McHugh and Krukonis, 1994) which have been tested and confirmed by the National Cancer Institute (NCI), U.S. to have growth inhibition properties. The extracts of these plants containing maytansine are used as chemotherapeutic agents in the treatment of leukemia. The results of the cytotoxicity assay of the samples indicated that all the SC CO_2 (275 bar, 35°C) extracts have positive cytotoxicity, sometimes more activity than the ethanolic extracts. As mentioned earlier, selectivity of recovery of the bioactive alkaloid improves with the progress of supercritical extraction, as triglycerides, fatty acids, and esters are preferentially removed in the beginning. This observation gave credence to the concept of simple fractionation with time. However, when the bioassay tests of some extracts with liquid CO_2 (275 bar, 20°C) and ethylene (276 bar, 20°C) were performed, they were found to have more potency than the SC CO_2 (275 bar, 35°C) extracts. Work on supercritical separation of *Vinca rosa* is in progress in many laboratories to selectively recover vincristine and vinblastine which are very effective in chemotherapy.

China uses an herbal medicine for the treatment of leukemia popularly known as *Danggui Longhui Wan*, which is a mixture of eleven different herbal products. The active ingredient is indirubin, a red colored relative of blue indigo dye. It has been established that indirubin stops the uncontrolled growth of tumor cells by inactivating the enzyme, kinases, responsible for cell division. Indirubin gets bound to enzymes and blocks their activity, thus stopping the growth and spreading of the cancer cells. It is believed that SC CO_2 extraction can yield a more concentrated extract, and an indirubin-enriched product can be obtained by SC CO_2 mixed with ethanol as the entrainer at a moderately high pressure of 300 bar at 40°C.

Taxol is an alkaloid known to have antileukemia and tumor-inhibitory activities and is used for the treatment of ovarian, breast, lung, and colon cancers. Taxol is available in the bark of yew tree (*Taxus brevifolia*) found in the northwestern Pacific Coast. Efforts are being made to increase the taxol level in the tree bark, which is the only available source of this unique molecule (mol wt 854). The taxol content varies between 50 to 160 mg/kg bark. Since the tree is very slow growing, and it

TABLE 7.11
Recovery of Taxol from Bark of *Taxus brevifolia* at 45°C with Supercritical CO_2 with and without Ethanol as Entrainer

SC CO_2 Pressure (bar)	Cut #	Taxol in Extract (%)	SC CO_2 + EtOH Pressure (bar)	Cut #	Mol% Ethanol in SC CO_2	Taxol in Extract (%)
257	1	—	257	1	1.3	0.271
257	2	0.865	257	2	2.0	0.861
257	3	1.137	257	3	2.5	1.645
257	4	0.663	257	4	2.4	1.541
213	1	0.579	257	5	2.8	1.185
213	2	0.838	208	1	0.7	0.335
213	3	1.131	208	2	3.4	0.976
183	1	0.431	208	3	3.9	1.719
183	2	0.618	208	4	3.6	1.822
183	3	0.918	208	5	3.8	1.682
183	4	0.891	181	1	0.2	0.371
			181	2	3.7	0.913
			181	3	5.7	1.559
			181	4	3.5	1.331
			181	5	4.8	1.051

Jennings et al., 1992.

reportedly requires two to three 60-year-old trees to yield enough taxol to treat one patient, its extraction process needs to be very efficient, recovering the entire content of taxol from the ground bark without any degradation and with relatively high concentration of that in the bark.

SC CO_2 extraction is then considered to be a promising technique for the separation of taxol. Studies made by Jennings et al. (1992) indicated that ethanol extracted mostly polar components with taxol concentration of 0.125%, whereas the taxol concentration in the SC CO_2 extracts ranged from 0.27 to 1.82%. That is, it increased 10- to 100-fold at pressures ranging from 180 to 260 bar at 45°C from initial taxol content of 0.016% in the bark. This showed that SC CO_2 extraction was more selective than solvent extraction with ethanol. SC CO_2 mixed with ethanol as an entrainer could recover double the amount of taxol that is recoverable by SC CO_2 alone. About 50 to 85% taxol present in the ground *Taxus brevifolia* bark could be recovered, depending on the pressure, ethanol concentration in the SC CO_2 solvent, and the time of extraction, as can be seen in Table 7.11. In general, a higher pressure, a longer time of extraction, and the addition of ethanol improved the recovery and purity of mildly polar taxol. SC CO_2 mixed with 3.6 mol% ethanol at 210 bar and 45°C was able to extract a higher amount of taxol in the extract (1.82%). The residence time for the solvent in the bed was approximately 25 min and the extraction time for each cut, excluding the first cut of SC CO_2 + ethanol extractant, was less than 6.0 to 6.5 h.

7.3.4 Anticarcinogenic Polyphenols

Polyphenols are organic substances having two or more hydroxyl groups or their functional derivatives, i.e., esters, methyl esters, and glycosides. They possess antioxidative, antitumorous, and antiviral activities. They occur naturally in leaves, flowers, rhizomes, fruits, and seeds of plants (Pratt and Hudson, 1990). They are regularly consumed in the form of food or medicines. Polyphenols have the ability to associate with other metabolites, e.g., protein, lipids, and carbohydrates, to form stable complexes and thus inhibit mutagenesis and carcinogenesis. Some medicinal plants rich in gallic acid, a triphenol, are *Geranii herba* (*Geranium maculatum*), Meadowsweet (*Filipendula ulmaria*), Raspberry (*Rubus idaeus*), and black currant (*Ribes rubrum*) (Haslam et al., 1992) and their extracts are popular as herbal medicines.

Proanthocyanidins belong to a group of polyphenols formed by two or more molecules of flavan-3-ol and are known to have therapeutic properties. They are generously found in natural products and their effects as antineoplastics and antioxidants have been largely studied. Separation of most of the proanthocyanidins from natural sources is very tedious and complex. SC CO_2 extraction of polyphenols from natural sources such as tamarind (*Tamarindus indica*) seed coat was studied by Tsuda et al. (1995). They reported that some flavan-3-ol molecules, such as epicatechin, was soluble in SC CO_2 and that its solubility was enhanced with the addition of an entrainer like ethanol. Murga et al. (1998) studied the extractability of three natural polyphenols, namely, gallic acid, (+)-catechin, and (–)-epicatechin from grape seeds by SC CO_2 extraction. Extractability of proanthocyanidins from grape seeds was explored by evaluating the solubilities of individual polyphenols in SC CO_2 as will be discussed in Chapter 8. In order to evaluate separability of these three components from grape seed extracts, solubility studies were carried out for a mixture of equal amounts of the individual components at 200 bar and 40°C with SC CO_2 mixed with 10% v/v ethanol. The results showed that (–)epicatechin was hardly extractable whereas gallic acid dominated the extract (85%).

7.3.5 Medicinal Constituents of Tea Extract

Current research on newer and more effective herbal medicines around the world reveals that regular intake of two to three cups of tea a day per person reduces the risk of several chronic ailments, like cardiovascular diseases, some cancers and tumors, and other diseases related to aging. It protects us from UV radiation, and prevents premature aging and wrinkle formation. It also lowers fatty acids in blood, thereby helping circulation. It has been established that antioxidative polyphenols present in tea are 20 times more effective than ascorbic acid (vitamin C). Polyphenols, such as catechins, flavonols, flavonol glycosides, theaflavins, thearubigens, and phenolic acids are found in tea in high concentrations (Ho, 1992), making tea a highly beneficial and healthy beverage. Upon being convinced about the cancer preventive attributes of tea, the National Cancer Institute of the U.S. has initiated a developmental project for using tea constituents for chemotherapy in human trials. Tea is the most widely consumed beverage in the world, with a per capita annual

TABLE 7.12
Composition of Black and Green Tea Leaves, and Dry Extracts from Green and Black Tea

Component	Black Tea (% dry wt)	Green Tea (% dry wt)	Green Tea (% dry extract)	Black Tea (% dry extract)
Polyphenols	30	39	—	—
Caffeine	0.4	0.4	6.9	7.1
Catechin	0.23	0.1	34.0	4.2
Epicatechin	0.41	0.9		
Epicatechin gallate	0.80	0.8		
Epigallocatechin	1.05	3.9		
Gallate	1.66	4.4		
Flavonol			0.4	
Flavonol glycosides	0.05		4.4	1.4
Theaflavin	0.25			1.8
Theaflavin gallate A	0.17			
Theaflavin gallate B	0.24			
Theaflavin digallate	0.25			
Thearubigens	5.94	—	—	17.0
Phenolic acids	—	1.5	9.5	11.0
Protein	15.0	15.0	7.6	10.7
Lipids		2.0	—	—
Fiber	30.0	0.5	—	—
Chlorophyll		0.5	—	—
Amino acids	4.0	4.0	5.3	4.8
Ash	5.0	5.0	—	—
Carbohydrate	7.0	25	12.5	13.5
Lignin	—	6.5		

Balentine, 1992.

consumption exceeding 40 l. Like tea, coffee is also a stimulant and some people even believe that it is coffee which is more beneficial than tea in lowering the risk of heart diseases. Both tea and coffee have antioxidative and anticancerous properties, though most teas are known to have less caffeine content than coffee.

Tea plant is a native of India and China, the two most densely populated countries in the world. Two most common varieties of tea are *Camellia assamica* and *Camellia sinensis*, obviously associated with their places of origin, India and China, respectively. There are two most popular varieties of tea leaves consumed — the green (unfermented) tea and the black (fermented tea). The compositions of the tea leaves and the hot water extracts from both green and black tea leaves are compared in Table 7.12. There are numerous types of polyphenols present in green tea leaves, such as flavan-3-ols (catechins), flavonols, and their glycosides. Green tea contains 13.6% catechins (flavan-3-ols) out of 39% polyphenols, 3.5% methylxanthins, 1.5% phenolic acid, and 4.0% amino acids, besides 25% carbohydrates, 15% protein, and 6.5% lignin. The phenolic acids include gallic acid, chlorogenic acid, caffeic acid,

TABLE 7.13
Flavonol and Flavonol Glycosides in Green and Black Tea (mg/g Dry Leaf)

Favonol Group	Green Tea	Black Tea
Quercetin	0.40	—
Myricetin	0.34	
Kaemferol	0.52	
Rutin	1.58	1.32
Quercetin-G	1.00	0.76
Kaemferol-G	1.33	0.70
Quercetin-G2	3.17	
Kaemferol-G1	2.30	
Isoquercetin	1.82	
Kaemferol G2	4.30	
Kaemferol-G3	—	1.01
Theaflavin		6.8
Theaflavin gallate A		8.3
Theaflavin gallate B		2.5
Theaflavin digallate		0.7

Balentine, 1992.

quinnic acid, cinnamic acid, and coumeric acid (Balentine, 1992). Gallic acid is a triphenol and is present in esterified form in tea catechins. Current research has revealed that the antioxidative and cancer preventive properties of green tea are due to epicatechin gallate and epigallocatechin gallate. These compounds inhibit free radical chain reaction of cell membrane, lipids mutagenicity, DNA-damaging activity, tumorogenesis, and tumor growth (Pratt and Hudson, 1990). The most common biologically active flavonol in green tea is quercetin. Quercetin also inhibits both initiation and promotion of tumor. Quercetin and phenolic acids, such as chlorogenic acid, may act synergistically to reduce carcinogenesis responsible for colon cancer. The phenolic acids like gallic acid, caffeic acid, and chlorogenic acid, along with other polyphenols present in tea and coffee, may form stable complexes with caffeine. It is known that caffeine, being a good proton acceptor, has the ability to form stable caffeine-polyphenol complexes with large polyphenolic molecules which are proton donors. After complexation, these molecules are precipitated out in an aqueous medium, thus alleviating some of the adverse physiological effects of caffeine.

Green tea contains simple polyphenols whereas black tea contains complex polyphenols, as shown in Table 7.13. The oxidative change of polyphenols by the polyphenoloxidase (PPO) enzyme leads to the formation of theaflavins and thearubigens, the orange and red pigments of black tea. The process of fermentation oxidizes simple polyphenols into more complex condensed polyphenols which give black tea its color and flavor. After fermentation, tea leaves are, of course, roasted and dried whereby the enzymes become inactivated. It is reported that two of the oxidized forms of tea catechins, the theaflavin monogallate B and theaflavin digallate, are important polyphenols of black tea, because they have even stronger inhibitory bioactivities due to lipoxygenase enzyme than the catechins prior to oxidation (Balentine, 1992).

Recently SC CO_2 extraction of black tea was performed using the 10 l pilot plant at IIT, Bombay, which resulted in natural tea extract with most of the valuable constituents of black tea. This tea extract is evenly soluble in cold water and contains aroma, caffeine, polyphenols, phenolic acids, xanthines, and amino acids, though the detailed analysis is not immediately available. This may prove to be a good health drink of instant tea which does not require hot water for brewing.

Coffee contains 3.5 to 5.0% caffeine and tea contains 1.0 to 3.5% caffeine, depending on the soil, altitude, and climate conditions. Caffeine is an alkaloid which stimulates the central nervous, muscular, and circulatory systems. It is also a diuretic. Tea and coffee with too much caffeine are not good for elderly people and caffeine-sensitive patients. So most people in developed countries prefer to drink decaffeinated coffee with 0.4% or less caffeine in it. Most of the caffeine produced as a byproduct from the decaffeination process is used in cola soft drinks and the pharmaceutical industries.

The SC CO_2 extraction process for decaffeination of coffee beans was commercialized two decades ago. Nearly 90% of the coffee consumed in the U.S. is now decaffeinated by this process. SC CO_2 extraction for decaffeination (McHugh and Krukonis, 1994) is performed on green coffee beans. Coffee aroma is subsequently developed by a simple roasting process of the decaffeinated green coffee beans (Lack and Seidlitz, 1993). For decaffeination of tea, the SC CO_2 extraction process is conventionally performed on black tea. However, SC CO_2 decaffeination of green tea has also been investigated to find out whether the enzymes (PPO) are inactivated in the high pressure CO_2 environment, since these enzymes are needed for the development of flavor and color during the subsequent fermentation process of the decaffeinated green tea. SC CO_2 extraction for decaffeination of green tea is also feasible.

Apart from caffeine, coffee also contains chlorogenic acid and trigonelline, which are also pharmaceuticals, having high physiological effects. SC CO_2 extraction and separation of these compounds may be carried out to obtain them as byproducts of the caffeine recovery process. However, due to very low solubilities in SC CO_2 by 3 to 4 orders of magnitudes less compared to that of caffeine (Saldana et al., 1997), their recovery may be inefficient.

7.3.6 Fat Regulating Agent

Onion (*Allium Cepa L.*) oleoresin is known to have biologically active sulfur compounds to correct for glandular imbalance and obesity, to purify blood, and to keep the skin clear. SC CO_2 extraction of dried onion powder at 300 bar and 65°C yielded 0.9% oleoresin, whereas steam distillation yielded only 0.044% of essential oil (aroma) (Gao et al., 1997). On the other hand, at a lower extraction pressure (90 to 100 bar), SC CO_2 could recover only onion flavor. However, it is possible to fractionate the extract from dried onion powder to essential oil and oleoresin fractions,

TABLE 7.14
Comparison of the Yield and Sulfur Recovery Efficiency from Dried Onion by Different Methods

Recovery Method	Yield (%)	Sulfur Content (%)	Sulfur Extraction Efficiency (%)
Steam distillation	0.44	41.98	3.37
Hexane extraction	1.11	0.81	1.75
Alcohol extraction	12.62	0.86	18.04
SC CO_2 extraction (300 bar, 65°C)	0.90	2.37	4.14

Gao et al., 1997.

if SC CO_2 extraction is carried out at a high pressure of 300 bar. A comparison of the sensory characteristics of onion extracts obtained by different methods revealed that SC CO_2-extracted oleoresin had better sensory characteristics and bioactivity, though it has a lesser amount of the more volatile aroma components. Sulfur components present in the onion oleoresin are different from those present in the onion essence. The yield of onion oleoresin by SC CO_2 extraction was similar to that by hexane extraction, but was only 7% of that by alcohol extraction, although sulfur recovering efficiency was the lowest with hexane and the highest with alcohol, as can be seen in Table 7.14. Accordingly SC CO_2 extraction seems to be a better method for recovering onion oleoresin devoid of aroma from dried onion powder.

The fruit kokam (*Garcinia indica*) and tamarind (*Garcinia cambogia*) are known for their acidic flavor which is attributed to α-hydroxy citric acid. In addition, it has considerable quantities of red and yellow pigments. Its red color is due to two anthocyanins, namely, cyanidin-3-glucoside and cyanidin-3-sambubiaside, and the yellow pigment is due to β-carotene. The red pigments are polar whereas yellow pigments are nonpolar. This fruit, kokam, has two very "well sought for" pharmaceutical properties, namely, fat regulating and antitumor activities. The fruit is also used in formulations for remedies of piles, pain, heart diseases, etc. The α-hydroxy citric acid is present in the bud to the extent of 16 to 20%. The acid can be recovered by dissolving it in water and solvent extraction followed by concentration and purification. Anthocyanin pigments are present to the extent of 1.0 to 1.5% in the rind portions and can be recovered together, whereas the yellow pigments can be recovered by hexane extraction. However, all these steps can be minimized by using SC CO_2 extraction with ethanol as an entrainer at 250 to 300 bar pressure. Kokam also contains 25 to 30% oil, which has a high therapeutic value and can be recovered and fractionated by SC CO_2 extraction. Hydroxy citric acid recovered by this method is of higher concentration and safer than traditional methods of recovery. Hydroxy citric acid helps to block the accumulation of unwanted fat in the body and to accelerate the burning of excess fat already accumulated in the body. Presently there is a huge demand for this natural pharmaceutical product all over the world for slimming and keeping fit.

7.3.7 Therapeutic Oils and Fatty Acids

Gamma-linolenic acid (GLA) is a very useful ingredient due to its efficacy in preventing and curing several human ailments. It is used for the treatment of skin disorders, e.g., atopic eczema, as dietary complements, or in the treatment of premenstrual tension. It is also a viable source of energy and precursor to prostaglandins which affect the nervous system and blood circulation, regulate hormone and metabolism, and inhibit the production of gastric juice.

The most common source of GLA is evening primrose seed (*Oenothera blennis* and *O. lamariciana*) oils, black currant seed oil, or borage oil. It is also found in Mortierella fungi. The content of γ-linolenic acid (GLA) in the polyunsaturated fatty acids (PUFA) obtained by hydrolysis of oil from both sources of oil is about 10% by weight of PUFA. The oil contains more than 80% PUFA. SC CO_2 with or without an entrainer like ethanol can be used for recovering the neutral oil at 300 bar and 58 to 60°C with 10% v/v alcohol. The oil sacs need to be broken while grinding, and grinding with alcohol was found to release most of the oil. The oil extracted with SC CO_2 showed significantly less phosphorus content than the hexane-extracted oils (15 vs. 800 ppm). The highest yield of extraction of evening primrose oil was about 21% by wt at 50°C and 300 bar. The fatty acids obtained by hydrolysis of the neutral oil were recovered and purified by supercritical fluid chromatography (SFC) and the results (Sugeta et al., 1998) indicated that γ-linolenic acid could be recovered with purity in the range 94 to 99% by SC CO_2 from both sources without esterification of the neutral oil.

The oil extracted from borage seed with SC CO_2 contains approximately 80% mono- and poly-unsaturated fatty acids (oleic-, linoleic-, and linolenic-) in the form of triglycerides, in which GLA content is 18 to 19% which considerably increases the therapeutic value of the oil. There is a cross-over pressure at 230 bar in the oil solubility vs. temperature behavior. The borage seed extraction with CO_2 at 250 bar and 35°C resulted in 24% oil yield which was 80% of the total oil available, i.e., 30% of the seeds (Illes et al., 1994).

Tea seed oil is known for its high therapeutic value and it is different from the tea tree oil mentioned earlier (Section 7.3.1). Tea seed oil is rich in mono-unsaturated fatty acids (MUFA) like olive or canola oil and is recommended for maintaining low levels of blood cholesterol. It is also used as a salad or frying oil. It has important pharmaceutical applications as a carrier for long-acting injection, such as antimalarial injection containing a mixture of tea seed oil and artemisinin. It is commercially produced from *Camellia sasanqua* in China and is widely used in Europe and Asia. Tea seed oil may be produced from *Camellia sinensis* as in Australia. The sasanqua species contains 45 to 50% oil, whereas the other species, sinensis, which is mainly used for tea leaves, contains a much lower amount (20 to 25%) of fixed oil. It is reported by Swift et al. (1994) that SC CO_2 extraction of triglycerides from tea seeds was investigated at pressures ranging from 85 to 350 bar and at temperatures of 25 to 70°C. It was demonstrated that tea seed oil could be extracted and fractionated into three fractions, namely low, medium, and high molecular weight triglycerides. Similar observation was reported earlier by Bharath et al. (1992) on fractionation of palm oil into three fractions of low, medium, and high molecular weight mixed

TABLE 7.15
Comparison of Hiprose (*Rosa canina L.)* Extracts from Different Sources

Material	Extractant	Carotenoid	Tocopherol	Fatty Acid Comp. (Area %)			
				C18:0	C18:1	C18:2	C18:3
Seed	SC CO_2 (250 bar, 35°C)	68	—	2.9	16.7	53.0	23.7
Seed	$C_3H_8 + CO_2$ (100 bar, 28°C)	148	176	2.6	15.4	54.0	23.6
Fruit	SC CO_2 (250 bar, 35°C)	827	—	2.6	17.1	44.0	31.6
Fruit	$C_3H_8 + CO_2$ (100 bar, 28°C)	1483	—	2.6	17.4	52.8	23.6
Peel	SC CO_2 (250 bar, 35°C)	1499	503	2.7	19.5	14.7	60.1
Peel	$C_3H_8 + CO_2$ (120 bar, 28°C)	—	194	3.7	16.3	70.4	1.7

Illes et al., 1997.

triglycerides (Bharath et al., 1992). But individual triglycerides were not separated. They were however unable to fractionate sesame oil containing fatty acid radicals of molecular weight >48.

The reddish orange fruit of hiprose (*Rosa canina*) is usually used as a natural source of vitamin C, besides pectin, flavonoids, tocopherols, carotenoids, and mono-unsaturated fatty acid (MUFA) oil. It is thus used as an additive to tea, jam, syrup, or soft drinks. Illes et al. (1997) studied supercritical extraction of hiprose fruit by using CO_2, propane, and a propane + CO_2 (3:2) mixture. SC CO_2 at 250 bar and 35°C yielded 5.71%, 3.03%, and 0.37% oil (extracts) from seeds, whole fruit, and peel, respectively. Using pure propane and propane + CO_2 mixture, the yields were slightly higher. Analysis of the extracted oils revealed that the highest levels of biologically active components, such as carotenoids (lycopene, lutein, and β-carotene) and tocopherols (α, β, and γ forms) were present in the peels and the lowest in the seeds. The highest amounts of mono- and polyunsaturated fatty acids were present in the oil from the seeds, which contained 52 to 55% of linoleic acid (C 18:2), 23 to 24% linolenic acid (C 18:3), and 15 to 19% of oleic acid (C 18:1), as can be seen in Table 7.15.

Jojoba (*Simmondsia california*), a native of Mexico and California, has now acquired worldwide attention due to its manifold commercial potential. It is also known as *Simmondsia chinensis*. Jojoba seeds contain 45 to 55% oil which has chemical properties similar to those of whale oil. This oil is not a triglyceride, but is composed mostly of esters of mono-unsaturated C_{20}-C_{22} fatty acids (MUFA) and long chain mono-unsaturated alcohols. It is not a fat but a liquid wax (mpt: 7°C). The oil has high thermal stability (up to 315°C) and finds wide-ranging applications, such as in cosmetics, lubricants, pharmaceuticals, vegetable cooking oil, disinfectants, surfactants, personal hygiene care, etc. Naturally derived jojoba oil can be

TABLE 7.16
Composition of Medicinal Oils from Black Currant Seeds by Different Methods

	SC CO_2 Extraction (300 bar, 35°C)	Solvent Extraction (Hexane)
$C_{16:0}$	4.7	5.1
$C_{18:0}$	1.5	1.3
$C_{18:1}$	19.6	19.0
$C_{18:2}$	41.0	50.8
α-$C_{18:3}$	23.2	14.2
γ-$C_{18:3}$	3.7	1.4
Carotene, μg/g	7–10	7
Tocopherol, μg/g	17	35.6

Then et al., 1998.

used as a base in "oil-free" cosmetics. More than 2000 t jojoba oil are produced each year for uses in a variety of chemical products marketed worldwide, out of which 750 t are used in cosmetics. The total world market potential for jojoba oil has been estimated to be 64,000 t/year. The solubility of jojoba oil in SC CO_2 was measured over a pressure range of 100 to 2600 bar and a temperature range of 20 to 80°C. The oil solubility sharply increases with pressure beyond 300 bar. For example, at 80°C, the solubility of jojoba oil in SC CO_2 increases from 1.8 to 5.0% by weight on increasing pressure from 300 to 500 bar (Schultz et al., 1991). SC CO_2 extraction of ground jojoba seeds in the 10-l pilot plant (10 l) at IIT Bombay at 330 bar and 60°C yielded 45% of high-quality oil, which has great potential for applications in cosmetics and pharmaceutical industries.

Raspberry, blackberry, and (black and red) currant seeds are known to contain fatty oils rich in linoleic and linolenic acids, besides tocopherol and carotenoids. SC CO_2 extraction of raspberry seeds at 300 bar and 35°C yielded 10.7% oil; blackberry seeds 10.3% oil; and black/red currant seeds 6.9% oil (Then et al., 1998). The analysis of black currant seed oil, as shown in Table 7.16, reveals that the oil is rich in linoleic and linolenic acids, which are better recovered by SC CO_2 extraction.

REFERENCES

Airy Shaw, H. K., *A Dictionary of Flowering Plants and Ferns*, 7th ed., Wills, J. C., Ed., Cambridge University Press, Cambridge, 1966.

Anthony, J. I. X., Garcinia Indica — a natural fat regulating functional beverage, *Bev. Food World*, 24–25, May, 1997.

Balentine, D. A., Manufacturing and chemistry of tea, in *Phenolic Compounds in Food and Their Effects on Health I*, Ho, C. T., Lee, C. Y., and Huang, M. T., Eds., ACS Symp. Series, No. 506, 102, 1992.

Bharath, R., Inomata, H., Adschiri, T., and Arai, K., Fractionation of palm oil using SC CO_2, *Fluid Phase Equilibria*, 81, 307, 1992.

Barth, D., Pop, E., Hubert, N., Mihaiescu, D., and Gainer, I., Supercritical CO_2 extraction from some Romanian lamiaceae, Proc. 5th Meet. SCFs, France, 2, 655, 1998.

Business Line (India), Agri-Business, Bangalore, India, February 20, 1999.

Calame, J. P. and Steiner, R., Supercritical extraction of flavours, in *Theory and Practice of Supercritical Fluid Technology*, Hirata, M. and Ishikawa, T., Eds., Tokyo, Metropolitan Univ., 227–318, 1987.

Choudhury, R. D., Chief Ed., *Herbal Drugs Industry*, Eastern Publishers, New Delhi, India, 1986.

Della Porta, G., Reverchon, E., and Ambruosi, A., Pilot plant isolation of celery and parsley essential oil by supercritical CO_2, Proc. 5th Meet. Supercritical Fluids, France, 2, 613, 1998.

Deora, R. S., *Selected Medicinal Plants of India*, CHEMEXCIL (publisher), compiled by Bharatiya Vidya Bhavan's Swami Prakashananda Ayurveda Research Centre, Bombay, India, 1992.

Dobelis, I. N., Project Ed., *Magic and Medicine of Plants*, the Reader's Digest Assoc., Pleasantville, New York, 1990.

Gao, Y., Simandi, B., Sass-Kiss, A., Stefanovits, P., Zs, F., and Czukor, B., Supercritical CO_2 extraction of oleoresin, Proc. 4th Intl. Symp. SCFs, Sendai, Japan, May, 1997.

Garau, R. and Pittau, B., Essential oils contained in the Myrtaceae family, extraction and fractination, Proc. 5th Meet. Supercritical Fluids, France, 2, 621, 1998.

Gasper, F., Guibert, D., Santos, R., and King, M. B., Extraction of essential oils from *Origanum virens L.*: a comparative study of hydrodistillation with SC CO_2, Proc. 5th Meet. SCFs, France, 2, 705, 1998.

Ho, Chi-Tang, Phenolic compounds in food, in *Phenolic Compounds in Food and Their Effects on Health I*, Ho, C. T., Lee, C. Y., and Huang, M. T., Eds., ACS Symp. Series, No. 506, 2, 1992.

Haslam, E., Lilley, T. H., Warminski, E., Liao, H., Cai, Y., and Martin, R., Polyphenol complexation, in *Phenolic Compounds in Food and Their Effects on Health I*, Ho, C., Lee, C. Y., and Huang, M. T., Ed., ACS Symp. Series, No. 506, 15, 1992.

Illes, V., Szalai, O., Szebenyi, N. I., Grosz, M., and Hethelyi, I., Oil recovery from borage seed with carbon dioxide and propane solvents, Proc. 3rd Intl. Symp. Supercritical Fluids, France, 2, 511, 1994.

Illes, V., Szalai, O., Then, M., Daood, H., and Perneczki, S., Extraction of hiprose fruit by supercritical CO_2 and propane, *J. Supercritical Fluids*, 10, 209, 1997.

Jain, S. K., *Medicinal Plants*, National Book Trust of India, New Delhi, 1996.

Jennings, D. W., Howard, M. D., Zalkow, L. H., and Teja, A. S., Supercritical extraction of taxol from the bark of *Taxus brevifolia*, Spec. Symp. Issue *J. Supercritical Fluids*, 5, No. 1, 1, March, 1992.

Kerey, A., Simandi, B., Ronyai, E., and Lemberkovics, E., Supercritical fluid extraction of some nonvolatile bioactive terpenoids, Proc. 5th Meet. Supercritical Fluids, France, 2, 561, 1998.

Lack, E. and Seidlitz, H., Commercial scale decaffeination of coffee and tea using SC CO_2, in *Extraction of Natural Products Using NC Solvents*, King, M. B. and Bott, T. R., Eds., Blackie Academic & Professional, an imprint of Chapman & Hall, Glasgow, 101, 1993.

Lashkari, Z., Ed., A story of resurgence of natural products, in *Finechem from Natural Products*, 1, No. 4, 1, Sept., 1999.

Lawless, J., *The Illustrated Encyclopedia of Essential Oils*, Barnes & Noble Books, New York, 1995.

Lemberkovics, E., Kery, A., Simandi, B., Marczal, G., and Then, M., Influence of SCFE and other facts on composition of volatile oils, Proc. 5th Meet. SCFs, France, 2, 567, 1998.

Mahindru, S. N., Indian plant perfumes, *Metropolitan*, New Delhi, India, 1992.

McHugh, M. and Krukonis, V., *Supercritical Fluid Extraction*, 2nd ed., Butterworth-Heinemann, Stoneham, MA, 304, 1994.

Murga, R., Beltran, S., and Cabezas, L., Study of the extraction of natural polypenols from grape seeds by using supercritical carbon dioxide, Proc. 5th Meet. Supercritical Fluids, France, 2, 529–532, 1998.

Peterson, N., *Herbal Remedies*, Blitz editions, Amazon Publishing Ltd., Middlesex, U.K., 1995.

Peplonski, R., Kwiatkowski, J., Jarosz, M., and Lisicky, Z., Extraction of some components for flavoring, coloring, and preservation of food using supercritical carbon dioxide, Proc. 3rd Intl. Symp. Supercritical Fluids, France, 2, 435–440, 1994.

Pratt, D. E. and Hudson, B. J. F., Natural antioxidants not exploited commercially, in *Food Antioxidants*, Hudson, B. J. F., Ed., Elsevier Applied Science, England, 1990.

Ronyai, E., Simandi, B., Deak, A., Kery, A., Lemberkovics, E., Reve, T., and Lack, E., Production of marigold extracts with CO_2 supercritical fluid extraction, Proc. 5th Meeting SCFs, France, 2, 607, 1996.

Saldana, M. A., Mazzafera, P., and Mohamed, R. S., Extraction of caffeine, trigonelline and chlorogenic acid from Brazilian coffee beans with SCFs, Proc. 4th Intl. Symp. SCFs, Vol. A, 219, 1997.

Sugeta, T., Sako, T., Nakazawa, N., Sakaki, K., and Sato, M., Extraction of oil containing γ-linolenic acid in fungi cells and purification of the acid using supercritical CO_2, 2, Proc. 5th Meet. Supercritical Fluids, 463–464, 1998.

Swift, D. A., Kallis, S. G., Longmore, R. B., Smith, T. N., and Trengove, R. D., Supercritical carbon dioxide extraction of oils from *camellia sinensis*, 2, Proc. 3rd Intl. Symp. Supercritical Fluids, 487, 1994.

Simandi, B., Ronyai, E., Hajdu, V., Kemeny, S., Domokos, J., Hethelyi, E., Oszagyan, M., Palinkas, J., Kery, A., and Veress, T., Supercritical fluid extraction of medicinal and aromatic plants for use in cosmetics, Proc. 5th Meet. Supercritical Fluids, France, 2, 601–606, 1998.

Schultz, K., Martinelli, E. E., and Mansoori, G. A., Supercritical fluid extraction and retrograde condensation, applications in biotechnology, in *Supercritical Fluid Technology*, Bruno, T. J. and Ely, J. F., Ed., CRC Press, Boca Raton, Florida, chap. 13, 464, 1991.

Schaeffer, S. T., Zalkow, L. H., and Teja, A. S., *Crotalaria spectabalis* using supercritical carbon dioxide-ethanol mixtures, *Biotechnol. Bioeng.*, 34, 1357–1365, 1989.

Then, M., Daood, H., Illes, V., Simandi, B., Szentmihalyi, K., Pernetczki, S., and Bertalan, L., Investigation of biologically active compounds in plant oils extracted by different extraction methods, Proc. 5th Meet. Supercritical Fluids, France, 2, 555–559, 1998.

Tsuda, T., Mizuno, K., Ohshima, K., Kawakishi, S., and Osawa, T., *J. Agric. Food Chem.*, 43, 2803–2806, 1995.

Zhimin, L., Souqui, Z., Renan, W., and Guanghua, Y., Separation of Flavonoids by Packed Column Supercritical Fluid Chromatography, Proc. 4th Intl. Symp. SCFs, Sendai, Japan, Vol. A, 19, 1997.

8 Natural Antioxidants

8.1 IMPORTANCE OF RECOVERY

Natural antioxidants are those phenolic or polyphenolic compounds commonly occurring in plant materials, which interfere with the formation of free radicals (i.e., the initiation reactions) and also deter the propagation of oxidation or the free radical chain reactions, thus preventing formation of hydroperoxides. Free radicals damage the cells of the human body which undergo oxidation with natural oxygen. In general, antioxidants provide primary defense to the bodily system by eliminating free radicals which interfere with metabolism. It is well known that some antioxidants occur naturally in different amounts in all foods and herbal medicines. However, their further addition in small quantities facilitates the delay, retardation, or prevention of the development of rancidity caused by atmospheric oxidation and thus preserves fats, oils, and fat-soluble components like vitamins, carotenoids, and other nutritive ingredients of foods. At higher concentrations, antioxidants may act as pro-oxidants, since they themselves are susceptible to oxidation. Thus, in order to ensure the quality, safety, and shelf-life of food products, it is essential to strictly follow the standards that exist in different countries with respect to the synergy between different classes of food products and antioxidants.

There is definite scientific evidence that dietary supplementation of natural antioxidant nutrients, such as vitamins A, C, E, and flavonoids to foods may prevent many human diseases caused by oxidative damage, including aging, cataract, coronary heart diseases, cancer, etc. Their natural occurrence in foods, such as fruits, vegetables, leaves, nuts, and grains provides a valuable degree of protection against oxidative attack. If the onset of such diseases could be delayed by even a few years, the social and economical implications would be tremendous.

However, when food commodities are subjected to processing, the antioxidants naturally present in them often get depleted due to physical operation or by chemical degradation. Consequently such foods are fortified with antioxidants, preferably from natural sources, or blended with some natural extracts enriched with antioxidants. Antioxidants should be added to the product as early as possible. There are two main groups of antioxidants: synthetic and natural. Though the synthetic antioxidants are cheap, most natural antioxidants are also commercially viable. One of the important trends in the food industry today is the demand for natural antioxidants from plant material. The superiority of natural antioxidants has been proven over synthetic ones in terms of safety, tolerance, and nontoxicity, without any side effects, because these components occur naturally in foods which have been consumed for years. In general, synthetic food additives are subjected to pharmacological scrutiny and technical evaluation for mutagenic, carcinogenic, and pathogenic effects. Such toxicity concerns, in addition to consumers' preference for "all natural" ingredients and stringent regulations in developed countries, have resulted in increased interest in natural antioxidants. Most commonly used natural antioxidants are tocopherols, ascorbic acids, flavonoids, lecithin, citric acid, and polyphenols. These are added in minute, predetermined concentrations (0.01 to 0.02%) to oils, sterols, emulsifiers, fat-soluble vitamins, phospholipids, flavors, aroma, even carotenoids (color) that are susceptible to oxidation during storage and transportation. Bioflavonoids, which are included in foods as flavorings, have been recognized as effective in decreasing the erythrocite aggregation and sedimentation rate of blood *in vitro* and are useful in dietary control for problems related to coagulation of platelets. These high-molecular weight polyphenolic antioxidants possess anticarcinogenic, antiproliferation, antimetastatic, and prodifferentiation activities. An increased intake of these antioxidants can contribute to minimizing the problem of oxidation of dietary cholesterol, which is a cause of cardiac arrest. Hence, antioxidants are essential for maintaining a good state of health for our body system and for the preservation of foods and medicines. Vitamin E is listed as the most widely consumed nutritional supplement (400 I.U. in a day). Vitamin E, which maintains healthy cell membranes, also reduces the risk of prostate cancer in men, fights Alzheimer's disease, cardiovascular disease, and kidney disease, and strengthens the immune system in elderly people. Vitamin C is the second most consumed antioxidant in the world and *Ginkgo biloba* the third.

8.2 CLASSIFICATION

Broadly there are five major types of antioxidants (Kochhar and Rossell, 1990) as described below:

1. Primary antioxidants are those compounds, mainly phenolic substances, that terminate the free radical chains in lipid oxidation and function as electron donors, e.g., natural and synthetic tocopherols, alkyl gallates, butylated hydroxy toluene (BHT), butylated hydroxy anisole (BHA), tertiary butyl hydroquinone (TBHQ), etc.
2. Oxygen scavengers are those substances which react with oxygen and can thus remove it in a closed system, e.g., ascorbic acid (vitamin C), ascorbyl palmitate, erythorbic acid (D-isomer of ascorbic acid), etc.
3. Secondary antioxidants are those compounds which function by decomposing the lipid hydroperoxides into stable end products.
4. Enzymic antioxidants are those enzymes which function either by removing dissolved or head space oxygen, e.g., glucose oxidase, or by removing highly oxidative species, e.g., super oxide dismutase.
5. Chelating agents are synergistic substances which greatly enhance the action of phenolic antioxidants. Most of these synergists exhibit little or no antioxidant activity, for example citric acid, amino acid, and phospholipids such as cephalin.

8.3 BOTANICALS WITH ANTIOXIDATIVE ACTIVITY

Some botanicals which have long been known to exhibit antioxidant properties are listed in Table 8.1.

TABLE 8.1
Some Sources of Natural Antioxidants

Algae	Green leaves and green/yellow vegetables
Cereals	Protein hydrolysates
Cocoa products	Resins
Citrus products	Various peppers
Herbs and spices	Onion and garlic
Legumes	Olives
Oil seeds	Tea and Coffee

Spices like clove, ginger, garlic, mace, nutmeg, etc. and labiatae herbs like rosemary, sage, thyme, and oregano are currently used for extracting natural antioxidants commercially. A large number of antioxidants, mainly phenolics, have been identified in the extracts of spices and herbs. Compounds like carnosic acid, carnosol, rosemanol, rosmarinic acid, rosmaridiphenol, and rosmariquinone have been isolated from rosemary leaves. Gallic acid and eugenol are antioxidants identified in clove extracts. Certain oil seeds, such as sesame and cottonseeds, contain characteristic antioxidant components in addition to tocopherols. Sesame seed oil contains sesamol, sesamin, sesamolin, and γ-tocopherol which render the oil its strong antioxidant activity. Other natural antioxidants are β-carotene (present in carrot and green leafy vegetables), phenolic acids, e.g., gallic, caffeic, quinnic, and ferrulic acids (present in oil seeds and oil seed flours), and flavonoids, e.g., quercetin, myricetin, quericitrin, and rutin (present in soybean, tea, and coffee). Various vegetable oils, e.g., soya, oat, and wheatgerm extracts, also containing tocopherols and lecithin, are often included in food formulations to prevent oxidation. The antioxidant activity of oat extract is however due to the presence of caffeic and ferrulic acids. Savory oil, which contains 25 to 45% of carvacrol as the major component, is currently being used to prevent rancidity and improve the shelf-life of polyunsaturated fatty acids and oils (Esquivel et al., 1999). However, all natural antioxidants are not good. For example, gossypol present in crude cottonseed oil and nordihydroguaiaretic (NDGA) from the bush *Larrea divaricata* have toxic properties.

8.4 TOCOPHEROLS AS ANTIOXIDANTS

The most common natural antioxidants commercially exploited are tocopherols. Tocopherols are a group of monophenolic antioxidants, commonly known as vitamin E, which are used as food supplements because of their strong biological antioxidant activity. They belong to two families with generic names tocols and tocotrienols. The basic structure of the two families is a side chain joined to the chromane ring to which are also attached methyl groups. The side chain is saturated in the case of tocopherols, whereas it is unsaturated in tocotrienols. They are present naturally in 0.2 to 0.02% by weight in edible oils and cereals (Schuler, 1990) as listed in Table 8.2. Steam distillation for deodorization of crude vegetable oils usually produces deodorizer distillates that are enriched in tocopherols and the tocopherol

TABLE 8.2
Concentration (mg/kg) of Tocopherols and Tocotrienols in Common Vegetable Oil

	Tocopherols (mg/kg)			
	α	β	γ	δ
Coconut	5–10	—	5	5
Cottonseeds	40–560	—	270–410	0
Maize, grain	60–260	0	400–900	1–50
Maize, germ	300–430	1–20	450–790	5–60
Olive	1–240	0	0	0
Palm oil	180–260	trace	320	70
Peanut	80–330	trace	130–590	10–20
Canola	180–280	—	380–590	10–20
Safflower	340–450	—	70–190	230–240
Soybean	30–120	0–20	250–930	50–450
Sunflower	350–700	20–40	10–50	1–10
Wheat germ	560–1200	660–810	270	270
	Tocotrienols (mg/kg)			
	α	β	γ	δ
Coconut	5	trace	1–20	—
Cottonseeds	—	—	—	—
Maize, grain	—	0	0–240	0
Maize, germ	—	—	—	—
Olive	—	—	—	—
Palm oil	120–150	20–40	260–300	70
Canola	—	—	—	
Safflower	—	—	—	
Soybean	—	—	—	
Sunflower	—	0	0	
Wheat germ	20–90	80–190		

Schuler, 1990.

content in the refined oil gets reduced by 35 to 40%, even under good manufacturing practices (GMP).

There are four main tocopherols, namely α, β, γ, and δ type, according to the position and the extent of methyl substitution on the aromatic ring of tocol, 2-methyl-2-4′,8′,12′-trimethyl (tridecyl)-6-chromanol (Figure 8.1). Each tocopherol has three asymmetric centers. The natural stereoisomer 2R, 4′R, 8′R-α-tocopherol is the most potent of all tocopherols. The tocopherols are isoprenoid, light yellow colored, oily liquids belonging to the lipids. They are insoluble in water, but soluble in triglycerides (oils) and are readily extracted with them during the process of recovering oil from cereals or seeds. It is recommended to keep α-tocopherol content at a level of 50 to 500 mg/kg of the substrate, depending on the nature of the foods.

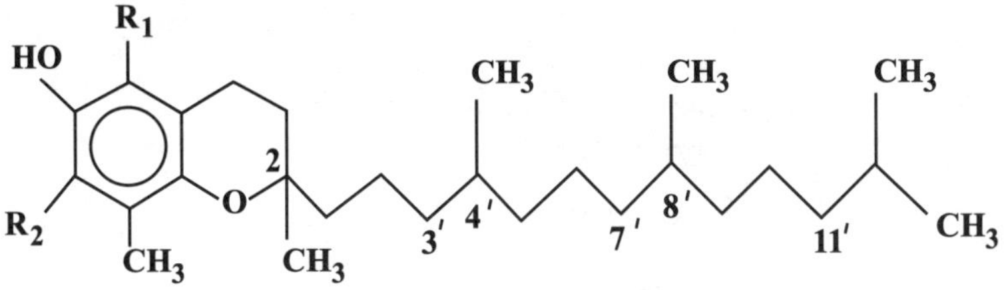

Substituents

R_1	R_2	Notation
CH_3	CH_3	α - Tocopherol
CH_3	H	β - Tocopherol
H	CH_3	γ - Tocopherol
H	H	δ - Tocopherol

FIGURE 8.1 Chemical structure of tocopherols.

TABLE 8.3
Distribution of Tocopherols and Sterols in Crude Oils

		Tocopherol/Sterol					
Components	**Mol Wt**	**Soybean**	**Corn**	**Cottonseed**	**Sunflower**	**Canola**	**Peanut**
			Tocopherols				
α	431	15	31	51	95	32	49
β	417	2	0	0	4	—	1
γ	417	52	65	48	1	66	47
δ	403	31	4	1	0	2	3
			Sterols				
Brassica sterol		0	0	0	0	9	0
Campesterol		21	19	17	8	37	15
Stigma sterol		21	4	0	7	0	8
Sitosterol		49	67	89	61	53	61
Other sterols		9	10	4	24	1	16

Wheat germ oil contains the highest level of tocopherols, followed by soybean oil, while coconut oil contains the lowest total content of tocopherol. Most oil, except coconut oil, has a sufficient tocopherol level to allow the use of distillates from these oils to be used as sources for vitamin E. Tocotrienols have very little vitamin E activity and cannot be economically converted to vitamin E. The distribution of tocopherols and sterols in crude vegetable oils is given in Table 8.3. The major components in the soybean oil deodorizer distillate are free fatty acids, squalane, tocopherols, and sterols, besides di- and triglycerides as given in Table 8.4.

The content of α-tocopherol in some of the Indian pulses is given in Table 8.5, which illustrates why pulses are included in most meals of the common man. Chevolleau et al. (1993) found that Mediterranean plant leaves such as *Quercus ilex*

TABLE 8.4
Composition of a Soybean Oil Deodorizer Distillate

Component	GC Peak Area (%)
C_{16} fatty acid	19.8
C_{18} fatty acid	56.1
Monoglycerides	9.2
Squalane	12.0
α-Tocopherol	2.2
Other tocopherols (γ, β, δ)	9.3
Sterols	14.8
Diglycerides	1.7
Triglycerides	1.7

Brunner et al., 1991.

TABLE 8.5
Tocopherol Content of Indian Pulses

Tocopherol	Bengal Gram (mg/kg)	Black Gram (mg/kg)	Green Gram (mg/kg)	Horse Gram (mg/kg)
α	16.8	0.3	0.9	0.3
β	0.9	—	0.1	—
γ	92.2	65.8	116.6	66.3
δ	4.1	1.5	7.8	6.9
Total	114.0	67.6	125.4	73.5

Gopala Krishna et al., 1997.

are good sources of α-tocopherol, as listed in Table 8.6, and contain up to 846 ppm of α-tocopherol on a dry basis.

8.4.1 Recovery by SC CO_2

There are various processes like solvent extraction, molecular distillation, and SC CO_2 extraction which are commercially used for the recovery of tocopherols, depending on the form of their natural sources. The first two processes have inherent drawbacks of thermal degradation and/or residual solvent, whereas the third method offers the advantage of selective recovery of a truly natural form of mixed tocopherols for vitamin E activity.

Conventionally, tocopherols are recovered from soybean sludge, which is a byproduct from the deodorization process of soybean oil. The tocopherol concentrate can be obtained by molecular or vacuum distillation after removing the sterols from

TABLE 8.6
α-Tocopherol Content of Dry Mediterranean Plant Leaves

Spices	α-Tocopherol Content Dry leaves (ppm)	α-Tocopherol Content Hexane Extract (%)
Centranthus ruber (*Valerianaceae*)	100	0.4
Cistus albidus (*Cistceae*)	33	0.1
Conium machulatum (*Apiaceae*)	210	0.6
Coronilla juncea (*Fabaceae*)	18	0.1
Eucalyptus globulus (*Myrtaceae*)	333	0.3
Ferula communis (*Apiaceae*)	75	0.3
Globularia alypum (*Globulariaceae*)	663	3.9
Hedera helix (*Araliaceae*)	75	1.3
Myrtus communis (*Myrtaceae*)	627	3.3
Phillyrea angustifolia (*Oleaceae*)	480	2.4
Pinus halepensis (*Pinaceae*)	210	0.2
Quercus ilex (*Fagaceae*)	846	4.7
Rhamnus alaternus (*Rhamnaceae*)	442	3.4
Smilax aspera (*Liliaceae*)	357	2.1
Staehelina dulia (*Asteraceae*)	138	0.3

Chevolleau et al., 1993.

the sludge via alcohol recrystallization. However, this process requires several steps involving large amounts of solvents and energy.

Lee et al. (1991) examined the feasibility of extraction of α-tocopherol from soybean sludge with SC CO_2 at a temperature ranging from 35 to 70°C and at a pressure ranging from 200 to 400 bar. They reported that a simple batch process could be utilized to recover tocopherols of 40% concentration from the esterified sludge initially containing 13 to 14 wt% tocopherols. The esterified soybean sludge was found to have 4 to 6 times higher solubility in SC CO_2 than the sterol-removed soybean sludge (Lee et al., 1991).

As mentioned earlier, wheat germ oil is a good source of tocopherols and tocopherols can be recovered from wheat germ by combining the SC CO_2 extraction process with supercritical fluid chromatography (SFC) in a preparative mode. Saito and Yamauchi (1990) reported that tocopherols could be enriched to 5% starting with a sample of SC CO_2 extracted wheat germ oil containing 0.05% tocopherols, i.e., by a factor of 100 in a single run of preparative SFC. Upon repeating the process, the final concentration was only 20% due to the low initial concentration of 0.05% in wheat germ oil. However, tocopherol-enriched fractions having compositions of 85% α-tocopherol and 70% β-tocopherol could be obtained from the wheat germ oil by recycle semipreparative SFC using two columns packed with silica gel operated at 250 and 200 bar, respectively. β-tocopherol could be collected after recycling twice, and α-tocopherol after two additional recycles prior to analysis by capillary gas chromatography.

For extraction of oil from the solid matrices of soybean flakes and rice bran, and fractionation of tocopherol from the extracted oil, King et al. (1996) combined

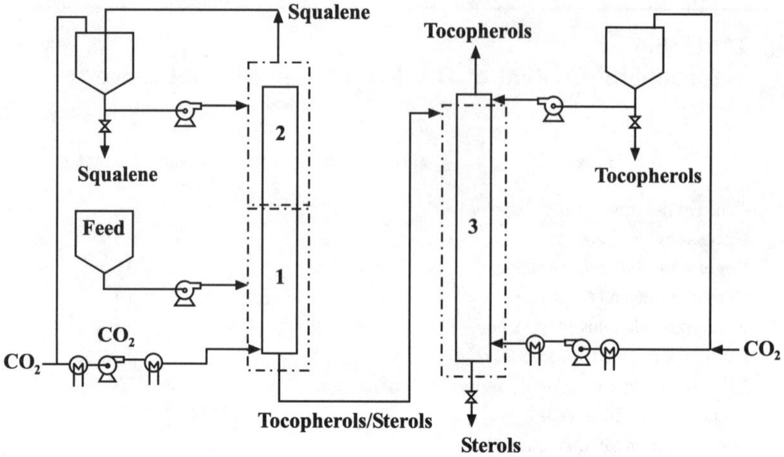

FIGURE 8.2 Separation sequence for recovering tocopherol.

SC CO_2 extraction with preparative SFC. The optimal recovery and enrichment of tocopherol could be obtained at 250 bar and 80°C for the SC CO_2 extraction stage and 250 bar and 40°C for the SFC stage.

Earlier Ohgaki and Katayama (1984) developed a process for extraction of tocopherols from palm oil and coconut oil using compressed-liquid CO_2 at 25°C and 85 to 100 bar pressure, in which tocopherols are separated from coextracted fatty acids at a higher temperature (supercritical) and 80% of the initial content (0.1 wt%) of tocopherols could be separated together with triglycerides. The fatty acids are subsequently separated from CO_2 in the second fraction at a lower pressure. However, if extraction is carried out at the supercritical condition, the tocopherol recovery could be only 40%.

Alternatively, tocopherol present in crude vegetable oils is concentrated in the deodorizer distillate (DOD) during the process of deodorization by the steam distillation of crude vegetable oils. The distillate is then separated into aqueous and organic layers. This organic layer contains most of the tocopherols transferred from the crude vegetable oil and is then subjected to vacuum distillation to obtain the tocopherol concentrate. This can then be further concentrated by SC CO_2 to obtain individual or mixed tocopherols. SC CO_2 processing can be thus easily integrated as a downstream process to the vegetable oil industry, either for fractionation of vegetable oils or for isolation of tocopherols from tocopherol concentrate.

Brunner et al. (1991) used SC CO_2 as well as SC CO_2 + ethanol as the solvent to obtain a higher enrichment of tocopherol product. Figure 8.2 represents a separation sequence for recovering tocopherols from a mixture of squalane, tocopherol, and sterols using SC CO_2 in two continuous sequential countercurrent fractionating columns. In the first column, the more volatile squalane is separated from tocopherol and sterols which are fed to the second column, enriching tocopherols from 56.1 wt% to 71.2% tocopherols. The nonvolatile sterols are removed from the bottom of the second column, whereas a fraction containing 95 to 85% tocopherol is removed from the top of the second column.

From a phase-equilibrium study for recovering α-tocopherol from a mixture of squalane-tocopherol-campesterol, it is reported by Brunner (1994) that the separation factor for squalane-tocopherol varies from a value of 4 at a low squalane content (0.5 wt%) to a value of 1 at a high squalane content (85 wt%) at the SC CO_2 density of 655 kg/m^3 for pressures in the range of 200 to 300 bar and temperatures 70 to 100°C. At similar operating conditions, the separation factor for α-tocopherol-campesterol is in the range of 2.6 to 1.5 for the tocopherol content in the pseudobinary mixture, varying in the range of 60 to 80 wt% tocopherol, but decreases rapidly with increasing α-tocopherol content. The higher values of separation factors are in general observed at lower pressures (e.g., 200 bar).

An experimental program was simulated for exploring the feasibility of a commercial-scale, separation process to recover tocopherol from mixtures of squalane, sterols, fatty acids, and triglycerides. In order to produce a tocopherol fraction having at least 85 wt% purity from tocopherol concentrate (55 wt% tocopherols) using SC CO_2 as the solvent at 295 bar and 100°C, in a column of 17 mm i.d. with 5 mm spirals as packing at a maximum throughput of SC CO_2 of about 20,000 kg/m^{2}h, the number of theoretical stages required was found to be merely three (Brunner, 1994). However, for recovering 99% pure tocopherol from a feed mixture of up to 60 wt% tocopherol using SC CO_2 mixed with 10 to 30% ethanol at 200 bar and 70°C, assumption of a constant mean separation factor for separation of more volatiles required 16 theoretical stages for the first column, and for separation of less volatiles from the tocopherol fraction required 24 theoretical stages in the second column (Brunner et al., 1991). Squalane and free fatty acids are enriched at the top of the first column, but the tocopherols, glycerides, and sterols are concentrated at the bottom. With increasing ethanol concentration in SC CO_2 there is a decrease in the separation factor of tocopherol, with respect to more volatiles and an increase in separation factor with respect to less volatiles. Thus continuous fractionation and recovery of tocopherol from DOD involves elaborate experimental investigation on the actual feed mixture and systematic optimization of process conditions.

For antioxidant applications, mixed tocopherols having various contents of α-, γ-, and/or δ-tocopherols (usually diluted in a vegetable oil) and synergistic mixtures composed of tocopherols, ascorbyl palmitate, or other antioxidants, synergists, e.g., lecithin, citric acid, and other carriers, are marketed in oily form.

Commercially, colorless and odorless mixed tocopherols of 50% strength is marketed as Tenox GT-1 by Eastman Chemicals, and mixed tocopherols of 70% strength as Tenox GT-2. Similarly Henkel Corporation has two commercial products called Covi-ox T50 and T70 having similar concentrations of tocopherols. They represent the largest group of commercial natural antioxidants currently being marketed (Nguyen et al., 1994).

8.5 SPICE AND HERBAL EXTRACTS AS ANTIOXIDANTS

Spices and herbs are used not only to enhance flavor but also the shelf-life of various foods. Chipault et al. (1952 and 1956) demonstrated that 32 spices could behave as antioxidants, out of which rosemary and sage, belonging to the Labiate family, have been identified as the most potent antioxidants. Besides, allspice, clove, mace, savory, oregano, nutmeg, turmeric, ginger, marjoram, spearmint, and thyme also possess significant antioxidant properties. Of all of these spices, oregano is found to be the most effective for lipid-containing foods. These spice extracts contain a number of components responsible for their antioxidant activities. For example, carnosic acid and rosmaric acid are the two phenolic compounds present in rosemary extract which are the most active components and it is believed that the antioxidative activity of the rosemary extract is due to its carnosic acid content.

However, carnosol, a phenolic diterpenic lactone, is believed to be actually responsible for the antioxidant activity of rosemary (*Rosemarinus officinalis*) as well as sage (*Salvia officinalis*) leaves (Brieskorn et al., 1964). Carnosol is produced by the oxidation of carnosic acid. The antioxidant activity of both carnosol and carnosic acid seems to be equivalent. Rosmanol and rosmaridiphenol are two other antioxidant compounds present in the rosemary extract. Rosmanol showed more antioxidative effectiveness than α-tocopherol and BHT, when applied to lard (Nakatani, 1994). The rosemary extract is found to be very effective in inhibiting oxidation in almost any type of food. The antioxidant activity of the rosemary extract may be further enhanced by the addition of ascorbic acid (200 ppm rosemary extract to 500 ppm ascorbic acid). The synergism between carnosic acid and ascorbyl palmitate was confirmed by Pongracz et al., (1978 to 1987). The synergistic behavior of a commercial antioxidant mixture containing 25% ascorbyl palmitate, 5% DL-α-tocopherol and 70% lecithin, was also established and it was found that the antioxidant activity of all mixtures is greater than the sum of the activities of the individual components.

It is reported by Masuda et al. (1995) that the rhizomes of several tropical gingers, e.g., *Curcuma domestica* and *Zingiber cassumunar* contain antioxidants which have antiinflammatory and antitumor promotion activities. *Curcuma domestica* contains two natural phenolic compounds besides four curcumonoids. These phenolics are found to have stronger antioxidative activity than the curcumonoids and can prevent cancer and inflammatory activities in mammals. Cassumunarins isolated from *Zingiber cassumunar* showed stronger antioxidant and antiinflammatory activities than those of curcumin, implying that they are more effective for cancer prevention.

It is already known that the structural feature required for antioxidant activity is a phenolic ring containing an electron-repelling group. Such characteristics are encountered in the constituent components of essential oil from all Labiatae herbs, e.g., rosemary and oregano. The antioxidant activity of essential oils obtained from other herbs belonging to the Labiatae family is attributed to carvone (caraway), eugenol (clove), cuminaldehyde (cumin), thujone (sage), and thymol (thyme). Thymol is always present in conjuction with its isomer carvacrol in Labiatae plants. For example, oregano, popularly known as the pizza herb, contains 80% carvacrol with 2% thymol in the essential oil extracted with SC CO_2 at 100 bar and 40°C (Calame and Steiner, 1987). The effectiveness of the oils in retarding oxidation of linoleic acid (control) in ascending order is caraway < sage < cumin < rosemary < thyme < clove (Tsimidou and Boskou, 1994). The phenomenon is concentration dependent, for example, the antioxidant activity of thymol and eugenol at 1200 ppm of concentration is nearly equivalent to 60 and 70% of the effectiveness of BHT at 200 ppm of the usage level, respectively. Sensory evaluation tests indicate that the addition of these

TABLE 8.7
Antioxidant Activity of Labiatae Oleoresin

Antioxidant	Peroxide Value (meq/kg)
Base	0.5
Control	33.7
BHA/BHT (1:1)	1.9
Rosemary officinallis extract	1.5
Sage officinallis extract	1.5
Sage triloba extract	2.1
Oregano vulgare extract	3.8
Thyme vulgaris extract	2.9

Nguyen et al., 1994.

essential oils at concentrations ranging from 50 to 1200 ppm does not affect the odor and color profiles of the substance to which they are added for stabilization.

Labex™ is a commercial antioxidant oleoresin fraction from the Labiatae herbs, rosemary and sage. An evaluation of the antioxidant activity, as given in Table 8.7 for a number of herbal extracts from Labiatae plants, reveals that they are fully effective at a level of 200 to 300 ppm and are at least as strong as synthetic antioxidants BHA + BHT (1:1) mixture. Figures 8.3a and 8.3b illustrate that Labex™ performs better and is more effective when compared to the 1000 ppm level required for mixed tocopherols and two commercial Rosemary Deodorized™ extracts A and B.

Labex™ shows excellent performance in food preservation. The usage levels are sufficiently low for Labex™ antioxidants and accordingly there is no change in the original aroma and flavor of the base food products. Labex™ has also been found to protect the color of paprika oleoresin during extended heating by preventing the oxidation of carotenoid pigments. It is GRAS and can be used at any level for any food applications.

Commercial antioxidants from spice oleoresins are normally in the form of fine powders. Spice extracts used as food antioxidants should be free from odor, color, and taste. Depending on their content of active substances, it is recommended to use them at levels between 200 to 1000 mg/kg of finished product to be stabilized. Due to increasing demand for chemical-free, all-natural food products, rosemary and sage oleoresins are finding ready market as commercial natural antioxidants. For example, Cal-Pfizer is marketing Rosemary Deodorized™, Nestle is marketing Spicer Extract AR™, Kalsec is marketing Herbalox™, and OM Ingredients is marketing Flavor Guard™ (Nguyen et al., 1994).

8.5.1 Recovery by SC CO_2

Various commercial processes are in practice for obtaining antioxidant spice extracts. The major requirements of the processes are to extract them with sufficient antioxidant activity to render the usage levels in the range of 0.01 to 0.10% of the substrate

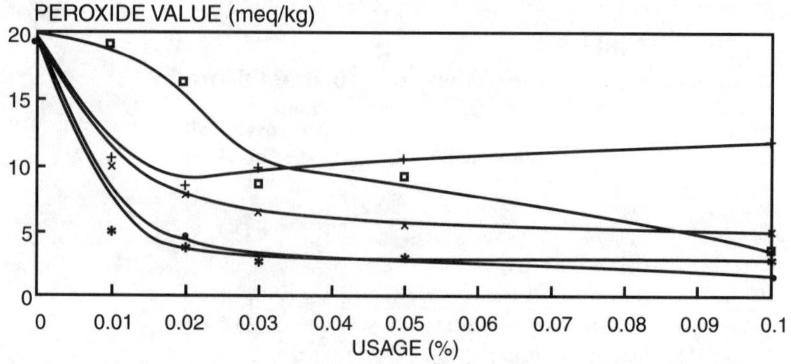

FIGURE 8.3a Comparison of antioxidant activity of Labex™ spice oleoresin for lard after 18 h at 100°C with (●) BHA/BHT (1:1); (+) tocopherol; (*) Labex™; (□) com-A; (x) com-B (Nguyen et al., 1994).

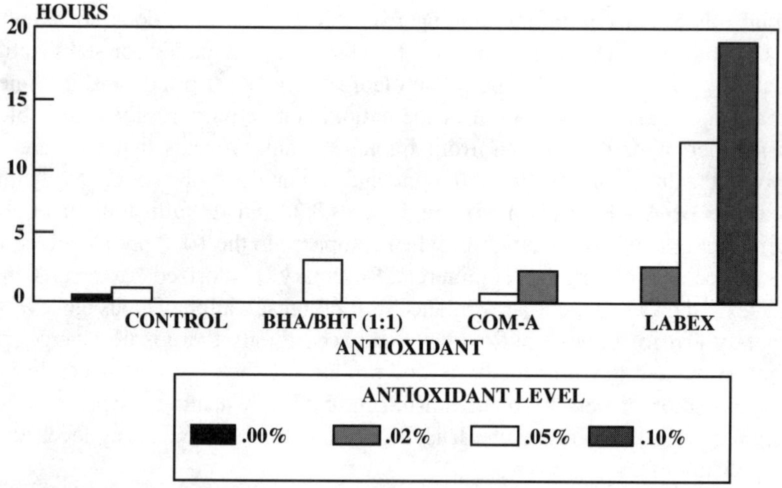

FIGURE 8.3b Labex™ spice oleoresin-Rancimat antioxidant comparison of induction time (hours) at 120°C using prime steam lard (Nguyen et al., 1994).

and to remove flavor, odor, and color components from the antioxidant extracts so that they are not detectable in the treated food products. The extraction processes that have been commercially employed for obtaining these extracts include (1) extraction with both polar and nonpolar solvents, e.g., ethanol, ethyl acetate, hexane, acetone, methyl chloride (Chang et al., 1977; Kimura and Kanamori, 1983; Aesbach and Philippossion, 1987; Todd, 1989); (2) aqueous alkaline extraction (Viani, 1977); (3) extraction with vegetable oils and/or monoglycerides or diglycerides (Berner and Jacobson, 1973); (4) steam distillation; (5) molecular distillation; and most recently (6) SC CO_2 extraction (Tateo and Fellin, 1988).

Most of these processes suffer from a number of drawbacks, such as the solvents used are not sufficiently selective for the active antioxidant ingredients, and they also may be left behind as unwanted residues, which is prohibited by food regulations. When molecular distillation is used to concentrate the active antioxidant ingredients, the process suffers from dilution with the carrier used in molecular distillation and it also may have an adverse effect on their solubilization in fat/oil.

SC CO_2 extraction, on the other hand, overcomes most of the above-mentioned drawbacks. Ground rosemary leaves were first extracted with SC CO_2 at a pressure of 300 bar and 35°C, followed by fractional separation of oleoresin from essential oil (Tateo and Fellin, 1988) and later the raffinate (the ground leaf residue) was further extracted with a more polar solvent, like ethanol, to recover the antioxidant principle. The commercial product, Rosemary Deodorized™, is produced by a similar method in which SC CO_2 is used to extract the oleoresin fraction which is then subjected to molecular distillation to improve its color and flavor. However the product contains 80% caprylic and capric triglyceride as the distillation carrier and accordingly it has less solubility in lipids and less antioxidant activity (Nguyen et al., 1994).

Labex™, the commercial Labiatae antioxidant, is, however, obtained by extraction with SC CO_2 at a higher pressure, in the vicinity of 500 bar, and a temperature in the range 80 to 100°C, followed by fractionation into two products (Nguyen et al., 1994). The first is an oil-soluble resin containing less than 2% essential oil, while the second is an aroma fraction containing more than 80% essential oil. This first fraction is used for the commercial Labiatae antioxidant.

The exceptional effectiveness of Labiatae extracts is attributed to carnosic acid, which usually gets converted by oxidation into carnosol and other unknown byproducts during the process of organic and/or aqueous solvent extraction. Carnosic acid is extremely stable in SC CO_2 and is fully extractable by SC CO_2 extraction without any degradation. As a result, the supercritically extracted product contains 5 to 6 times as much total phenolic antioxidant (TPA), and 80% of this is carnosic acid. This active antioxidant compound is stable up to cooking temperatures. Labiatae extracts thus produced successful natural alternatives to synthetic antioxidants for a variety of food products such as eggs, meat, seafood, etc.

8.6 PLANT LEAF EXTRACTS AS ANTIOXIDANTS

It is well known that plant leaves, like any dark green leafy vegetables, are rich sources of phenolic compounds which are lipid-soluble antioxidants. These include α-tocopherol (vitamin E) and β-carotene (precursor to vitamin A), since chlorophyll and carotenes are closely related. Their recovery from plant leaves could provide a valuable source of natural vitamins.

The antioxidant activity of several leaf extracts of Mediterranean plants has been reported by Chevolleau et al. (1993). α-tocopherol was identified as the main antioxidant in leaf extracts of 15 plant species. The tocopherol content could be enriched by hexane extraction up to 4.7% from dry leaves containing 846 ppm of tocopherol. The highest tocopherol content was earlier reported to be 1000 ppm in the leaves of *Vrtica dioica* (Booth, 1963). This level of tocopherol content is even higher than what most common vegetable oils contain and is much higher than the level of tocopherol content in oil seeds. This suggests that the leaves, both edible and nonedible types, are better sources of vitamins A and E than the seeds.

TABLE 8.8
Comparison of Yield and Concentration of Tocopherol or β-Carotene in the Supercritical CO_2 Extracts from Various Sources

Plant Leaf	SC CO_2 Condition	Dry Leaf Yield (%)		Extract (%)	
		α-Tocopherol	β-Carotene	α-Tocopherol	β-Carotene
Guava	200 bar/60°C	0.15	0.75	5.1	26.0
Neem	200 bar/60°C	0.46	0.001	15.5	0.41
Basil	200 bar/40°C	0.50	0.0005	15.0	0.015
Wheat grass	200 bar/60°C	0.28	0.15	19.5	10.8
Papaya	200 bar/60°C				0.80
Kholkhol	200 bar/60°C				0.36

Joshi, 1999.

Recently SC CO_2 extraction of several plant leaves, such as papaya, kholkhol, mango, guava, neem, basil, wheat grass, etc. was carried out in our laboratory-scale supercritical extraction apparatus using dried and finely ground powders at pressures in the range of 185 to 250 bar and at two temperatures, 40 and 60°C. The α-tocopherol and β-carotene contents were analyzed by a UV Spectrophotometer (Joshi, 1999). The maximum recovery was mostly possible at 200 bar and 40 or 60°C. Both α-tocopherol and β-carotene could be simultaneously extracted, as can be seen in Table 8.8.

Wheat grass, spinach, and guava leaves were found to be excellent sources of both vitamins E and A, whereas neem and basil leaves yielded SC CO_2 extracts enriched with α-tocopherol to an extent of 15%. It is interesting to note that leaf extracts are more enriched in vitamins than seed extracts and it is easier to recover them from dried ground leaves than seeds. Further investigations on SC CO_2 extraction of both vitamins from other green leaves are in progress. Preliminary pilot plant studies indicate that SC CO_2 extraction of ground leaves in a larger (by volume) 10 l extractor and a higher pressure in the range of 250 to 300 bar at 60°C, can yield much higher recovery and concentrated extracts.

SC CO_2 extraction was performed (Rebeiro et al., 1998) using dried leaves of savory (*Satureja hortensis*), thyme (*Thymus capitatus*), and coriander (*Coriandrum sativum*), and it was observed that the antioxidant activity was more prevalent in the leaves rather than in the seeds in the case of coriander. An analysis of antioxidant activity in terms of protection factors (PF) indicated that all of these leaves have significant antioxidant activity as PF was in the range of 1.4 (for thyme and coriander freeze-dried leaves) to 1.1 (for savory leaves).

Eucalyptus essential oils have been widely used for cosmetics and medicinal purposes. However, diketones with long alkyl side chains isolated from *Eucalyptus globulus* leaf waxes are found to show strong antioxidant activity in comparison with other diketone analogues. β-diketones and hydroxy-diketones are commonly present in leaf waxes of plants, such as acacia, rhododendron, and oat, and are found to show strong antioxidant activity (Osawa and Namiki, 1985).

TABLE 8.9
Relative Abundance of Flavonoids

Tisue	Relative Concentration
Fruit	Cinnamic acids > catechins > flavonols
Leaf	Flavonols (cinnamic acids) > catechins
Wood	Catechins > flavanols > cinnamic acids
Bark	As wood, but greater concentrations
Brussel sprouts	500 mg/kg quercetin
Lettuce	200 mg/kg quercetin
Apricots	50 mg/kg quercetin
Onions	10 g/kg (highest) quercetin

Pratt and Hudson, 1990.

8.7 FLAVONOIDS

The most common natural antioxidants are flavonoids, such as flavones, flavonols, isoflavones, catechins, flavanones, flavonones, and cinnamic acid derivatives. Flavonoids occur widely in leaves, flowering tissues, and pollens. They are also present in roots, fruits, seeds, stems, and all woody parts of the plant. Flavonoids are characteristic constituents of a large number of species of the Apiaceae family. They are present in cottonseed oils, oat extracts, seed coats, and several herbs. They represent a wide variety of compounds such as polyphenols which result in plants from photosynthesis. The term flavonoid refers to plant phenols characterized by the basic structure of two aromatic rings linked by a 3-carbon pyran or furan ring. The most effective flavonoids are quercetin, rhamnetin, kampferol, and rutin, or the vitamin P group. The flavonols are present in plant materials in the form of either glycosides or aglycones. High levels of antioxidant activity of certain plant materials are not confined to flavonoids alone, but also to closely related compounds, e.g., chalcones and phenolic acids. Chalcones which are natural precursors of the flavones and flavanones are also highly potent antioxidants. Dihydrochalcones are even more active than the corresponding chalcones. Flavonoids and related antioxidants are present in all parts of the plant, and their relative concentrations may vary depending on their sources, as shown in Table 8.9.

In addition to these sources, cereals, tea, coffee, and chocolates also provide these flavonoids to our daily meals and imbibe health benefits, such as prevention of heart diseases. Flavonoids did not find much commercial exploitation until a couple of years ago, possibly due to their poor solubilities in aqueous system. Quercetin is effective at 150 mg/kg in milk, lard, and butter, and at 3 mg/kg in methyl linoleate. But they find applications with synergists, such as citric acid, ascorbic acid, phospholipids, and even tocopherols in the presence of which flavonoids are more soluble. It is generally observed that there is synergism of flavonoids and related compounds with other food components. For example, in the case of leaves, tocopherols function as primary antioxidants, phospholipids as proton donors, ascorbic acid as an oxygen scavenger, and flavonoids as primary antioxidant and metal chelators, but all of them together provide a high degree of protection. Synergism between phospholipids, flavones, isoflavones, and phenolic acids is well documented in the case of seeds as well (Pratt and Hudson, 1990).

Phenolic acids, including chlorogenic, caffeic, and ferulic acids are present in soybeans, cottonseeds, and peanuts, of which, the first two in significant concentrations. Leafy material is well known as a rich source of both flavones and phenolic acids. Gallic acid and its esters are recognized as potent antioxidants and are present in onion and garlic.

A large number of flavonoids have been identified, though with varying metabolic properties and polarities. The following decreasing order of potency was found while testing 14 flavonoids in beef heart mitochondria: chalcone > flavone > flavonol > dihydroflavonol > anthocyanidin (Bermond, 1990). Readily oxidizable fat, such as lard, may be effectively stabilized by chalcones in the concentration range of 0.025 to 0.1%, more so than by the corresponding flavonones. Typically, lettuce contains 200 mg/kg and onion 10 g/kg of quercetin (flavone). Four flavanol glycones and one flavanonol aglycone have been identified in cottonseed. Quercetin, kaempferol, gossypetin, and heracetin are flavanols and dihydroquercetin is a flavanonol. Flavonoids, hydroxy cinnamic acids, carnosic acids, caffeic acid, gallic acid, coumarins, tocopherols, and polyfunctional phenolic acids are extracted together from plant material such as cotton seed, apricot, grape seed, soybean, and peanuts. However, the best potential source of flavonoids for food antioxidants is wood and whole bark. For example, Douglas fir bark contains about 5% of taxifolin, a dihydroquercetin, while the cork contains 22%. Quercetin, which can also be obtained from dihydroquercetin by oxidation, is the most potential commercial flavonoid antioxidant for food products.

There is an increasing interest in the antioxidative activity of different types of tea related to the formation of free radicals, carcinogenesis, and atherogenesis due to the presence of natural polyphenols. The total amount of flavonols present in fresh tea flush varies from 15 to 25% (on dry-weight basis). These flavonols are oxidized to theaflavin and thearubigin, together with other oxidation products by catechol (polyphenol) oxidase enzyme, endogeneous to the leaves of tea plants. The polyphenolic compounds constitute about half of the total solids (0.3% by weight) present in a cup of tea (Sanderson et al., 1992). Tea extracts from nonfermented tea, e.g., green, yellow, and white teas, are found to have stronger antioxidant activity to inhibit lipid oxidation compared to BHA and BHT (Chen et al., 1996) and are stable up to 100°C. These natural antioxidants of dietary origin are also believed to suppress atherosclerosis, thrombosis, and mutagenesis.

8.7.1 Recovery of Flavonoids by SC CO_2

Natural polyphenols like gallic acid, catechin, and epicatechin can be commercially extracted from grape seeds by using SC CO_2. Pro-anthocyanidins are polyphenols formed by condensation of flavan-3-ol molecules. They also find applications as antioxidants or antineoplastic agents. Solubilities of these monomers in SC CO_2 and SC CO_2 mixed with ethanol were studied to ascertain the possibility of their

TABLE 8.10
Effect of Ethanol as an Entrainer on the Solubilities of Phenols and Polyphenols in SC CO_2 at 200 bar and 40°C

	Solubility, ppm (Volume % Ethanol)			
Components	**0**	**2**	**5**	**10**
Gallic acid	0.005	0.11	0.55	7.48
(+)-Catechin	—	0.16	0.47	4.80
(–)-Epicatechin	—	0.05	0.45	2.23

Murga et al., 1998.

extraction from natural materials such as grape seeds. Table 8.10 gives the amount of solute dissolved in the amount of CO_2 used with increasing amounts of ethanol addition.

It is apparent from Table 8.10 that while (+)-catechin and (–)-epicatechin are not soluble at all in pure CO_2 at a pressure of 200 bar and 40°C, gallic acid is slightly soluble in pure CO_2, though its solubility increases 2500 times in SC CO_2 mixed with 10 vol% ethanol. The poor solubilities of catechin and epicatechin in pure SC CO_2 are attributed to their high polarity and the size of the molecules (mol wt = 290.3). Solubility of gallic acid (mol wt = 170.1) is the highest, whereas that of (–)-epicatechin is the lowest even from a equimolar mixture of these three flavanols, using SC CO_2 with 10% by volume ethanol. These results indicate that while carrying out SC CO_2 extraction from grape seeds, the extract will contain more gallic acid than epicatechin.

8.8 CAROTENOIDS AS ANTIOXIDANTS

Carotenoids are natural, fat-soluble substances of which more than 500 different molecules have been identified so far. The classification of carotenoids will be presented in the next chapter. The transformation of β-carotene to vitamin A occurs mainly by cleavage of the molecule at the central double bond by the action of the enzyme, carotene deoxygenase, as described in Figure 8.4. Carotenoids are known to deactivate free radicals and excite oxygen, both of which are implicated in a multitude of degenerative diseases for protection against uncontrolled oxidations (e.g., tumors, cardiovascular diseases, etc.).

Many vegetables, such as carrot, parsley, dill, red and yellow pepper, celery, pumpkin, and sweet potato, and fruits are natural sources of carotenoids and offer the possibility of obtaining natural beta carotene in the concentration range of 1 to 26% in extracts. As mentioned earlier, any dark green, leafy or yellow vegetables are a good source of both tocopherols and β-carotene. Of them, cultivated grasses of wheat, rice, corn, oats, rye, and barley are the richest sources of both β-carotene and tocopherol. For example, dehydrated wheat grass contains 23,136 (I.U.) of vitamin A and 51 mg of vitamin C per 100 g of dried wheat grass (Seibold, 1990).

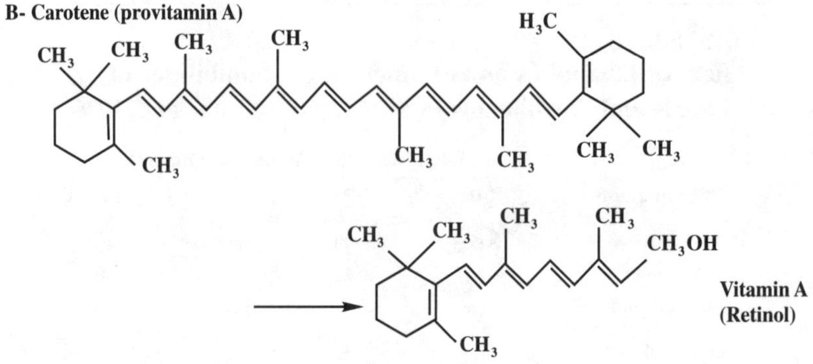

FIGURE 8.4 Structure of β-carotene and retinol (vitamin A).

Carotenoids are also one of the major groups of natural pigments that find widespread utilization in the food industry. β-Carotene, an orange color lipid and precursor to vitamin A, is often added to food products in order to give uniform coloring.

8.8.1 Recovery of β-Carotene by SC CO_2

SC CO_2 extraction of crude palm oil containing 550 ppm of β-carotene was carried out in a continuous countercurrent column. It was possible to concentrate the β-carotene level up to 1225 ppm at 300 bar and 65°C with SC CO_2 mixed with 7.5 mol% ethanol (Ooi et al., 1996). SC CO_2 extraction of sweet potatoes at 48°C and 414 bar yielded extract containing 94% β-carotene (Spanos et al., 1993).

Goto et al. (1994) studied the recovery of β-carotene from carrots by SC CO_2. They established that ethanol as an entrainer had a favorable effect on the recovery. Subra et al. (1994) also investigated the extraction of carotenoids from carrots with SC CO_2. The highest recovery of the two major carotenoids α- and β-carotene could be achieved at 57°C and 250 bar, which was merely 0.2 mg/g in 2 h of extraction time. The loading was found to be much less than the solubility, due to mass transfer resistance of the freeze-dried substrates. Since the extraction yield was very low, it was necessary to have pretreatment of solid matrices, such as freeze drying to ambient temperature, drying to remove moisture, and grinding to fine particle size to minimize the mass transfer resistance. The highest yield (10 mg/g) of β-carotene is obtained from partially dried raw carrot, rather than freeze-dried carrot, using SC CO_2 mixed with an entrainer, e.g., ethanol at 294 bar and 40°C (Goto et al., 1994). Carotenoids are susceptible to denaturation during freeze drying, which also increases mass transfer resistance.

8.9 ANTIOXIDANT SOLUBILITY IN SC CO_2

In order to obtain natural antioxidants from different raw materials using the SC CO_2 extraction process, it is useful to have knowledge of the solvent power and selectivity of separation from associated components. Solubilities of tocopherol and β-carotene in SC CO_2 are, respectively, given in Tables 8.11 and 8.12. Solubilities of flavone

TABLE 8.11a
Solubility y_2 of α-Tocopherol in CO_2 at 25 and 40°C

t (°C)	P (bar)	y × 10^4	t (°C)	P (bar)	y × 10^4
25	100	4.24	40	104	3.14
	105	4.32		104	3.71
	131	4.16		128	2.59
	252	4.47		130	2.85
	154	4.5		153	4.75
				154	4.81
				170	6.19
				179	7.31

Ohgaki et al., 1989.

TABLE 8.11b
Solubility S, g/kg CO_2 and Mole Fraction y_2 in SC CO_2

T (K)	P (bar)	ρ (kg/m³)	S (g/kg)	$y_2 \times 10^3$
40	199	839.3	13.7	1.42
	219	856.8	16.3	1.69
	239	872.3	18.1	1.88
	259	886.1	20.2	2.10
	279	898.5	23.6	2.46
	299	909.9	25.5	2.67
	319	920.5	25.7	2.69
	349	935.0	27.7	2.90
60	199	721.6	10.6	1.09
	219	750.8	13.6	1.41
	239	774.9	17.1	1.77
	259	795.4	19.8	2.06
	279	813.3	22.9	2.39
	299	829.2	25.1	2.62
	319	843.5	29.4	3.09
	349	862.7	33.9	3.57
80	199	591.6	6.2	0.64
	219	634.9	10.1	1.04
	239	669.7	14.4	1.49
	259	698.6	17.3	1.80
	279	723.2	19.9	2.07
	299	744.5	23.1	2.41
	319	763.4	28.4	2.98
	349	788.2	33.2	3.50

Johannsen and Brunner, 1997.

TABLE 8.12
Solubility in S in g/kg CO_2 in Mole Fraction y_2 of β-Carotene (2) in SC CO_2

7°C	P (bar)	ρ (kg/m³)	S (g/kg)	$y_2 \times 10^6$
40	200	840.2	0.11	0.09
	220	857.7	0.12	0.10
	240	873.6	0.13	0.11
	260	886.7	0.16	0.13
	280	899.1	0.17	0.14
	300	910.5	0.19	0.16
	320	921.0	0.29	0.24
	350	935.4	0.34	0.28
60	200	723.2	0.34	0.28
	220	752.1	0.43	0.35
	240	776.0	0.64	0.52
	260	796.4	0.72	0.59
	280	814.2	1.05	0.86
	300	830.0	1.26	1.03
	320	844.2	1.46	1.19
	350	863.3	1.44	1.18
80	200	594.2	0.42	0.35
	220	671.3	1.07	0.88
	240	699.9	1.17	0.96
	260	724.3	2.05	1.68
	280	745.5	2.45	2.01
	300	764.3	2.64	2.17
	320	788.9	3.95	3.24

Johannsen & Brunner, 1997.

TABLE 8.13
Mole Fraction Solubilities, y, of Flavone and 3-Hydroxy Flavone in SC CO_2

	Flavone		3-Hydroxy Flavone	
P (bar)	35°C 10^4 y_2	45°C 10^4 y_2	35°C 10^5 y_2	45°C 10^5 y_2
91	1.07	0.35	1.49	0.788
101	2.33	1.65	2.39	2.66
152	3.59	4.73	3.04	4.54
203	4.22	5.36	3.25	5.06
253	4.62	5.61	3.37	5.56

Uchiyama et al., 1997.

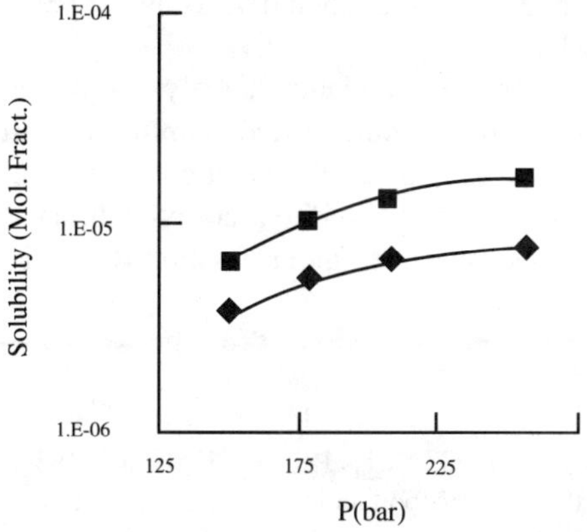

FIGURE 8.5 Solubility of propyl gallate in CO_2 40°C (♦) and 60°C (■) (Cortesi et al., 1997).

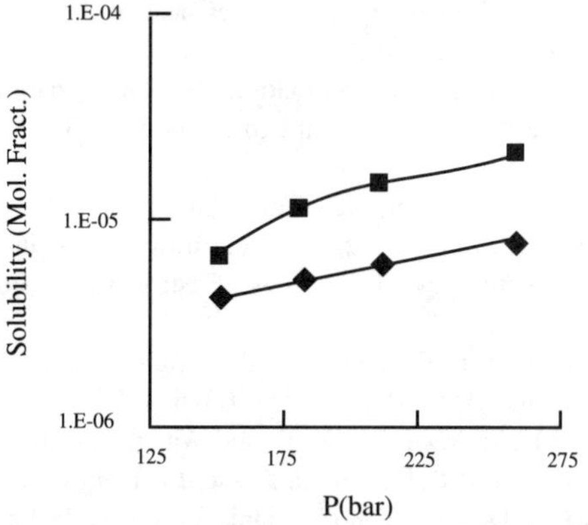

FIGURE 8.6 Solubility of dodecyl gallate in CO_2 at 40°C (♦) and 60°C (■) (Cortesi et al., 1997).

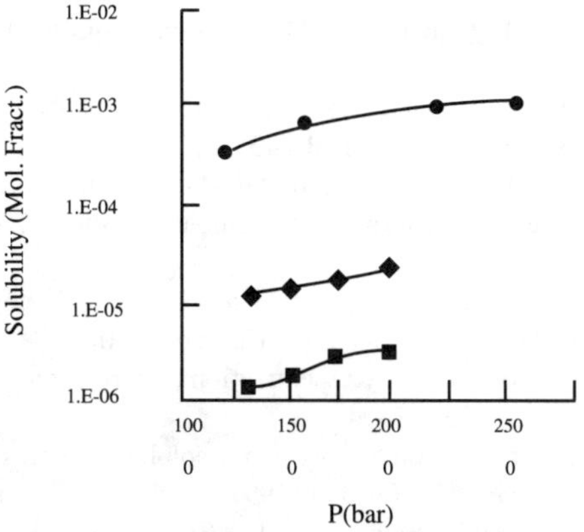

FIGURE 8.7 Solubility of ascorbic acid (♦), ascorbyl palmitate (■), and palmitic acid (●) in CO_2 at 40°C (Cortesi et al. 1997).

and 3-hydroxy flavone in SC CO_2 at 35 and 45°C and at pressures 91 to 253 bar are presented in Table 8.13.

The solubilities of other antioxidants, namely, propyl gallate, dodecyl (lauryl) gallate, ascorbic acid, ascorbyl palmitate, and palmitic acid, are presented in Figures 8.5 to 8.7 at pressures 130 to 250 bar at 40 and 60°C. It is observed that gallic acid esters have higher solubilities compared to ascorbyl palmitate in SC CO_2, and gallic acid has solubilities less than 10^{-6} in molar fraction (Cortesi et al., 1997).

REFERENCES

Aesbach, R. and Philippossion, G., Swiss Patent, 672,048,45, 1987.

Bermond, P., Biological effects of food antioxidants, in *Food Antioxidants*, Hudson, B. J. F., Ed., Elsevier Applied Science, U.K., chap. 6, 193, 1990.

Berner, D. L. and Jacobson, G. A., U.S. Patent, 3,732,111, 1973.

Booth, V. H., *Phytochemistry*, 2, 21, 1963.

Brieskorn, C. H., Fuchs, A., Bredenburg, J. B., McChesney, J. D., and Wankert, E., *J. Org. Chem.*, 29, 2293, 1964.

Brunner, G., Malchow, Th., Struken, K., and Gottaschu, Th., Separation of tocopherols from deodoriser condensate by countercurrent extraction with CO_2, *J. Supercritical Fluids*, 4, 72–80, 1991.

Brunner, G., *Gas Extraction*, Springer, New York, 290, 1994.

Calame, J. P. and Steiner, R., Supercritical extraction of flavors, in *Theory and Practice of Supercritical Fluid Technology*, Hirata, M. and Ishikawa, T., Eds., Tokyo Metropolitan Univ. Press, 301, 1987.

Chang, S. S., Ostric–Matijaseric, B., Hsieh, O. A. L., and Huang, C., Natural antioxidants from rosemary and sage, *J. Food Sci.*, 42, 1102–6, 1977.

Chen, Z. Y., Chen, P. T., Ma, H. M., Fung, K. P., and Wang, J., Antioxidative effect of ethanol tea extracts on oxidation of canola oil, *J. Am. Oil Chem. Soc.*, 73, No. 3, 375, 1996.

Chevolleau, S., Mallet, J. F., Debal, A., and Ucciani, E., Antioxidant activity of Mediterranean plant leaves: occurrence and antioxidative importance of α-tocopherol, *J. Am. Oil Chem. Soc.*, 70, No. 8, 807–809, 1993.

Chipault, J. R., Mizuno, G. R., Hawkins, J. M., and Lundberg, W. O., *Food Res.*, 17, 46, 1952.

Chipault, J. R., Mizuno, G. R., Hawkins, J. M., and Lundberg, W. O., *Food Technol.*, 10 (5), 209, 1956.

Cortesi, A., Allessi, P., Kikic, I., and Turtoi, G., Antioxidants solubility in SC CO_2, Proc. 4th Int. Symp. SCFs, Sendai, Japan, Vol. B., 435–437, 1997.

Esquivel, M. M., Ribeiro, M. A., and Bernardo-Gil, M. G., Supercritical extraction of savory oil: study of antioxidant activity and extract characterization, *J. Supercritical Fluids*, 14, 129–138, 1999.

Gopala Krishna, A. G., Prabhakar, J. V., and Aitzetmuller, K., Tocopherol content of some Indian pulses, *J. Am. Oil Chem. Soc.*, 74, 12, 1603–1606.

Goto, M., Sato, M., and Hirose, T., SC CO_2 extraction of carotenoids from carrots, Proc. Intl. Cong. Foods, No. 2, 835–837, 1994.

Johannsen, M. and Brunner, G., Solubilities of fat soluble vitamins A, D, E, and K in SC CO_2, *J. Chem. Eng. Data*, 42, 106–111, 1997.

Joshi, N. R., Supercritical Fluid Extraction of α-Tocopherol and β-Carotene from Plant Materials, M. Tech. dissertation, Indian Institute of Technology, Bombay, India, 1999.

Kimura, Y. and Kanamori, T., U.S. Patent, 4,380,506, 1983.

King, J. W., Favati, F., and Taylor, S. L., Production of tocopherol concentrates by supercritical fluid extraction and chromatography, *Sept. Sci. Tech.*, 31, 13, 1843–1857, 1994.

Kochhar, S. P. and Rossell, J. B., Detection, estimation and evaluation of antioxidants in food systems, in *Food Antioxidants*, Hudson, B. J. F., Ed., Elsevier Applied Science, England, 19, 1990.

Lee, H., Chung, B. H., and Park, Y. H., Concentration of tocopherol from soybean sludge supercritical carbon dioxide, *J. Am. Oil Chem. Soc.*, 68, 8, 571–573, 1991.

Masuda, T., Jitoe, A., and Mabry, T. J., Isolation of structure determination of cassumunarins, *J. Am. Oil Chem. Soc.*, 72, No. 9, 1053, 1995.

Murga, R., Beltran, S., and Cabezas, J. L., Study of the extraction of natural polyphenols from grape seed by using SC CO_2, Proc. 5th Meet. France, 2, 529, 1998.

Nakatani, N, Antioxidative and antimicrobial constituents of herbs and spices, in *Spices, Herbs and Edible Fungi*, Charalambous, G., Ed., Elsevier Science, Amsterdam, 251, 1994.

Nguyen, U., Evans, D. A., and Frakman, G., U.S. Patent, 5,017,397, 1991.

Nguyen, U., Evans, D. A., and Frakman, G., Natural antioxidants produced by supercritical extraction, in *Supercritical Fluid Processing of Food and Biomaterials*, Rizvi, S. H. H., Ed., Blackie Academic & Professional, An imprint of Chapman & Hall, chap. 8, 103, 1994.

Ohgaki, K. and Katayama, T., German Patent 3424 614, 1984; as reported by Brunner et al., in *J. Supercritical Fluids*, 4, 72–80, 1991.

Ohgaki, K., Tsukahera, I., Semba, K., and Katayama, T., A fundamental study of extraction with a supercritical fluid, solubilities of α-tocopherol, palmitic acid and tripalmitin in compressed CO_2 at 25°C and 40°C, *Int. Chem. Eng.*, 29, 302–308, 1989.

Ooi, K., Bhaskar, A., Yener, M. S., Tuan, D. Q., Hsu, J., and Rizvi, S. S. H., Continuous SC CO_2 processing of palm oil, *J. Am. Oil Chem. Soc.*, 73, No. 2, 233–237, 1996.

Osawa, T. and Namiki, M., Natural antioxidants isolated from eucalyptus leaf waxes, *J. Agric. Food Chem.*, 33, No. 5, 777, Sept./Oct., 1985.

Pongracz, G., Kracher, F., and Schuler, P., 1978–1987, as reported in Natural Antioxidants Exploited Commercially, by Schuler, P., in chap. 4, *Food Antioxidants*, Hudson, B. J. F., Elsevier Applied Science, 136, 1990.

Pratt, D. E. and Hudson, B. J. F., Natural antioxidant not exploited commercially, in *Food Antioxidants*, Hudson, B. J. F., Ed., chap. 5, 171, 1990.

Ribeiro, M. A., Lopes, I., Esquivel, M. M., Bernardo-Gil, M. G., and Empis, J. A., Comparison of antioxidant activity of plant extracts using SC CO_2 and classical methods, Proc. Meet. SCFs, France, 2, 669, 1998.

Saito, M. and Yamauchi, Y., Isolation of tocopherols from wheat germ oil by supercritical fluid chromatography, *J. Chromatogr.*, 257–271, 1990.

Sanderson, G. W., Ranadive, A. S., Eisenberg, L. S., Farrel, F. J., and Coggon, P., Contribution of polyphenolic compounds to the taste of tea, chap. 2 in *Phenolic, Sulfur and Nitrogen Compounds in Food Flavors*, Charalambous, G., Ed., Elsevier Science, B.V., Amsterdam, 1992.

Schuler, P., Natural antioxidants exploited commercially, in *Food Antioxidants*, Hudson, B. J. F., Ed., Elsevier Applied Science, England, chap. 4, 99, 1990.

Siebold, R. L., Cereal Grass, chap. 1, Wilderness Community Education Foundation, Lawrence, KS, 1990.

Spanos, G. A., Ghen, H., and Schwartz, S. J., SC CO_2 extraction of β-carotene from sweet potatoes, *J. Food Sci.*, 58, No. 4, 817–820, 1993.

Stahl, E., Quirin, K. W., and Gerard, D. C., *Dense Gases for Extraction and Refining*, Springer-Vorgler, Berlin, 1987.

Subra, P., Castellani, S., and Garrabos, Y., Supercritical CO_2 extraction of carotenoids from carrots, Proc. 3rd Intl. Symp. Supercritical Fluids, France, 2, 447, 1994.

Tateo, F. and Fellin, M., *Perfumer Flavorist*, 13, 48, 1988.

Todd, P. H., U.S. Patent, 4,877,635, 1989.

Tsimbidou, H. and Boskou, D., Antioxidant activity of essential oil from the plants of Lamiceae family, in *Spices, Herbs and Edible Fungi*, Charalambous, G., Ed., Elsevier, Amsterdam, 251, 1994.

Uchiyama, H., Mishima, K., Oka, S., Ezawa, M., Ide, M., Takai, T., and Park, P. W., Solubilities of flavone and 3-hydroxy flavone in SC CO_2, *J. Chem. Eng. Data*, 42, 570–573, 1997.

Viani, R., Process for Extracting Antioxidants, U.S. Patent, 4,012,531, 1977.

9 Natural Food Colors

9.1 CAROTENOIDS AS FOOD COLORS

Natural colors present in a wide variety of plant sources, such as roots, seeds, leaves, fruits, and flowers, are principally due to the occurrence of one or more of the groups of color compounds, such as carotenoids, betacyanins, anthocyanins, and other flavonoids. Increasing consumer demand for natural food colors has led not only to their recovery from natural sources but also to the synthesis of nature-identical carotenoids, important to the food industry. Natural colors and their formulations find wide-ranging applications from foods to pharmaceuticals and from dyes to cosmetics. Global consumption of natural food colors is estimated to be more than $1 billion (U.S.) (Lashkari, 1999).

A major part of natural food colors constitute the extracted carotenoids or the yellow and orange pigments which are widely distributed in plants and animals. Carotenoids are isoprenoid polyenes ($C_{40}H_{56}$) which are formed by C_5 isoprene units. So far more than 500 different molecules of carotenoids have been identified. Carotenoids are generally regarded as lipids and therefore are also known as lipochromes or chromolipids. They are soluble in lipids and in solvents which dissolve fat, viz., acetone, alcohol, diethyl ether, and chloroform. Some carotenoids are soluble in water. They are known as lychromes.

Hydrocarbon carotenoids are known as carotenes and are soluble in nonpolar solvents, such as petroleum ether and hexane. Other carotenoids are oxygenated derivatives of carotenes. These are xanthophylls. Xanthophylls are best soluble in polar solvents like ethanol and methanol. There are also acids and esters, e.g., carotenoid acids and esters, and xanthophyll esters, etc., as elaborated in Table 9.1.

Carotenoids are unstable pigments that are particularly sensitive to light, oxygen, and peroxide. Chlorophyll is mostly associated with carotenoids. There are three types of pigments belonging to the carotenoids: red pigment (capsanthins), yellow pigment (xanthophylls), and orange pigment (carotenes). Different natural carotenoids with wavelengths at maximum absorbence in saturated solutions and solvents are listed in Table 9.2.

Carotene ($C_{40}H_{56}$) was first isolated from carrot. There are three isomers, e.g. α-carotene (mp 187-187.5°C), β-carotene (mp 184.5°C), and γ-carotene (mp 176.5°C). β-carotene is precursor to vitamin A ($C_{20}H_{30}O$). The carotenoid content and carotenoid distribution in several fruits and vegetables are shown in Table 9.3.

9.2 RECOVERY OF CAROTENOIDS BY SC CO_2

In recent years there has been an upsurge in consumer demand for natural food colors from natural sources. This has led not only to large-scale commercial production of natural colors using conventional solvent extraction processes, but also to serious investigation into newer processes for improvement of recovery efficiency

TABLE 9.1
Classification of Carotenoids

On the Basis of Molecular Structures

Hydrocarbon
- Lycopene ($C_{40}H_{56}$)
- Rhodopurpurine ($C_{40}H_{56}$ or $C_{40}H_{58}$)
- α, β, γ Carotene ($C_{40}H_{56}$)
- Leprotene ($C_{40}H_{54}$)

Xanthophylls
- Containing 1–OH group
- Lycoxanthin ($C_{40}H_{56}O$)
- Kryptoxanthin ($C_{40}H_{56}O$)
- Rubixanthin ($C_{40}H_{56}O$)
- Gazaniaxanthin ($C_{40}H_{56-58}O$)
- Rhodopin ($C_{40}H_{56-58}O$)

- Containing 2–OH groups
- Lutein ($C_{40}H_{56}O_2$)
- Lycophyll ($C_{40}H_{56}O_2$)
- Zeaxanthin ($C_{40}H_{56}O_2$)
- Rhodoviolascin ($C_{42}H_{60}O_2$)
- Eschischolzanthin ($C_{40}H_{54+-2}O_2$)

- Containing 3–OH groups
- Flavoxanthin ($C_{40}H_{56}O_3$)
- Antheroxanthin ($C_{40}H_{56}O_3$)
- Petaloxanthin ($C_{40}H_{58(-2)}O_3$)
- Eloxanthin ($C_{40}H_{56}O_3$)

- Containing 4–OH groups
- Violaxanthin ($C_{40}H_{56}O_4$)
- Taraxanthin ($C_{40}H_{56}O_4$)

Ketones
- Myxoxanthin ($C_{40}H_{54}O$)
- Alphanin ($C_{40}H_{54}O$)
- Rhodoxanthin ($C_{40}H_{50}O_2$)
- Echinenone ($H_{40}H_{58(-2)}O$)

Hydroxyl carbonyl compounds
- Capsanthin ($C_{40}H_{58}O_3$)
- β-citraurin ($C_{30}H_{40}O_2$)
- Capsorubin ($C_{40}H_{60}O_4$)
- Astaxanthin ($C_{40}H_{52}O_4$)
- Fucoxanthin ($C_{40}H_{60}O_4$)

Carboxylic compounds
- Azafrin ($C_{27}H_{38}O_4$)
- Bixin ($C_{25}H_{30}O_4$)
- Crocetin ($C_{20}H_{24}O_4$)

TABLE 9.1 (continued)
Classification of Carotenoids

On the Basis of Property Variations

Property	Lychrome	Lipochrome
Solubility in water	Soluble	Insoluble
Color of solution	Yellow orange	Yellow red
Floroscence	Strong green	Weak yellow green
Acid medium	Stable	Very sensitive
Alkaline medium	Sensitive	Stable
Oxidizing medium	Very stable	Very sensitive
Biological activity	Vitamin B_2 and oxidation ferments	Vitamin A

Mayer, 1943.

TABLE 9.2
Different Carotenoids with λ_{max} in Saturated Solutions and Specific Extinctin Coefficients

Carotenoid	Solvent	λ (nm)	E_1 (%)
Astaxanthin	Acetone	473	1900
Capsanthin	Benzene	483	2072
Capsorubin	Benzene	489	2200
Lycopene	Benzene	487	8370
Lutein	Benzene	458	2236
α-Carotene	Petroleum ether	444	2800
β-Carotene	Cyclohexane	457	2505
α-Kryptoxanthin	Hexane	446	2636
Rubixanthin	Benzene	462	2909
Violaxanthin	Acetone	454	2240
Zeaxanthin	Acetone	452	2340

Eder, 1996.

and stability of colors, since these have been serious problems with natural colors. Due to impending regulatory constraints and process limitations in conventional solvent extraction processes, the feasibility of recovering natural pigments using supercritical and liquid CO_2 is being seriously explored all over the world.

There are a variety of natural food colors which can be extracted from botanicals, such as orange peel, marigold flowers, paprika, grass, annatto, turmeric, carrot, and seeds of grape, currant, tamarind, etc. It is not surprising that the solubilities of the hydrophobic substances, such as xanthophyll, carotenes, chlorophyll, and curcumin

TABLE 9.3
Carotenoid Content and Carotenoid Distribution in Fruits and Vegetables

	Fruits/Vegetables				
Carotenoids (%)	**"Valencia" Orange**	**Papaya**	**"Yolo Wonder A" Paprika**	**"B-6274" Carrot**	**"Red Chief" Tomato**
Phytofluene	4.2	0.1		—	8.4
Phytoene	2.0	0.1		—	19.7
α-Carotene	1.0	—		34.0	2.3
β-Carotene	2.0	29.6	15.4	55.3	
Kryptoxanthin	10.6	48.2	12.3	—	
Lutein	9.0	—		—	
Isolutein	2.0	—		—	
Violaxanthin	1.8	3.4	7.1	—	
Zeaxanthin	10.2	—	3.1	—	
Luteoxanthin	11.5	—	—	—	
Neoxanthin	2.5	0.2	2.0	—	
Lycopene	—	—	—	—	66.7
Capsanthin	—	—	33.3	—	
Capsorubin	—	—	10.3	—	
Total carotenoids content (mg/kg)	12	13.8	1755.9	99	950

Eder, 1996.

in liquid CO_2 (–20 to 20°C) and SC CO_2 at less than 300 bar, are quite low in view of their high molecular weights, in spite of their low polarity. However solubilities can be improved by adding a food-grade entrainer, e.g., ethanol or ethyl acetate. Table 9.4 gives approximate values of the solubilities of color compounds in liquid CO_2 with and without different polar entrainers. Components may be dissolved differently in the presence of natural oil, which may act as entrainers. It can be noted that the carotenoids and other oil-soluble materials are only slightly soluble in liquid CO_2 at near its vapor pressure at 20°C, but the solubilities of all components increase significantly in the presence of an entrainer. Ethanol is a better entrainer than acetic acid for most of the oil-soluble components. However, more polar, water-soluble color compounds are not soluble at all, as expected. In view of the low solubilities of color compounds in liquid CO_2, even with an entrainer, it is clear that the extraction of food colors from vegetable matter with liquid CO_2 is not viable at low pressures.

Figure 9.1 indicates the pressure dependencies of solubilities of different carotenoids in SC CO_2 at 15, 35, and 55°C. The solubilities of different carotenoids, as were measured by spectrophotometric method, were found to increase by more than two orders of magnitude at high pressures (≥500 bar) and temperatures (≥55°C) (Jay et al., 1991). The solubility of β-carotene was found to be 1.7×10^{-3}% by weight at 500 bar and 55°C, which was an order of magnitude less than that reported by Johannsen and Brunner (1997) by quasistatic measurement with an on-line analysis system to which supercritical fluid chromatography was coupled. Johannsen and

TABLE 9.4
Approximate Solubilities, (% w/v) × 10^4 of Pure Constituents of Food Coloring Materials in Liquid CO_2 at 60 bar, 20°C

Color Compounds	CO_2	CO_2 + 5% ethanol	CO_2 + 5% acetone	CO_2 + 1% acetic acid
Carotenoids				
β-apo-8-Carotenal	2	20	20	5
Bixin (95%)	0.2	3	3	2
Canthaxanthin	0	1	2	0.5
Capsanthin	1	5	2	2
β-Carotene	1	2	2	1
Lutein	0.1	2	1	2
Others				
Anthocyanins (3% aq.)	0	0	0	0
Betalaines (0.5% aq.)	0	0	0	0
Chlorophyll (10% in oil)	4	10	30	10
Copper chlorophyllins	0	30	0	10
Curcumin (96%)	0.1	10	50	10
Indigo (crystals)	0	0	0	0
Riboflavin (95%)	0	0.1	—	1

Jay et al., 1991.

Brunner measured the solubility of β-carotene over a pressure range of 200 to 350 bar at three temperatures. It was found to increase from 0.019 to 0.034% at 40°C on increasing pressure from 300 to 350 bar and from 0.264 to 0.395% by weight at 80°C on increasing pressure from 300 to 320 bar only. The discrepencies in the reported values may be attributed to the uncertainties in sampling or light sensitivity of β-carotene. Jay et al. (1991), however, found the solubility of β-carotene to increase by six times with 6% ethanol as the entrainer in SC CO_2, at 500 bar and 55°C. The solubility of β-carotene in liquid CO_2 was earlier found by Hyatt (1984) to be in the range of 0.01 to 0.05% by weight.

Though the solubilities further increase with an entrainer above 500 bar and 80°C, the direct extraction of food colors using SC CO_2 is presumably economically viable, provided another product can be extracted at a lower pressure from the same natural source, for example, supercritical extraction and simultaneous fractionation of essential oil, oleoresin, and pigment from red pepper or marigold flowers.

In general, sequential extraction and selective fractionation of essential oil, oleoresins, and certain colors are possible with the progress of the extraction by suitably changing the SC CO_2 extraction condition after each fraction is exhausted. For complete extraction and separation of colors, a scheme of stagewise pretreatment and extraction sequence can be followed in a single supercritical CO_2 extraction plant to get a solvent-free natural color concentrate of the desired specification. The following subsections will deal with various sources of natural food colors and how they can be recovered by SC CO_2 extraction.

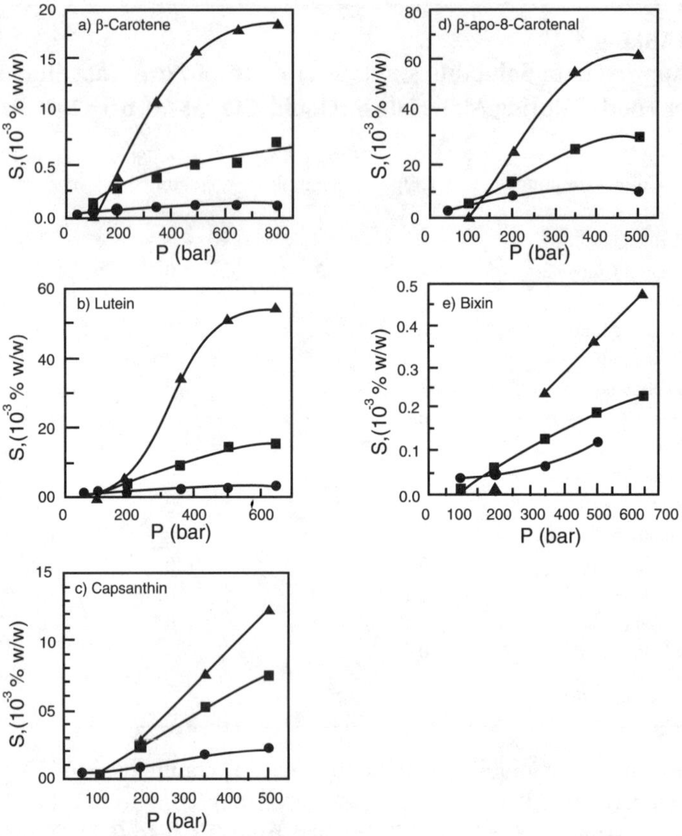

FIGURE 9.1a–e Solubility (S) of carotenoids vs. pressure (P) in SC CO_2; at 15°C (●), 35°C (■), and 55°C (▲) (Jay et al., 1991); (a) β-carotene; (b) lutein; (c) capsanthin; (d) β-apo-8-carotenal; (e) Bixin.

9.2.1 GRASS

Grass has been providing food for animals from prehistoric times. It can grow easily on a wide variety of soils and climatic conditions. Importance of chlorophyll as a therapeutic agent has been investigated in depth, which reveals that this green plant substance has many benefits for humans. Hence there is a recent interest in cereal grass cultivation for humans. The principal cultivated grasses are from wheat, rice, corn, barley, oats, rye, and millet, out of which wheat grass is the most common. Wheat grass contains 0.543% dry weight of chlorophyll and is considered to be a good source of chlorophyll like ordinary grass. Chlorophyll is an oil-soluble color and is mostly used in cosmetics and toiletries, and very little in foods, due to its poor stability. Chlorophyll is not permitted as a food additive in the U.S., though it is used in the U.K. in small quantities.

Supercritical CO_2 extraction of dried grass was carried out by Jay et al. (1988) at pressures in the range of 300-500 bar and temperatures in the range of 50 to 60°C in

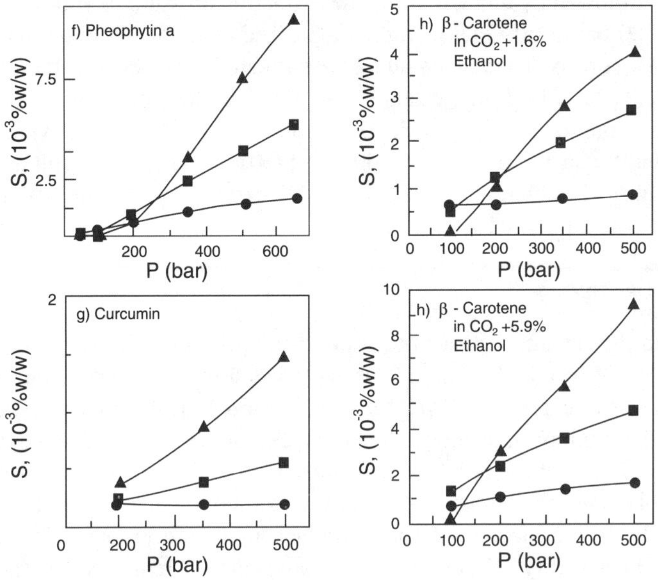

FIGURE 9.1f–i Solubility (S) of carotenoids vs. pressure (P) in SC CO_2; at 15°C (●), 35°C (■), and 55°C (▲) (Jay et al., 1991); (f): pheophytin a; (g) curcumin; (h) β-carotene in CO_2 + 1.6% ethanol; (i) β-carotene in CO_2 + 5.9% ethanol.

a sequential manner. After the recovery of essential oils and β-carotene in the first extract at the beginning of extraction at 35°C and a lower pressure of 100 bar, water was subsequently added as an entrainer to SC CO_2. The second phase of extraction was carried out at a higher temperature (50 to 60°C) and at a higher pressure (300 to 500 bar) to facilitate extraction of lutein diesters and chlorophyll. In the final stage of extraction, the remaining lutein diesters and some dark green solid matter (chlorophyll) were collected. At 500 bar and 60°C in the presence of water, the value of pH became 3.0 which probably converted chlorophylls to pheophytins. Pheophytins are green and luteins are yellow in color. The total yield was 1.56% by weight (Jay et al., 1988). The solubility of chlorophyll-A was, however, found to be quite low, for example, 10^{-3}% in SC CO_2 at 200 bar and 35°C in the presence of 0.5% hexane (Subra and Tufeu, 1990).

9.2.2 Orange Peel

Orange peel is a good source of a natural orange coloring agent which contains a large number of xanthophylls besides α- and β-carotenes (Table 9.3). Orange peel also contains higher levels of many constituents than the corresponding juice, such as cinnamic acids, which may be responsible for off flavors being developed on storage. As the coloring substance should be free from undesirable flavors, it is preferable to use supercritical CO_2 extraction for selective recovery of better quality xanthophylls from orange peel. SC CO_2 fractional extraction of orange peel was also carried out in stages starting with 100 bar pressure at 35°C in the beginning of the extraction, followed by an increase in the extraction pressure to 300 bar at 45°C and finally to 500 bar at 60°C (Jay et al., 1988). The outer rind of the orange contains 2% terpenes, mostly d-limonene and 45% carotenoids (violaxanthin). The pressure is sequentially increased first to remove oil and then the more "difficult to dissolve" carotenoids from the ground powder of dried orange peel. The total yield was about 6% w/w and the last two fractions contained 0.01% carotenoids with a recovery of 1.6 to 3.0 μg/g. The orange colored oil soluble extract is predominantly used in soft drinks and confectioneries.

9.2.3 Turmeric

There is an increasing interest in the application of curcumin as a natural coloring agent in food, drugs, and cosmetics. It is extracted from turmeric (*Curcuma longa*) which consists of curcumin ($C_{21}H_{20}O_6$, mol wt: 368.4) as the main constituent along with two curcuminoids, namely, dimethoxycurcumin and bisdimethoxycurcumin, amounting to 70% of the total curcuminoids. Curcumin and curcuminoids are sparingly soluble in water. Curcumin is dissolved in water by increasing pH or by adding an emulsifier, since most food and pharmaceuticals are prepared in water as the medium. The stability of curcumin also poses a problem, because it undergoes degradation or dissociation on storage under alkaline conditions and with exposure to light. The color of curcumin depends on its medium, because its intensity is different when an organic solvent is used as the medium compared to that for an aqueous medium. In acidic medium (pH 1 to 7) the color is yellow, but it changes to brownish or deep red by changing the pH in the range of 7 to 9.

SC CO_2 extraction of fresh, chopped turmeric rhizome was carried out with pure SC CO_2 at 500 bar and 80°C which gave a yield of 0.35% w/w (Jay et al., 1988). However, the addition of ethanol as an entrainer to SC CO_2 results in at least an order of magnitude higher yield. Pure (95%) curcumin is crystallized from the extracted oleoresin, which normally contains 37 to 55% curcumin. The yellow and orange commercial product of turmeric (*Curcuma longa*) contains 4 to 10% curcumin which is dissolved in a good grade solvent with an emulsifier.

9.2.4 Paprika

Paprika spice (*Capsicum annum*) or sweet red peppers is characterized by a deep red color (due to polar ketocarotenoids) and a minimum pungency (burning sensation). There are two varieties of paprika-type chilies produced in India, the color value of which varies from 125 to 175 ASTA. Since paprika is primarily utilized for its red color, its quality is decided by its carotenoid pigment content. The highest concentration of carotenoid pigments is present in the pericarp of the fruit, hence the spice which contains only pericarp will have the highest pigment content. Paprika oleoresin is a dark red colored oil obtained by extraction from the dry fruit (without pungency). Paprika pigment contains primarily mono- and difatty acid esters of carotenoid alcohols and diols. Six diesters present are zeaxanthin, antheroxanthin, capsanthin, capsorubin, violaxanthin, and capsorubin-capsanthin epoxide. Three monoesters are kryptocapsin, kryptoxanthin, and capsolutein. Normally paprika pigment is a mixture of red, yellow,

TABLE 9.5
Constituents in Paprika Oleoresin Pigment

Keto Carotenoids (red)	Xanthophylls (Yellow)	Carotene (Orange)
Capsanthin + Capsorubin (40–60%)	Zeaxanthin (7–9%)	β-Carotene (8–12%)
	Violaxanthin (4–6%)	
Bixin (2–4%)	β-Kryptoxanthin (6–8%)	
	Antheroxanthin	

and orange colors having the compositions as given in Table 9.5 and it is desirable to separate red pigments from yellow xanthophylls and orange carotenes.

When paprika red color is used as a food additive, it must be free from aroma. Conventional solvent extraction process by means of n-hexane, dichloromethane, dichloroethane, etc., is used to recover oleoresins. This process may lead to oxidation of aroma and coloring compounds, especially in the presence of air. Aroma and color compounds are next separated from paprika oleoresin by molecular distillation under high vacuum. However, there may be a thermal degradation due to high temperature involved. SC CO_2, on the other hand, overcomes the disadvantages of conventional solvent for extraction of paprika. Thus for production of natural red color from paprika, a two-step process is involved, namely separation of color substances from aroma by SC CO_2 extraction and fractionation and then separation of the red color from the yellow by preparative supercritical fluid chromatography. Alternatively, the nonpolar yellow pigments can be first extracted along with essential oils using SC CO_2 at a relatively lower pressure and a lower temperature condition. Subsequently, an entrainer mixed SC CO_2 at a higher pressure and a higher temperature condition may be used to recover the polar red pigments in the second stage as in the case of SC CO_2 extraction from dried grass as discussed earlier.

For the extraction of paprika by SC CO_2, the extraction pressure was varied between 90 and 400 bar, the extraction temperature between 20 and 60°C, and solvent ratios between 20 and 200 kg of CO_2/h/kg of dry solid (Knez et al., 1991). The highest color concentration is obtained when the extraction is carried out at both high temperature and high pressure. Recently Nguyen et al. (1998) showed that SC CO_2 extraction followed by fractionation into two products could lead to a color value of 7200 ASTA for the heavy fraction of the paprika extract.

9.2.5 Red Chili

When red chili is extracted with SC CO_2 and subsequently fractionated, both essential oil and pungent principles (i.e., compounds responsible for hotness) are collected in the light fraction and the lipids and color compounds are collected in the heavy fraction. The red chili extract thus constitutes two fractions, namely, (1) the hot fraction (pungent or burning principle), which amounts to 90% of the extract and is enriched with two main active ingredients capsaicin and dihydrocapsicin, and (2) the

TABLE 9.6
Compositions of Red Pepper Oleoresin Fraction

Component	Raw Spice (%)	Commercial Oleoresin (%)	Heavy Fraction (%)	Light Fraction (%)
Capsaicin	0.21	1.83	0.57	8.10
Dihydro-capsaicin	0.14	1.52	0.31	4.05
Nordihydro-capsaicin	0.04	0.58	0.07	1.35
Total capsaicin	0.39	3.93	0.95	13.5

Nguyen et al., 1998.

colored carotenoid fraction having very little hotness, as decided by the capsaicin concentration. However, the recent trend is to follow a three-stage extraction scheme for sequential recovery of essential oil at 100 bar, then pungent principles of oleoresin along with nonpolar (yellow/orange) pigments at an intermediate pressure, and finally the polar red colored fraction at 500 bar, i.e., in three subsequent stages, while increasing pressure fivefold from 100 to 500 bar.

Recently Nguyen et al. (1998) carried out SC CO_2 extraction of red pepper at 500 bar or higher pressures and at temperatures in the range of 80 to 100°C, with subsequent fractionation into light and heavy fractions and the compositions of the red pepper fractions are given in Table 9.6. It could be seen that most of the capsaicins could be collected in the light fraction and the ratio of the capsaicin concentration in the heavy to the light fraction could be optimized in order that the heavy fraction is used for color with very little hotness. Total yield of extract was 10% and only 3.93% of this extract was total capsaicin.

9.2.6 CARROT

Carotenoids are one of the major natural colors. Carotenoids, particularly β-carotene, are important for the food and pharmaceutical industries, not only as an antioxidant and a precursor to vitamin A, but also as a coloring material. Carotenoids can be extracted from raw or freeze-dried carrots using SC CO_2 with ethanol as the entrainer. The effects of pressure and entrainer on the extraction yields are shown in Figure 9.2 (Goto et al., 1994) for both freeze-dried and raw carrot containing 7 and 20 g/kg (dry weight) of carotenoid, respectively. For extraction from raw carrot, more carotenoids were extracted and an increase in pressure and the addition of entrainer enhanced the extraction rate and total yield. On the other hand, only 70% of the carotenoids could be extracted from the freeze-dried carrot due to mass transfer hindrance from the cell membranes, which was unaffected by the addition of entrainer.

9.2.7 MARIGOLD FLOWERS

Marigold (*Calendula officinalis*) flowers contain flavonoids, carotenes, and luteins. Marigold extracts have lately found applications in being blended with chicken feed

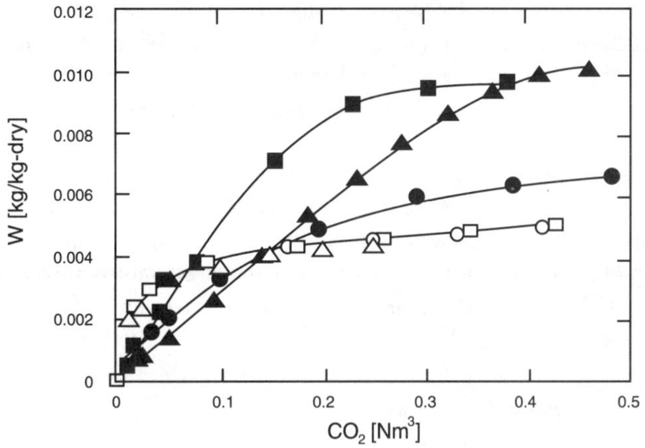

FIGURE 9.2 Effect of pressure and entrainer on the cumulative extraction from raw and freeze-dried carrots (Dp < 0.1mm); raw carrots: (●) 40°C, 147 bar; (▲) 40°C, 294 bar, (■) 40°C, 147 bar, 3wt% EtOH; freeze-dried carrots: (○) 40°C, 147 bar; (Δ) 40°C, 294 bar, (□) 40°C, 147 bar, 3 wt% EtOH (Goto et al., 1994).

to induce attractive yellow color to the poultry eggs. Marigold extract is also used as natural color in different food applications. SC CO_2 extraction of color compounds from marigold flower requires relatively high pressures in the range of 300 to 450 bar and 60 to 80°C from ground pretreated dried flower. The yield of essential oil is quite low at such high pressures, compared to color and oleoresin. However, extraction followed by fractionation yields both products simultaneously. Addition of ethanol to SC CO_2 allows a faster rate of extraction and a higher recovery. Yields range from 5.3 to 11%, depending on the extraction pressure and the amount of entrainer in SC CO_2. The yield by alcohol extraction is much more than that by SC CO_2 extraction, as in the latter case, selective separation of more desirable components may be achieved. A scheme of simultaneous extraction and fractionation, followed by the addition of ethanol to the substrate may be devised for the recovery of marigold color, oleoresin, and essential oil in three different fractions.

9.2.8 Annatto

Annatto is a vegetable yellow-orange color recovered from the thinly deposited substance around the seeds of the plant *Bixa orellana*. The main carotenoid present in this pigment is bixin, which constitutes 70 to 80% of the total pigment mass. Annatto is mainly used in dairy products, such as cheese, butter, margarine, etc. or in the cosmetics industry, leather, furniture, shoe polishes, etc. Annatto color is marketed as (Choudhury, 1996) one of the following products:

1. Suspension or solution in vegetable oil having 0.53 or 1.06% strength. Super strong solution contains 3% or more bixin.
2. Solution in water of similar strengths as above.
3. Water soluble annatto dry powder.

4. Bixin crystals, powder having 50 to 99% bixin.
5. Annatto mix: the red pigment from annatto can be mixed with curcumin, the mixture is suspended in oil to produce a color mix. This color mix can be used in edible oils, butter, salad oil, etc.

SC CO_2 extraction of bixin from annato seeds was performed by Jay et al. (1988) at pressures in the range of 300 to 500 bar and at temperatures of 35 to 55°C. The solubility of bixin in SC CO_2 was found to be two times more than that of β-carotene at 500 bar and 55°C, which increases further when ethanol is mixed with SC CO_2 solvent (Jay et al., 1991). Hence bixin-enriched product can be obtained by SC CO_2 extraction from annato powder.

9.2.9 Other Natural Colors

Supercritical CO_2 at high densities can co-extract minute quantities of chlorophyll from green colored seeds, e.g., fennel and parsley, along with lycopene and pheopytin, rendering a light yellow color to the extracted oil, as can be seen from Table 9.7.

Alfalfa leaves extract from lucerne grass contains yellow carotenoids such as 40% lutein, 34% violaxanthin, 2% zeaxanthin, 19% neoxanthin, and 4% kryptoxanthin besides β-carotene. This extract could be green if saponification is not done. Tomato extracts contain lycopene as the main carotenoid pigment (to the extent of about 80 to 90% of the total carotenoids). The SC CO_2 solvent at high densities and modified polarity with an entrainer can extract these natural colors, since these compounds have reasonable solubilities, as mentioned earlier.

Carthamus tinctorius contains two coloring matters, yellow and red, the latter being most valued for dyeing silk. It contains kaemferol and quercetin, which are flavonoids and are soluble in SC CO_2 solvent.

The flowers of saffron (*Crocus sativus*) have their characteristic yellowish orange pigment owing to the presence of crocin, a digentiobioside of crocetin. The original pigment crocin is a glycoside and is hydrolyzed followed by extraction with ethanol and acetone in the conventional method. However, SC CO_2 extraction can produce a better and stable extract when it is mixed with ethanol as entrainer.

Black tea has theaflavins and thearubigins which are formed by oxidative polymerization assisted by polyphenol oxidase (PPO) enzyme. These compounds are responsible for the orange and red pigments of tea, which can be extracted by SC CO_2 and spray dried to form a powder.

Echoides root or *Arnebia nobilis* (root based) gives purple-blue natural pigment. Red sandalwood-based amaranth (*Pterocarpus santaninus*), *Rheum emodi*, and *Rubia cordifolia* root give natural red pigments. Further work is needed to establish the recovery efficiency of these natural food colors using SC CO_2 extraction.

9.3 ANTHOCYANINS AS FOOD COLORS

Most of spectacular natural colors are attributed to anthocyanins. Anthocyanins are among the most important groups of plant pigments. The term anthocyanin, though scientifically designates the blue pigments of flowers, also signifies purple, violet,

TABLE 9.7
Coloring Compounds of Different Fatty Oils Obtained by Supercritical CO_2 Extraction at 300 bar, 35°C

	Fennel	Parsley
Pheophytin (mg/g)	69–1026	32–811.2
Tocopherol (mg/g)	13.0–15.2	10–12.3
Umbellipheron (mg/g)	0.09	0.13
Chlorophyll (mg/g)	10–26	12–17

Then et al., 1998.

magenta, and nearly all red hues occurring in flowers, fruits, leaves, stems, and roots. However, there are two exceptions: tomatoes owe their red color to lycopene and red beets to betanin, which are pigments not belonging to the anthocyanin group.

Anthocyanins have come into prominence due to the opposition to and official delisting of artificial food dyes, mainly those of the azo type. As a result, anthocyanins are highly desirable substitutes for synthetic food colors and are being considered as replacements for the banned dyes. Anthocyanins are water soluble, which simplifies their incorporation into aqueous food systems. But they also have disadvantages in that their tinctorial power and stability in foods are generally low as compared to the azo dyes. The stability of anthocyanins depends on several factors, such as concentration, temperature, pH, and their environment.

9.3.1 Classification of Anthocyanins

A red anthocyanin could be isolated and crystallized from red cornflower petals by Saito et al. (1964), using 70% methanol for extraction, followed by mild treatment. The pigment was presumed to be pelargonin-OH. Another red pigment extracted with neutral solvents from red rose petals by Saito et al. (1964) was understood to be cyanin-OH. From pansy petals, the violet colored anhydrobase of delphinidin 3-p-coumaroyl rutinoside 5 glucoside, was crystallized, which is named violanin. Patyconin is another plant pigment obtained as blue crystals from the purplish-blue petals of Chinese bell flowers. Cinerarin is another blue pigment isolated from the blue petals of cineraria (Yoshitama and Hayashi, 1974 and Yoshitama et al., 1975). Both of these blue pigments are very stable in neutral solutions. From red petals of cineraria, a red pigment, rubrocinerarin, can be isolated which is stable like cinerarin.

A new anthocyanin called peonidin was isolated by Asen et al. (1977) from morning glory flowers (*Ipomea tricolor*) and is known as "Heavenly Blue." This pigment is very stable in the pH range of 2 to 8 and Asen et al. (1979) were granted a patent for the possible use of this pigment as a food colorant (U.S. Patent 4,172,902). Commelinin is a blue anthocyanin metal complex extracted from the petals of *Commelina communis*, which was found to contain magnesium. Protocyanin is also a blue-colored complex of cyanin, polysaccharide, Fe, and Al, extracted from the petals of *Centaurea cyanus*. From the same plant, a blue pigment that contained cyanin, bisflavone, and Fe, was extracted and was named cyanocentaurin (Markakis, 1981).

The color of isolated anthocyanins can be altered by the presence of other substances by the process of what is known as copigmentation. Copigments do not have any or have very little visible color as such, but when they are added to an anthocyanin solution, they may greatly enhance the color of the solution or change it towards higher wavelengths for maximum absorption. Various groups of compounds including flavonoids, chlorogenic acid, and polyphenols related to gallic acid, caffeine, and theophylene may act as copigments.

9.4 RECOVERY OF ANTHOCYANINS

Besides flowers and leaves, anthocyanins are naturally present in fruits and vegetables. For example, anthocyanins are important constituents of grapes and wines. They are responsible for the red and white varieties of grapes and wines. High-performance liquid chromatography (HPLC) or SC fluid chromatography can be used for the separation of anthocyanins and other phenolic compounds (e.g., tannins) extracted from fruit and vegetable materials. Anthocyanins are, compared to tannins, more easily extractable from vegetable tissue, but on the other hand, they are more sensitive to oxidation.

The extent of the processing of raw plant materials depends on the degree of purity of the desired anthocyanin product. Sometimes anthocyanin-colored fruit juices may be used as carriers of these pigments or the purification may be carried out to obtain crystalline anthocyanins. Extraction of the pigment from the raw material constitutes the major purification step. Anthocyanins are water-soluble pigments. Conventionally water, water-containing SO_2, or acidified alcohols are used for extraction. SO_2 enhances the extraction of color severalfold (creating a purer extract) over plain water, and so it is dissolved in water or ethanol to the extent of 200 to 2000 ppm (SO_2 in water) or 200 to 3000 ppm (SO_2 in alcohol), respectively. Aqueous SO_2 extract is more stable as a soft drink colorant than the water-extracted colorant (Markakis, 1981). The anthocyanins along with flavonoids can be eluted with alcohol containing 1% HCl or less, and the eluate is then freeze dried. The extracts obtained by the above-mentioned methods are usually concentrated, minimizing exposure to oxygen and high temperatures. Accordingly, SC CO_2 mixed with an entrainer (water or alcohol) has great potential in the separation and purification steps, obviating both of these problems. SC CO_2 extraction followed by SC fluid chromatography can be employed for concentrating the extract and for isolating a purer product of anthocyanins which can be used in the formulation of a marketable liquid, paste, or spray-dried food colorant powder.

9.5 COMMERCIAL ANTHOCYANIN-BASED FOOD COLORS

The oldest, more than a century old, commercial red color, based on anthocyanin, was probably produced as a byproduct of the red grape processing industry. This is not surprising since the red variety of grapes are produced, even today, in large quantities, perhaps as one fourth of the total fruit production in the world. Towards the beginning, this deep red colorant used to be utilized for enhancing the intensity of the color of red wines. However, now it has found applications as a common food colorant. Aqueous sulfur dioxide solution is normally used as the extractant. Seeds of several fruits such as berry, currant, and tamarind have bright colors, which can be extracted with alcohol, chloroform, or SC CO_2 mixed with these polar solvents as entrainers.

The anthocyanin pigments present in the rind portions of many fruits are commercially attractive. For example, fruit of kokam (*Garcinia indica*) has a dense thick purple-brown rind which contains up to 1.5% coloring matter comprising anthocyanins and tannins. The anthocyanins consist of two red pigments distinctly separable with different concentrations and intensities, as cyanidin-3-glucoside and cyanidin-3-sophoroside, of which the latter is present in major quantity.

The color of the leaves and fruit skins of cherry-plum (*Prunus ceracifera*) is presently extracted with acidified ethanol. The extract is a good food colorant and is also found to inhibit mutagenesis. The red colored flower *Hibiscus rosa-sinensis* is found in many backyard flower gardens. The flower petals contain cyanidin diglycoside and delphinidin mono- and biosides which can be extracted with hot water. The alcoholic extract of the flower is used in India and China as a natural stable black hair dye and also as an oral contraceptive.

Even the anthocyanins present in the skin of olives are extracted and used as a food colorant. There are a host of flowers, fruits, seed coats, and leaves which can be extracted using a modified SC extraction process along with an aqueous or alcohol SO_2 solution; the stability of the anthocyanins in such extracts is very good over a long period of time.

9.6 BETACYANINS

Beet root has been known to mankind for hundreds of years for its intense red color and high sucrose content. It is an excellent source of natural red color. The pigments present in beet root are collectively termed as betalains, which can be classified under red betacyanins and yellow betaxanthins. Beet root contains up to 0.2% by weight of betacyanins and may yield 2% solid extract. Of the total color, 75-90% is due to the specific pigment called betanin ($C_{24}H_{26}N_2O_3$, mol wt = 550.5), which is the major single compound among the betacyanins, and vulgaxanthin I and II are among the betaxanthins. The concentrated beet root juice containing 0.5 to 1.0% betanin is commercially used as a food additive. The juice is spray-dried to obtain a powder, but for drying, it is required to add maltodextrin to the juice in view of the high (70%) sucrose content of the concentrated juice (Henry, 1992). Betanin has such an intense color that the dosage level of betanin for addition to food product is kept very low at 5 to 20 ppm. However, due to high sensitivity for heat, light, oxygen, SO_2, and water activity, betanin has restricted applications as a red colorant. No studies have been reported on SC CO_2 processing in the production of betanin color, since betanin is water soluble and can be easily obtained by pressing or diffusion techniques followed by centrifugation, pasteurization, concentration, and spray-drying of the juice.

REFERENCES

Asen, S., Stewart, R. N., and Norris, K. N., *Phytochemistry*, 16, 1118, 1977.

Choudhury, R. D., Ed., *Herbal Drugs Industry*, Eastern Publishers, New Delhi, India, 1996.

Eder, R., Pigments, in *Handbook of Food Analysis, Natural Colors for Foods and Other Purposes*, Nollet, L. M. L., Ed., Marcel Dekker, New York, 1, 940–942, 1996.

Goto, M., Sato, M., and Hirose, T., Supercritical CO_2 extraction of carotenoids from carrots, in *Developments in Food Engineering*, Part 2, Yano, T., Matsuno, R., and Nakamure, K., Eds., Blackie Academic & Professional, Glasgow, 835, 1994.

Hyatt, J. A., *J. Org. Chem.*, 49, 5097, 1984.

Henry, B. S., Natural food colors, in *Natural Food Colorants*, Hendry, G. A. F. and Houghton, J. D., Eds., Blackie & Sons Ltd., Glasgow, chap. 2, 39, 1992.

Jay, A. J., Smith, T. W., and Richmond, P., The extraction of food colors using supercritical CO_2, Proc. Intl. Symp. Supercritical Fluids, Nice, France, 2, 821, 1988.

Jay, A. J., Steytler, D. C., and Knights, M., Spectrophotometric studies of food colors in near critical CO_2, *J. Supercritical Fluids*, 4, 131–141, 1991.

Johannsen, M. and Brunner, G., Solubilities of fat soluble vitamins A, D, E, and K in SC CO_2, *J. Chem. Eng. Data*, 42, 106–111, 1997.

Knez, Z., Posel, F., Hunek, J., and Golal, J., Proc. 2nd Symp. Supercritical Fluids, Boston, MA, 101, 1991.

Lashkari, Z., Ed., A story of resurgence of natural products, in *Finechem from Natural Products*, 1, No. 4, 1, Sept., 1999.

Mayer, F., The constituents, properties and biological relations of the important natural pigments, in *The Chemistry of Natural Coloring Matters*, Revised and Transl. by A. H. Cook, Reinhold Publishing, New York, chap. 1, 18–19, 1943.

Nguyen, V., Anstee, M., and Evans, D. A., Extraction and fractionation of spices using supercritical fluid carbon dioxide, in Proc. 5th Meet. Supercritical Fluids, France, 2, 523, 1998.

Markakis, P., Anthocyanins and food additives, in *Anthocyanins as Food Colors*, Markakis, P., Ed., Academic Press, NY, chap. 9, 245, 1981.

Saito, N., Hirata, K., Hotta, R., and Hayashi, K., *Proc. Jpn. Acad.*, 40, 516, 1964.

Subra, P. and Tufeu, R., *J. Supercritical Fluids*, No. 3, 20, 1990.

Then, M., Daood, H., Illes, V., Simandi, B., Szentmihalyi, K., Pernetezki, S., and Bertalan, L., Investigation in biological active compounds in plant oils extracted by different extraction methods, Proc. 5th Meet. SCFs, held in Tome, 2, 555, 1998.

Yoshitama, K. and Hayashi, K., *Bot. Mag.*, 87, 33, 1974.

Yoshitama, K., Hyashi, K., Abe, K., and Kakisawa, H., *Bot. Mag.*, 88, 213, 1975.

10 Plant and Animal Lipids

10.1 IMPORTANCE OF RECOVERY

Lipids are a large group of fatty organic compounds present in living organisms. Lipids form an important food and energy source in plant and animal cells. These include animal fats, fish oils, vegetable seed oils, natural waxes, and natural oils from sea weeds, squid organs, and fungi cells. Plant lipids comprise a complex mixture of monoglycerides (MG), diglycerides (DG), triglycerides (TG), and free fatty acids (FFA) associated with some minor constituents, such as squalane, tocopherols, sterols, phosphatides (gums), alkaloids, flavonoids, waxy materials, color compounds, and volatiles that provide the taste and odor of the oils. The oils are normally refined prior to making them suitable for human consumption and for enhancing their nutritional and market value. The conventional refining of vegetable oils involves a number of sequential processes, namely, degumming, deacidification, deodorization, clarification, and stabilization. These processes invariably require strong chemicals and harsh operating conditions, causing decomposition and thermal degradation of valuable chemicals and oils. These losses have significant socioeconomic implications. As a result, there have been continuous efforts all over the world to develop newer processes for production of better-quality edible oils in higher quantity and to establish a variety in edible oils, with simultaneous value addition by recovering valuable nutrients, oleochemicals, and pharmaceuticals, in order to compensate for the overall cost of vegetable oil processing.

Vegetable oils are essentially triglycerides or triesters of glycerol and monocarboxylic fatty acids of varying chain lengths. Nomenclature and distribution of fatty acids in various vegetable oils are given in Appendix B. In addition to edible oils, the plant and animal lipids find several important applications in foods, pharmaceuticals, and cosmetics.

For example, polyunsaturated fatty acids (PUFA) are important for their medicinal and therapeutic values and for their role as precursor to prostaglandins. While unsaturated fatty acids like linolenic acid (18:3), linoleic acid (18:2), and oleic acid (18:1) are in great demand, saturated fatty acids such as stearic (18:0) and palmitic acids (16:0) are also of immense health benefit. While the pharmaceutical industry requires the polyunsaturated fatty acids in a high degree of purity, the food industry requires triglycerides with certain degree of unsaturation to be present in high concentration in cooking oils.

In recent years there has been a growing interest in value-added specialty oils enriched with different omega-3 (ω-3) polyunsaturated fatty acids such as docosohexanoic acid (DHA, 22:6 ω-3) and eicosapentanoic acid (EPA, 20:5 ω-3) for human nutrition and long-term health benefits. Because they have direct effect on plasma lipids they are beneficial in the treatment of inflammatory diseases, e.g., atherosclerosis, asthma, arthritis, and cancer. The ω-3 fatty acids occur naturally in marine fish oils, some seaweeds fungi, and algae from which their recovery is highly beneficial.

10.2 RECOVERY METHODS

Fatty acids and triglycerides are the predominant lipids present in fats and oils. Their separation is very important for many applications in the food, pharmaceutical, plastics, lubricant, cosmetic, surfactants, soap, and detergent industries. Most fatty acids occur in nature in the form of triglycerides.

Free fatty acids (FFA) are obtained by the hydrolysis of triglycerides, fatty acids methyl esters (FAME) by alcoholysis with methanol, and monoacylglycerides (MAG) by alcoholysis with glycerol, followed by fractionation and purification. It is advantageous to transesterify triglycerides with methanol to form methyl esters which can be subsequently fractionated and then hydrolyzed to free the individual fatty acids. The transesterification can be achieved by using either an alkaline or acidic catalyst. For separation of fatty acids, or their methyl-esters, and glycerides, there are two conventional processes, such as vacuum distillation and solvent extraction. Vacuum distillation requires relatively high temperature which causes degradation of the fatty acids. Solvent extraction poses the problem of residual solvent. The vapor pressures of methyl esters are about four orders of magnitude higher than those of triglycerides and about one order of magnitude higher than those of free fatty acids. Compared to high vacuum distillation, molecular distillation, or crystallization, SC CO_2 extraction is more advantageous. Higher vapor pressures give higher solubilities in SC CO_2. SC CO_2 has a great potential for the separation of these fatty acids and their methyl esters, even with the same number of carbon atoms, but with a varying degree of unsaturation. The solvent power and selectivity of separation may be easily enhanced by addition of small amounts of cosolvent or entrainer to SC CO_2. Separation and fractionation of fatty acids using SC CO_2 has been studied by several workers (Eisenbach, 1984; Nisson et al., 1989 and 1991; Rizvi et al., 1988) for two important reasons: (1) it has emerged as an effective selective solvent for food and pharmaceutical applications, and (2) the importance of these fatty acids as life-savings drugs has been realized to increase with time.

In recent years several researchers have established that SC CO_2 can be used as a good alternative solvent for the following lipid processing operations:

1. Separation of free fatty acids (FFA) from vegetable oil
2. Separation of polyunsaturated fatty acids (PUFA) from animal lipids
3. Refining and deodorization of vegetable oil
4. Fractionation of glycerides
5. Recovery of oil from oil-bearing materials
6. Deoiling of lecithin
7. Decholesterolization and delipidation of food products

10.3 SEPARATION OF FFA FROM VEGETABLE OIL BY SC CO_2

In recent years, there has been increasing interest in separating and rearranging the fatty acids from food materials to formulate new products. SC CO_2 is found to be a good selective solvent for separation of FFA as a byproduct from crude vegetable oils and for removal of undesirable flavor. The role of CO_2 as the extractant and fractionating agent has long been established (Brunner and Peter, 1982; Brunetti et al., 1989; and Krukonis, 1989). Solubilities of fatty acids and triglycerides measured by several workers reported in Table 10.1 clearly ratify the potential of SC CO_2 in fractionating vegetable oils and animal fats.

Chrastil's (1982) solubility data for five triglycerides and two fatty acids in SC CO_2 in the pressure range of 80 to 250 bar and temperatures between 40 and 80°C suggest that fatty acids are much more soluble than the corresponding triglycerides. The high pressure phase equilibrium data for oleic acid-triolein-CO_2 system was studied at three pressures, 200, 250, and 300 bar, at 40°C. The distribution coefficients of both oleic acid and triolein (Bharath et al., 1991) were found to be both pressure and concentration dependent. The distribution coefficient of oleic acid was found to be an order of magnitude higher than that of triolein. This indicates the selective solubilization of FFA even at high pressures up to 300 bar. Thus, the selectivity of SC CO_2 for separation of oleic acid from triolein was established as a function of extraction pressure. At higher pressures a large amount of CO_2 gets dissolved in the liquid phase, necessitating determination of high-pressure-phase equilibria.

However, Bamberger et al. (1988) observed that solubilities of saturated triglycerides (trilaurin, trimyristin, and tripalmitin) at 40°C and pressures between 80 to 300 bar were similar to those from their mixtures (except for the less soluble component) and are not affected by dissolution of CO_2 in liquid phase. Futher, the solubilities of trilaurin and palmitic acid in SC CO_2 were about the same order of magnitude, though the ratio of their molecular weights is 2.5. This is attributed to the higher polarity of free fatty acids compared to the saturated triglycerides.

The operating conditions for processing vegetable oils with SC CO_2 are selected on the consideration whether deacidification or fractionation of fatty acids is the objective. For deacidification of vegetable oils, higher pressures are usually preferred because of higher solvent power. For fractionation of fatty acids, however, lower pressures and higher temperatures are usually selected for better selectivity. Alternatively, the fatty acids are first esterified with methanol or ethanol and then fractionation is carried out from the methyl or ethyl esters of the fatty acids. Methyl esters of oleic, linoleic, and linolenic acids could be selectively fractionated with high purity from their mixture using SC CO_2 as the extractant, followed by chromatographic separation with SC CO_2 mixed with 4% entrainer as the eluent (Ikushima et al., 1989). For selective separation of the dissolved methyl esters, the separation chamber was packed with $AgNO_3$-doped silica gel, which was placed immediately after the extraction column, both maintained at the same temperature and pressure conditions. The alternate addition of hydrocarbons and ethyl acetate as an entrainer to CO_2 affected the improvement in both solvent power and selectivity of the separation of the methyl esters, even at relatively low pressure of 108 bar and 35°C (Ikushima et al., 1989).

SC CO_2 mixed with ethanol as the entrainer was used (Brunner and Peter, 1981) in a multistage countercurrent extraction column at 137 bar and 80°C to reduce the FFA content of palm oil from 3 to 0.1 wt%. Brunetti et al. (1989) deacidified olive oil at 60°C and 200 bar, and observed that SC CO_2-deacidification of olive oil was feasible. Solubilities of four fatty acids and two triglycerides were measured at 200 and 300 bar in the temperature range from 35 to 60°C, which indicated that

TABLE 10.1
Literature on Phase Equilibrium Data of Fatty Acids, Fatty Acid Methyl Esters, and Triglycerides in SC CO_2

System	Pressure Range (bar)	Temperature Range (°C)	Ref.
A large number of fatty acids and triglycerides - CO_2	80–250	40–80	Chrastil, *J. Phys. Chem.*, 86, 3016–3021, 1982.
Oleic acid - CO_2	50–200	40–80	Peter et al., Proc. Intl. Symp. SCF, 1, 99, 1988.
Methyl myristate - CO_2 Methyl stearate - CO_2 Methyl oleate - CO_2 Methyl linoleate - CO_2 Methyl laurate - CO_2 Methyl oleate - CO_2 Methyl linoleate - CO_2	0–200	40–60	Wu et al., Proc. Intl. Symp. SCF, 1, 107, 1988.
Methyl palmitate - CO_2 Methyl laurate - CO_2 Methyl caprate - CO_2 Methyl caprylate - CO_2 Methyl caproate - CO_2	0–160	60	Ashour, Proc. Intl. Symp. SCF, 1, 115, 1988.
Trilaurin - CO_2 Trimyristin - CO_2 Tripalmitin - CO_2	80–300	35–55	Bamberger et al., *J. Chem. Eng. Data*, 33, 327, 1988.
Tri-stearin - CO_2 Lauric acid - CO_2 Myristic acid - CO_2 Palmitic acid - CO_2	80–300	35–55	Bamberger et al., *J. Chem. Eng. Data*, 33, 327, 1988.
Oleic/linoleic/linolenic acid/ methyl ester - CO_2	93–198	28–40	Ikushima et al., *Ind. Eng. Chem. Res.*, 28, 1364, 1989.
Palmitic acid - CO_2 Tripalmitin - CO_2	80–160	25–40	Ohgaki et al., *Intl. Chem. Eng.*, 29, 302, 1989.
Stearic acid + CO_2	140–467	45–65	Kramer & Thodos, *J. Chem. Eng. Data*, 34, 184, 1989.
Oleic acid/linoleic acid/ methyl oleate/methyl linoleate - CO_2	38–300	40–60	Zou et al., *J. Supercritical Fluids*, 3, 23–28, 1990.
Myristic acid Palmitic acid	81–218	35	Iwal et al., *J. Chem. Eng. Data*, 36, No. 4, 431, 1991.
Oleic acid - CO_2 Triolein - CO_2	0–300	40–80	Bharath et al., Proc. 2nd Intl. Symp. SCFs, 288, 1991.

TABLE 10.1 (continued)
Literature on Phase Equilibrium Data of Fatty Acids, Fatty Acid Methyl Esters, and Triglycerides in SC CO_2

System	Pressure Range (bar)	Temperature Range (°C)	Ref.
Palm oil - CO_2 Sesame oil - CO_2			
Methyl oleate - CO_2 Oleic acid - CO_2 Oleyl glycerol - CO_2 Mono olein - CO_2 Diolein - CO_2 Triolein - CO_2	172–309	50–60	Nisson et al., *J. Am. Oil Chem. Soc.*, 68, No. 2, 87, 1991.
Ethyl esters of DHA, arachiodonic acid, eicosatrienoic acid, and oleic acid - CO_2	90–250	40–100	Liong et al., *Ind. Eng. Chem. Res.*, 31, 400, 1992.
Methyl ester/ethyl ester/fatty acids/TG/fats/oils - CO_2	70–350	40–60	Yu et al., *J. Supercritical Fluids*, 7, 1, 51, 1994.
Oleic acid - CO_2 Methyl oleate - CO_2 Methyl oleate + ethanol - CO_2 Oleic acid + ethanol - CO_2	0–200		Navaro et al., Proc. 3rd Intl. Symp. SCFs, 2, 257, 1994.
Stearic acid - CO_2	80–160	35	Guan et al., Proc. 4th Symp. SCFs, Vol. B, 409, 1997.
Rice bran oil - CO_2 Cotton seed oil - CO_2	80–155	40–60	Mukhopadhyay and Nath., *Ind. Chem. Engr.*, 37, 53, 1995.
Squalane - CO_2	100–250	40–60	Catchpole et al., Proc. 4th Symp. SCFs, 175, 1995.

SC CO_2 had a higher selectivity for fatty acids than triglycerides at 60°C and 200 bar. However, their results were based on the solubility of pure components, i.e., the fluid-phase composition as liquid phase compositions were not measured. Goncalves et al. (1991) investigated the feasibility of separating FFA from olive husk oil containing very high (56%) content FFA using SC CO_2 at 35 and 40°C and at pressures up to 220 bar. But there were discrepencies in their observations. Subsequently, Carmelo et al. (1994) varied the initial FFA content (0.7 to 7% oleic acid) of the commercial olive oil fed to a 2-m long countercurrent supercritical extraction column operated at pressures up to 220 bar and temperatures between 40 and 50°C. Their results also indicated very high selectivity (for initial FFA content of 7.1%) varying between 11.8 and 14.7 at 210 bar on changing temperature from 40 to 50°C.

The selectivity further increased with decreasing the FFA content at 180 bar and it was found to increase with pressure as well, indicating greater efficiency for deacidification of olive oil with low FFA content. At a higher pressure and at a higher temperature, the selectivity further increased. However, the solvent-to-feed ratio used was very high (44), though the extraction yield was 70% at 210 bar and 50°C.

SC CO_2 was demonstrated to be a potential solvent for the removal of FFA from rice bran and cotton seed oils (Mukhopadhyay and Nath, 1995) with and without addition of an entrainer like isopropanol. SC fluid–liquid phase equilibria of the oils with CO_2 at pressures in the range of 90 to 155 bar at 40 and 60°C indicated the solubility of CO_2 in the liquid oil phase to be in the range of 13 to 33 wt%, depending on the pressure, temperature, and FFA content, which marginally increased with the addition of isopropanol in CO_2. The solubility of the lipids in SC CO_2 was in the range of 0.012 to 0.135 wt%, which increased to 0.08 to 0.30 wt% due to the addition of isopropanol to the oil in the pressure range of 110 to 145 bar at 40°C. The FFA content in the extract increased with pressure, for example, to 74 wt% at 130 bar for rice bran oil containing 7.6 wt% FFA in the feed. The selectivity of separation was accordingly found to be very high, in the range of 80 to 200 for cotton seed oil and 10 to 40 for rice bran oil, at 40°C, which further improved with the entrainer. It was observed that selectivity increased with lowering of temperature and at any temperature there is an optimum pressure beyond which selectivity decreases with increasing pressure. The lower values of selectivity in the case of the rice bran oil is due to higher (7.6%) FFA content in the oil compared to 0.28% FFA in the cotton seed oil.

There are wide variations and discrepencies in the solubility data reported in the literature, and accordingly, it is preferable to generate phase equilibrium data for the particular oil having the specific composition, i.e., the multicomponent system in question, in order to arrive at the feasibility of separation and the number of stages required.

10.4 FRACTIONATION OF PUFA FROM ANIMAL LIPIDS

Fish and animal lipids contain mainly triglycerides in the range of molecular weight from 500 to 1200 Da, but also contain other components, like valuable free fatty acids, retinol, squalane, tocopherols, cholecalciferol, cholesterol, cholesterol esters, wax esters, diglycerides, and phospholipids in minor quantities (Borch-Jensen et al., 1998). Certain polyunsaturated fatty acids (PUFA) present in animal lipids are highly beneficial for human nutrition and metabolism. The role of ω-3 PUFA, such as EPA (20:5) and DHA (22:6), in lowering serum lipids and blood cholesterol has been known for a long time. These PUFA have the ability to enhance the antiaggregatory activity and to reduce ischemic heart diseases and thrombosis. This is believed to be the reason why there are very low incidents of ischemic heart diseases among Greenland eskimos who consume mainly seafoods rich in ω-3 PUFA. Fish oils as well as lipids from other aquatic and terrestrial animals are the principal sources of EPA and DHA.

Further, an important factor affecting our health is the oil balance of PUFA, i.e., the ratio of the ω-3 to ω-6 fatty acids in foods. For example, a shift of the oil balance to the ω-6 side, due to a higher intake of meat rather than fish, increases the incidence of such diseases as allergies, myocardial infarction, hypertension, thrombosis, etc. The ω-3 fatty acids, such as γ-linolenic (18:3) acid, EPA, and DHA, are considered to have desirable effects in preventing or curing the adverse medical condition caused by the excess intake of ω-6 PUFA. Therefore healthy food ought to contain these ω-3 PUFA. Certain fish oils naturally contain ω-3 fatty acids, as high as 25%.

These polyunsaturated fatty acids are not stable and are difficult to separate even by molecular distillation, which requires elevated temperatures up to 180°C, resulting in undesirable degradation and polymerization. In purities above 90%, it is easier to fractionate them by using SC CO_2 from their methyl or ethyl esters, preferably from ethyl esters, because methanol released on hydrolysis is not suitable for food and pharmaceutical applications. These fatty acid esters can be fractionated using SC CO_2 at temperatures less than 80°C. For example, EPA and DHA in purities above 96 and 98%, respectively, could be obtained by Eisenbach (1984), using SC CO_2 extraction at 150 bar and 50°C. The EPA concentration in menhaden oil esters could be enhanced to 96% purity by employing a combined process of urea crystallization and SC CO_2 extraction. Urea crystallization removes less unsaturated species, such as C(20:0) and C(20:1) from the other C_{20} fatty acids. The distribution coefficients of some of the ethyl esters of menhaden oil in SC CO_2 were measured by Nilsson et al. (1988) and the selectivities of C_{16}/C_{18}, C_{18}/C_{20}, and C_{20}/C_{22} fatty acids were respectively found to be 1.63, 1.86, and 2.24 at 150 bar and 60°C (Krukonis, 1988), which are sufficient for fractionation of high purity EPA. Berger et al. (1988) employed preparative supercritical fluid chromatography (SFC) to obtain highly pure EPA and DHA.

The separation and fractionation of ω-3 PUFA using SC CO_2 has also been studied by several other workers (Yamaguchi et al., 1986; Rizvi et al., 1988). They established that the concentration of ω-3 fatty acid esters could be enriched up to 88 to 98% from fish oil depending on the initial EPA/DHA content in the feed, using SC CO_2 up to a maximum temperature of 80°C at 150 bar pressure, with temperature or pressure programming along the fractionating column to provide reflux. SC CO_2 fractionation based on the principle of retrograde solubility behavior was utilized for effective separation and purification of the most useful components of fish oil esters. When a temperature gradient along the column was maintained to provide internal reflux, the fractionation was primarily in accordance with the carbon numbers. An incremental pressure programming was utilized to obtain EPA and DHA exceeding 76% purity (Zhu et al., 1994). Riha and Brunner (1994) could achieve fractionation of methyl- and ethyl esters of fatty acids by chain length and not so much by degree of saturation.

10.5 REFINING AND DEODORIZATION OF VEGETABLE OIL

Four major vegetable oils that are mostly consumed and traded worldwide are soybean oil, palm oil, canola oil, and sunflower oil (see Appendix C). Further, safflower, corn, peanut, olive, and cottonseed oils are recent additions to world trade. Apart from being used as edible oil, they may be utilized to recover valuable

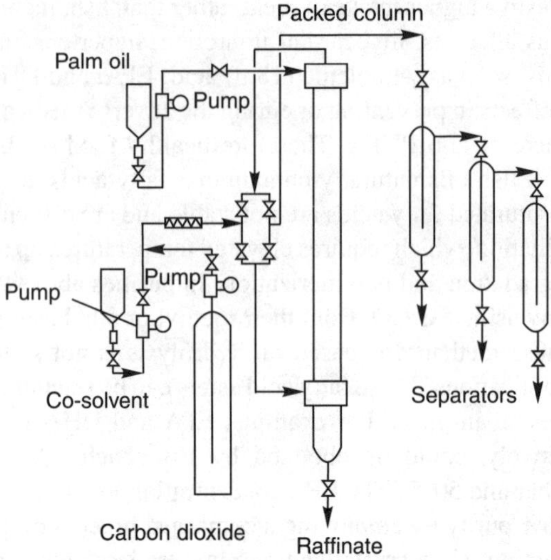

FIGURE 10.1 Schematic diagram of continuous supercritical CO_2 processing of palm oil (Ooi et al., 1996).

vitamins, sterols, lecithin, and fatty acids for application in food and pharmaceutical industries. As mentioned earlier, refining of crude vegetable oil is carried out to remove free fatty acids and objectionable odor. It is conventionally accomplished by a series of physical or chemical processes, such as degumming, alkali treatment, and vacuum distillation at a relatively high temperature, leading to degradation and decomposition of valuable nutrients and vitamins, including loss of the neutral oil. However, continuous SC CO_2 processing can combine all these steps and can sequentially reduce the contents of FFA, squalane, sterols, DG, MG, certain TG, and some carotenes in the edible oils.

Ooi et al. (1996) investigated refining crude Malaysian palm oil in a continuous countercurrent packed column of 61 cm length and 1.75 cm in diameter as shown in Figure 10.1. The FFA content could be reduced from 2.35 to 0.19% at 240 bar and 50°C, with solvent-to-feed ratio of 58.2. However, the addition of 3.7 mol% ethanol as a cosolvent to SC CO_2 was required to reduce the FFA level to less than 0.1%, achieved at 206 bar and 50°C. With a further increase of ethanol to 6.3 mol% in SC CO_2 solvent at 171 bar and 50°C and solvent-to-feed ratio of 40, the FFA content could be reduced to 0.04% with very little MG and much reduced DG. The crude palm oil contains 500 to 800 ppm of carotene which is also coextracted with FFA and the extract contains 100 to 322 ppm of carotene. The low DG content of the SC CO_2 processed oil helps in improving the crystallization characteristics of palm oil.

Penedo and Coelho (1997) proposed a combined process entailing deacidification, deodorization, and clarification — all in one stage, using SC CO_2 mixed with ethanol (5%), in which 65.3% of the FFA in the crude soybean oil and 55.8% of FFA in the Brazilian peanut oil could be extracted out at 140 bar and 80°C.

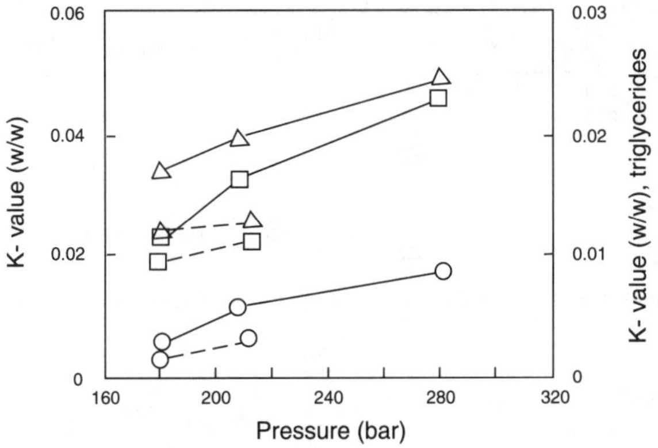

FIGURE 10.2 Distribution coefficients of FFA (□), triglycerides (○), and squalane (△) as a function of pressure for an initial concentration of 7 wt% FFA. The right axis scale refers to the K-factor of triglycerides: (-) 40°C, (---) 50°C (Simoes et al., 1994).

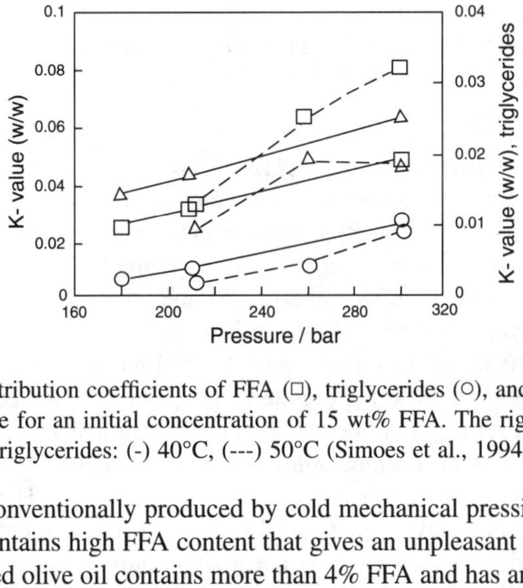

FIGURE 10.3 Distribution coefficients of FFA (□), triglycerides (○), and squalane (△) as a function of pressure for an initial concentration of 15 wt% FFA. The right axis scale refers to the K-factor of triglycerides: (-) 40°C, (---) 50°C (Simoes et al., 1994).

Olive oil is conventionally produced by cold mechanical pressing from the olive fruit and often contains high FFA content that gives an unpleasant odor. Nearly 50% of the cold-pressed olive oil contains more than 4% FFA and has an unpleasant odor, making it unsuitable for immediate human consumption, so it must be refined. SC CO_2 may be considered for crude olive oil deodorization with simultaneous improvement of organoleptic characteristics of the oil. Phase equilibrium data of such an oil with SC CO_2 with varying amounts of FFA (3 to 15 wt%) indicate that the vapor phase is at least 10 times more enriched with squalane and FFA (up to 60 wt% on solvent-free basis) than that in the liquid phase (Simoes et al., 1994). This implies that squalane and FFA can be selectively removed at 40°C and 300 bar. Squalane has higher volatility than FFA. The distribution coefficients of squalane, FFA, and TG are shown in Figures 10.2 and 10.3 for 7 and 15% by weight of FFA, respectively, in the initial-feed olive oil. An increase in pressure increases the distribution coefficients of all

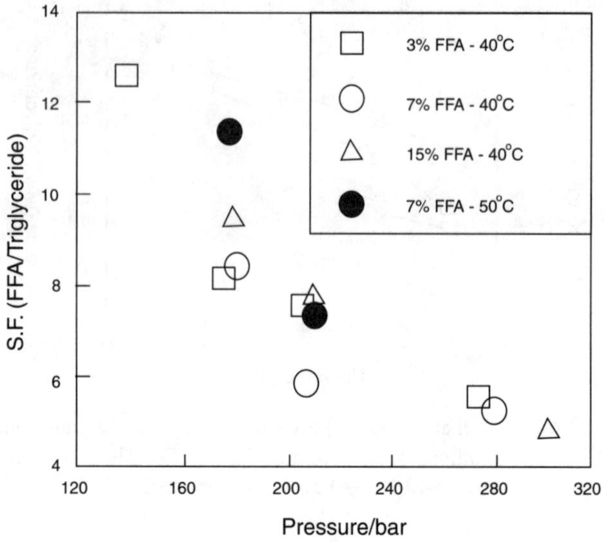

FIGURE 10.4 Separation factor of free fatty acids with respect to triglycerides at 40 and 50°C, and three different initial compositions of FFA in the oil (Simoes et al., 1994).

components for two different FFA concentrations in the feed. But the distribution coefficient of FFA decreases with temperature up to 50°C and then increases with temperature up to 80°C. However, for other components, the distribution coefficient decreases with temperature. The separation factors decrease with pressure at this level of pressure and initial concentration, as seen in Figure 10.4. The best condition for separation was found to be 260 bar at 80°C (Simoes et al., 1994) for 7% FFA in the feed which was reduced to 0.33% in the refined oil.

On the other hand, SC CO_2 refining of three different varieties of olive oils by Bondioli et al. (1992) at different pressures, temperatures, and solvent-feed ratios, indicated that best results could be obtained at 130 bar and solvent-to-feed ratio of 100 kg CO_2/kg oil with a temperature gradient 50/40/30°C, respectively, at the top/middle/bottom of the continuous countercurrent column. The compositions of the crude and refined olive oils from the three different sources are shown in Table 10.2. From data on the average concentration ratios of the extract to the refined oil (last column of Table 10.2), it can be noticed that the extract is highly enriched in squalane, and next, the fatty acids are extracted in preference to triglycerides, even though fatty acids have high polarity. SC CO_2 processed and deodorized oil thus compare similar to the conventionally refined oil, but with much less loss.

Pure (95%) squalane finds applications both as a health food tonic and in pharmaceutical preparations. Squalane is present in 40 to 70% by weight concentrations in shark liver oil. The shark liver oil is mostly sold as a crude oil, which also contains triglycerides, squalane, glyceryl ethers, pristane, and FFA which cause an unpleasant odor. Continuous SC CO_2 extraction of squalane from shark liver oil containing 50% squalane was performed at 250 bar and 60°C (Catchpole et al., 1997) and the squalane

TABLE 10.2
Composition Differences Before and After Deodorization with SC CO_2 at 130 bar

	Italian Olive Oil		Spanish Olive Oil		Tunisian Olive Oil		Average Enrichment Ratio
	Crude	Refined	Crude	Refined	Crude	Refined	
FFA (%)	1.73	0.20	3.38	0.38	3.93	0.92	85
MG (%)	0.04	0.03	0.05	0.05	0.06	0.03	22
DG (%)	2.90	2.22	4.22	3.17	4.52	3.97	4
Squalane (ppm)	5200	<400	2500	<200	3200	<200	168
Aliphatic alcohol (ppm)	76.1	12.1	88.7	23.2	80.4	22.7	60
Tocopherol (ppm)	—	—	144	57	43	15	9
Sterols (ppm)	1104	811	1545	1185	1096	887	7
Terpenes (ppm)	819	611	948	726	848	706	10
Waxes (ppm)	150	132	233	193	323	280	2
Sterol esters (ppm)	632	787	1081	1244	1117	1248	0.4
Terpene esters (ppm)	514	601	749	798	723	736	0.2

Bondioli et al., 1992.

purity of 99.5% could be achieved with 5% mass of the raffinate as reflux, for a high solvent-to-feed ratio.

10.6 FRACTIONATION OF GLYCERIDES

Utilization of surface-active lipids, such as monoacylglycerides (MAG), as emulsifiers, defoaming agents, and oil stabilizers has led to their wide ranging applications in the food and pharmaceutical industries. Accordingly they are used in bread, biscuits, margarines, and instant powders. Monoacylglycerides of saturated and unsaturated fatty acids are partial esters of glycerol containing high-molecular weight fatty acids. The MAG having molecular weights in the range of 274 to 354, diacylglycerides (DAG) in the range of 456 to 616, and triacylglycerides (TAG) in the range of 639 to 878 are normally separated by molecular distillation from the reaction product of esterification reaction of fatty acids and glycerol. However, SC CO_2 extraction obviates the two main drawbacks of this process, namely, the requirements of high vacuum and high temperature, normally more than 200°C, leading to the formation of undesirable components. The vapor pressures of MAG are three to five orders of magnitude higher than those of DAG and TAG, and can be thus selectively separated. SC CO_2 at 200 bar combined with an upward temperature gradient of 50 to 100°C on a packed column was shown to be effective for fractionation of the mixtures of MAG, DAG, and TAG with enrichment of MAG to as high as 90 wt%. Cosolvent addition to SC CO_2 could enhance the selectivity of separation. On doubling the column pressure from 172 to 344 bar, the concentration of MAG in the top product decreased from 94 to 56 wt%, although the yield increased (Sahle-Demessie et al., 1997). However, at 207 bar, a linear temperature gradient from the bottom to the top (65 to 95°C) gave the best selectivity of separation, and all three components, MAG, DAG, and TAG, could be obtained sequentially on a semibatch operation mode of the packed column.

Bharath et al. (1991) studied fractionation of mixtures of triglycerides present in palm oil and sesame oil using SC CO_2 in the pressure range of 150 to 350 bar and at a temperature in the range of 40 to 80°C. It was observed that triglycerides having a lower number of C-atoms in the three fatty acid parts, were more soluble and had higher K-values at lower pressures, but the difference in these K-values between the C-numbers decreased at higher pressures. This shows that the palm oil can be fractionated at 60°C and 200 bar. However, sesame oil, having higher (>48) C-number triglycerides, has all lower K-values and cannot be fractionated.

Crude palm oil contains more than 90% triglycerides (TG), 2.7% DG, less than 1% MG, 3 to 5% FFA, and 1.0% minor constituents, such as carotenoids (500 to 700 ppm), tocopherol and tocotrienols (600 to 1000 ppm), and sterols (250 to 620 ppm). Choo et al. (1996) successfully demonstrated that these five groups of components could be separated by preparative supercritical fluid chromatography (SFC) with a combination of a C-18 and a silica gel column, based on the functional groups using SC CO_2 mixed with 3 to 6 vol% ethanol at 180 bar. The first fraction was FFA containing 0.9% myristic acid, 47.9% palmitic acid, 37.1% oleic acid, and 10.7% linoleic acid, whereas the last fraction contained carotenes.

A complex mixture of triglycerides, such as butter (anhydrous milk fat), can be fractionated into a number of fractions as a function of the molecular weight and unsaturation using SC CO_2, which can be reconstituted later to produce a higher melting point (32 to 35°C) butter with good spreadability and stability. This product has the advantage of not requiring refrigeration for storage (Castera, 1994). The short-chain triglycerides in the carbon number range of C_{24} to C_{34} can be selectively separated at a lower pressure range of 100 to 150 bar and a temperature in the range of 40 to 50°C. The medium-chain length triglycerides in the carbon number range of C_{36} to C_{42} are more enriched at the medium pressure range of 200 to 250 bar, and lastly, the higher carbon-numbered triglycerides (more than C_{46}) are separated at about 350 bar and 70°C.

10.7 EXTRACTION OF OIL FROM OIL-BEARING MATERIALS

Triglycerides are the major source of energy since they are metabolized by the body to form essential fatty acids required for growth. Necessary dietary triglycerides are obtained from cooking oil, margarine, and butter. Oil from oil seeds mainly consist of neutral lipids, such as triglycerides of C_{16} to C_{20} fatty acids. To enhance functional properties, triglycerides are separated from protein present in the seeds by extraction or leaching.

Seed oil is traditionally cold-expelled or -extracted using organic solvents. But it needs to be further processed to attain its required quality, stability, and market value. SC CO_2 extraction can be employed to produce edible or pharmaceutical oils of the desired quality by sequentially or simultaneously fractionating different groups of compounds present in the seeds. Despite all its positive attributes, SC CO_2 extraction technique has not been employed on a commercial scale to a great extent due to widespread apprehension about the high investment cost. Reverchon and Osseo

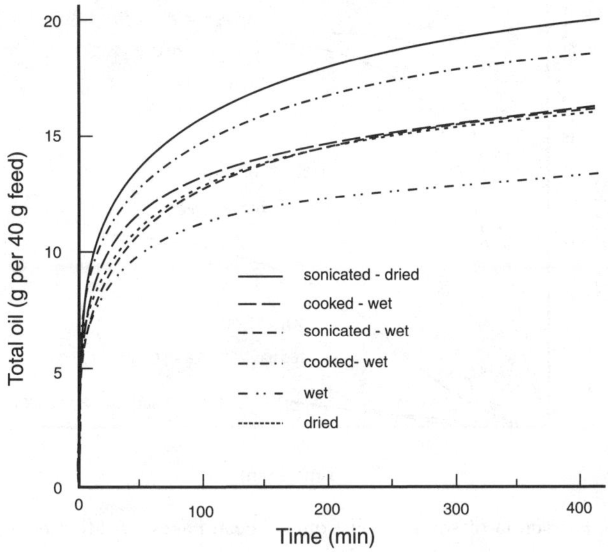

FIGURE 10.5 Extraction of oil (triglycerides) from pretreated palm fruits. Influence of pretreatment method (Brunner, 1994).

(1994) analyzed three different process alternatives for the production of soybean oil by SC CO_2 on an industrial scale, compared the operating cost of production with that for the hexane-extracted product, and inferred that the operating costs were in the same range as conventional commercial extraction plants. However, with proper selection of process parameters and process integration, it is possible to reduce the operating cost to a level much less than that by solvent extraction technique. Further, the high initial investment cost can be compensated for by much less batch time and significant value additions due to the high quality oil and byproducts recovered.

SC CO_2 has been used for extraction of several ingredients from oil-bearing materials ranging from oil seeds to algae. The oil extracted with SC CO_2 from oil seeds is known to have less polar lipids such as glycolipids and phospholipids, facilitating the elimination of the need for degumming and deodorization steps involved in the conventional oil refining process. The SC CO_2-extracted oil however contains tocopherol and free fatty acid levels comparable to hexane-extracted oil. Triglycerides are fairly soluble in SC CO_2, and more so in the presence of an entrainer. Pretreatment of oil seeds either by flaking or by cooking with a polar solvent prior to extraction is highly beneficial in enhancing oil recovery, as can be seen from Figure 10.5 in the case of palm fruit. The rate of extraction increases with temperature, pressure of extraction, and cosolvent (e.g., propane, butane, or ethanol) addition to SC CO_2, as can be seen from Figure 10.6 for soybean oil. Fractionation of the extract with respect to time of extraction as well as depressurization at different conditions of separation, results in better quality of oil and enrichment of desired constituents in the product. For example, the content of β-carotene in the first fraction of the extract is enhanced by a factor of 10, against the net content in the extract (Brunner, 1994), as can be seen in Figure 10.7.

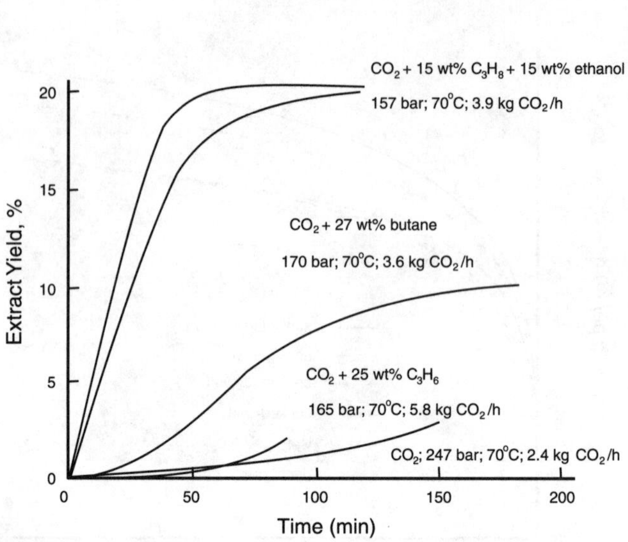

FIGURE 10.6 Extraction of soybean oil from soybean flakes by different supercritical solvents and different entrainers (Brunner, 1994).

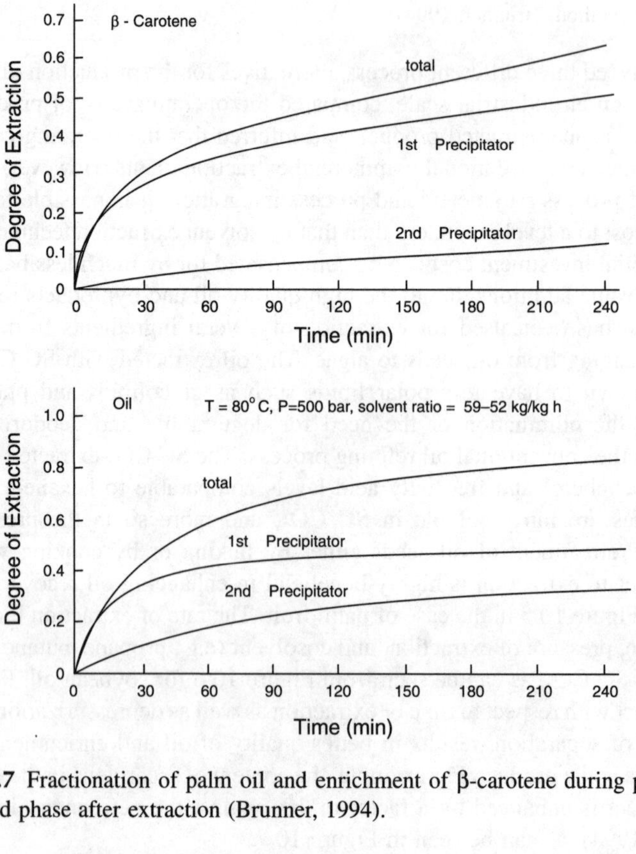

FIGURE 10.7 Fractionation of palm oil and enrichment of β-carotene during precipitation from the fluid phase after extraction (Brunner, 1994).

TABLE 10.3
Comparison of Major Fatty Acids in SC CO_2 (150 bar and 40°C) and Hexane-Extracted Oils

Oil	Recovery (%)	Myristic C14:0	Palmitic C16:0	Stearic C18:0	Oleic C18:1	Linoleic C18:2
Corn germ (SC CO_2)	50.1	—	31.28	6.03	53.27	7.32
Corn germ (hexane)	100	—	10.57	1.72	26.02	55.92
Sunflower (SC CO_2)	63.8	0.4	30.46	20.51	47.24	1.38
Sunflower (hexane)	100	0.06	6.43	4.16	31.08	56.06

Esquivel et al., 1994.

The oil extracted from sunflower and soybean seeds, corn germ and wheat germ, olive husk and rice bran with SC CO_2 are equivalent to the oil refined by the conventional process, in terms of the amounts of triglycerides and tocopherol content (List and Friedrich, 1985). Extraction of olive husk oil, corn germ oil, and sunflower oil was carried out by Esquivel et al. (1994) using SC CO_2 at 120, 150, and 180 bar at 35, 40, and 45°C. Table 10.3 presents the composition of these partially extracted oils obtained by SC CO_2 extraction at 150 bar and 40°C, compared with total oil extracted by hexane. For corn germ oil and sunflower oil, it can be noted that the linoleic acid content in these oils is lower than that in the hexane-extracted oil, whereas the content of oleic acid or the monounsaturated fatty acids (MUFA) is higher. This was, however, not observed in the case of olive husk oil, probably due to much less linoleic acid content (8%) in the olive husk as compared to 50 and 64% in the sunflower and corn germ oil, respectively.

The free fatty acids are, in general, preferentially recovered by SC CO_2. The selectivity is greater for free fatty acids than for triglycerides due to smaller molecular weights. Hence a scheme involving simultaneous extraction and fractionation of the oil can result in deacidified oil (triglycerides), eliminating the need for additional steps for the removal of free fatty acids, off flavor, and gummy materials, as in the case of solvent-extracted crude oil. Performance of supercritical CO_2 extraction of oil from different oil bearing materials is elaborated in the following subsections.

10.7.1 Sunflower Oil

Sunflower seeds contain up to 43.4% of good-quality, highly nutritive oil, rich in monounsaturated fatty acid (MUFA) esters. The feasibility of SC CO_2 extraction of this oil from sunflower seeds was explored by Calvo et al. (1998). The loading of sunflower oil in SC CO_2 increases with pressure and decreases with temperature up to the crossover pressure of 400 bar (Cocero and Calvo, 1996). The loading also increases with addition of ethanol in SC CO_2 at the rate of 4.5 g/kg CO_2 per 1 wt% increase in ethanol content. For example, the loading of sunflower oil in SC CO_2 at 300 bar and 42°C is about 0.5% by weight, which increases by more than an order of magnitude with the addition of 10% ethanol in SC CO_2. The effect of pressure on enhancing the solvent power is even greater when a greater quantity of ethanol is added. Phospholipids are hardly soluble in SC CO_2. However, addition of a little amount of ethanol causes phospholipids to be extracted in small concentrations. Since phospholipids are not desirable in the edible oil, the addition of ethanol is restricted to not more than 10% by weight in SC CO_2. Ethanol addition also shifts the cross-over pressure from 400 to 375 bar for 5% ethanol and to 350 bar for 10% ethanol in SC CO_2 (Cocero and Calvo, 1996). However, Favati et al. (1994) reported 350 bar as the cross-over pressure for SC CO_2 with no cosolvent. Phospholipids are coextracted only with triglycerides when SC CO_2 mixed with ethanol is used as the solvent, and when triglycerides are exhausted from the ground sunflower seeds, no increase is detected in the phospholipid content of the extract, even though the seed still contains residual phospholipids. It is interesting to note that sunflower oil can be extracted at 200 bar and 42°C with SC CO_2 mixed with 10% ethanol to have a loading of 2.6% and the dissolved oil can be separated isobarically by increasing the temperature to 80°C, as the loading reduces to about 0.5%. Alternatively, the oil extraction can be performed at 400 bar and 80°C with 10% ethanol for which loading is 12%, and the oil separation may be affected at 100 bar and 80°C at which there is hardly any dissolving capacity of SC CO_2 for oil, even with 10% ethanol in it. The minor constituents coextracted with oil like flavor and fatty acids can be subsequently removed from the SC CO_2 phase by further reducing the pressure to 60 bar and 40°C.

In order to find the optimal condition (Favati et al., 1994) for SC CO_2 extraction of sunflower oil, experiments were conducted over a wide range of pressure (200 to 700 bar) and temperature conditions. The highest loading of sunflower oil in neat SC CO_2 was reported by Favati et al. (1994) to be 18% at 700 bar and 80°C, which was three times more than that at 500 bar, 80°C with neat SC CO_2. The color of the oil depended on the extraction pressure and temperature. Up to 500 to 700 bar pressure levels, the oil color appeared bright yellow at 60°C, whereas it became turbid and dark yellowish brown at 80°C.

Addition of an alcohol as an entrainer to SC CO_2, no doubt enhances the solubility of oil but decreases selectivity, resulting in coextraction of other constituents. The total content of phospholipid in the oil extracted from sunflower seeds was found to be less with 5% butanol as the cosolvent with SC CO_2 at 300 bar and 40°C than with ethanol as the cosolvent, because the polarity of CO_2-ethanol mixture is higher than that with butanol (Calvo et al., 1998). With 20% butanol in SC CO_2, the solubility of sunflower oil in SC CO_2 is forty times more, i.e., 24.4% by weight as compared to 0.6% in neat SC CO_2, whereas with 5% butanol, the solubility increases only four times, as can be seen in Figure 10.8. The oil solubility is higher in SC CO_2 mixed with 10% butanol than that in SC CO_2 mixed with the same amount of ethanol at 300 bar and 40°C. It also depends on the operating conditions of pressure and temperature, as shown in Figure 10.9. Further, addition of butanol decreases the co-recovery of phospholipid in the oil as compared to ethanol, as can be seen from Figure 10.10, for the operating conditions of 300 bar and 40°C, as the polarity of ethanol is higher (μ = 25.7 D) than that of butanol (μ = 17.7 D).

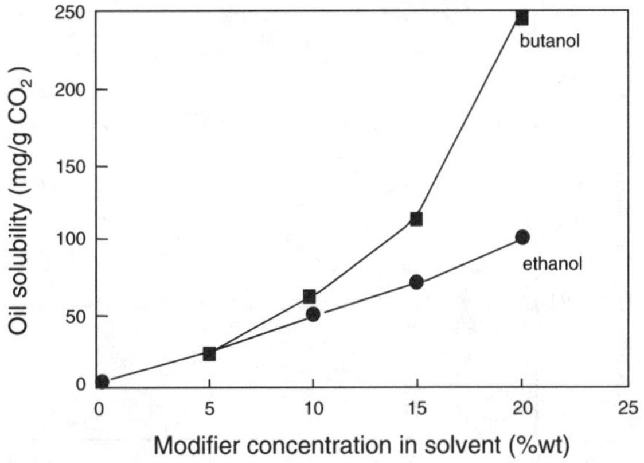

FIGURE 10.8 Solubility of sunflower oil in CO_2-alcohol mixtures of different alcohol concentrations at operating conditions: 300 bar at 40°C (Calvo et al., 1998).

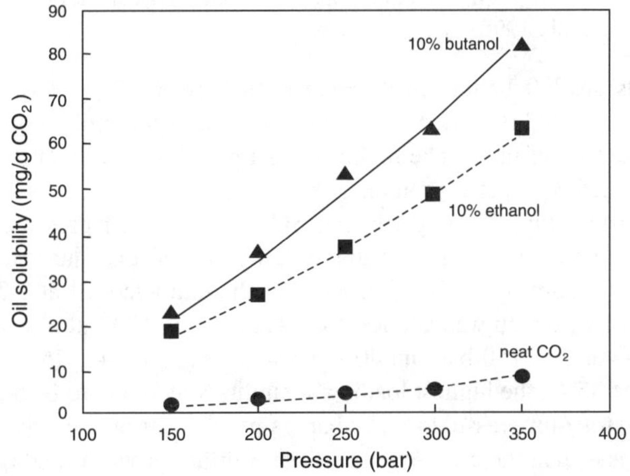

FIGURE 10.9 Effect of operating pressure on oil solubility using neat SC CO_2 and mixtures of SC CO_2 + 10% modifier as solvent at 40°C (Calvo et al., 1998).

10.7.2 Corn Germ Oil

Corn germ oil is considered a valuable oil due to its high content of PUFA and high nutrition value. Crude corn oil is traditionally obtained by cold pressing and/or solvent extraction with hexane. Corn germ contains about 45 to 50% oil.

SC CO_2 extraction of corn germ has been investigated by several authors (Christiansen, 1985; List, 1985; Fontan et al., 1994; Esquivel et al., 1994). Corn germ oil obtained by SC CO_2 is lighter colored and has lower phosphorus content than the hexane-extracted oil. The level of tocopherols found in the SC CO_2-extracted oils is more or less the same as that in the crude oil obtained with hexane. Further, the raffinate, the corn germ protein meal after SC CO_2 extraction, shows better nutritive properties than the hexane-defatted protein, because most of the

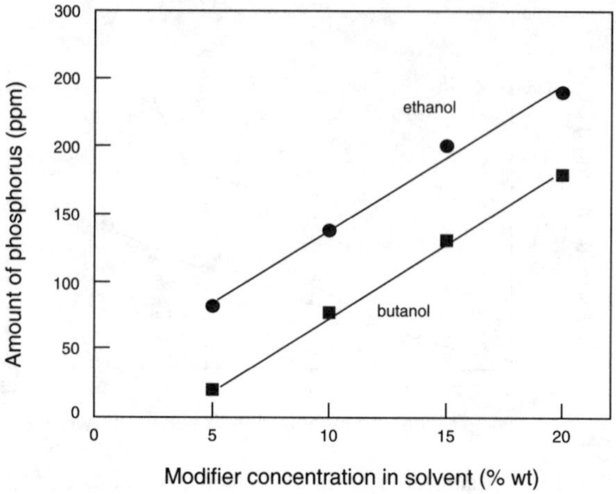

FIGURE 10.10 Phospholipids concentration in the extracted oil expressed as μg phosphorus per gram of oil and per gram of modifier vs. the modifier concentration in SC CO_2 at 300 bar, 40°C (Calvo et al., 1998).

phospholipids are left behind in the meals after the SC CO_2 extraction. SC CO_2 does not, or only slightly, dissolves polar lipids (phospholipids or glycolipids), so they remain in the raffinate. The higher polar lipid content influences the physico-chemical properties of germ protein.

The highest loading of corn germ oil in SC CO_2 varies with pressure, temperature, cosolvent, and particle size. For example, it was found that the loading remained more or less constant up to 30% recovery of the extracted oil at 2.2 g/kg CO_2 at 180 bar and 40°C, which was about 40% less (1.3 g/kg CO_2) at a marginally lower extraction pressure of 150 bar and 40°C (Fontan et al., 1994). However, at a lower temperature of 35°C, the highest loading/solubility was found to be higher (1.8 g/kg CO_2) at the extraction pressure of 150 bar, as per the retrograde solubility behavior at a pressure less than the cross-over pressure. Further, it was found (Ronyai, 1997) that SC CO_2 extraction carried out at a much higher pressure of 300 bar and 42°C with ethanol as an entrainer resulted in 48 to 50% yield of good quality, clear oil with a pleasant flavor, the color of which varied from straw yellow to pale red with variation of the alcohol content in the solvent. The phospholipid content of the oil, however, increased with the amount of alcohol present in the solvent as entrainer, as can be seen from Table 10.4. The time of extraction and solvent requirement decreased with ethanol concentration. The nutritive and emulsifying properties of the SC CO_2-extracted corn germ meals with 2.5% ethanol in SC CO_2 showed better results than the rest of the other meals extracted by pure SC CO_2 or SC CO_2 mixed with higher amounts of ethanol as entrainer.

10.7.3 Soybean Oil

There is a huge demand for soybean oil production and it is the most-consumed vegetable oil in the world. It is estimated that more than 20 million tonnes of soybean

TABLE 10.4
Phospholipids (%) in Corn Germ Oil with Variation in Alcohol Content in SC CO_2 at 300 Bar, 42°C

Ethanol, % in SC CO_2	Phosplipid in Oil, %	Solubility of Oil, %
0	0.026	0.747
2.5	0.048	1.027
5.0	0.115	1.746
7.5	0.772	5.701
10.0	0.756	
Hexane (solvent)	0.920	

Ronyai et al., 1997.

oil are being produced in the world, and more than 7 million tons of it in the U.S. every year, consuming large amounts of organic solvents. Accordingly, there is a need to find alternative solvents in view of the socioeconomic and regulatory constraints currently being faced in some developed countries.

Friedrich et al. (1982) investigated SC CO_2 extraction of soybean flakes at 335 bar and 50°C. Soybean oil extracted with SC CO_2 was found to be lighter in color, since it contained less iron and one tenth of the phospholipids extractable by hexane, thus obviating the need for any additional steps of dugumming and deacidification. SC CO_2-extracted oil is equivalent to the refined oil obtained from the hexane-extracted crude soybean oil in terms of odor, color, and stability. Hexane-extracted oil contains 1.5% phospholipids, whereas the SC CO_2-extracted oil contains 0.13%.

The solubility of soybean lipids in SC CO_2 is strongly influenced by temperature, pressure, and cosolvent. There is a retrograde solubility behavior at a cross-over pressure of 400 bar with neat SC CO_2, beyond which solubility increases with temperature. Pretreatment of soybeans is very important in improving the quality and yield of the oil, as mentioned earlier. It is reported that all pretreatments, including oven toasting, microwave heating, and live steam treatment improve the quality and yield of the oil from soybeans, which contain 15 to 18% moisture levels (Chu, 1995).

10.7.4 Olive Husk Oil

Olive husk oil is rich in MUFA and is extracted from the residual fat contained in the olive fruit. Olive husk is the residual material after olive oil is removed from the pulp of the fruit of olive trees by cold pressing. The husk contains up to 21.7% oil (Esquivel, 1994). This residual oil is normally extracted with hexane and subsequently refined to make it fit for human consumption. Alternatively, SC CO_2 extraction can be used to recover the husk oil. Both the yield and the rate of extraction increased with an increase in pressure. For example, SC CO_2 at 350 bar and 60°C was found to give the highest recovery (90%) at the lowest time of extraction (3 to 4 h) for a particle size of 0.17 mm by de Lucas et al. (1998). Mass transfer limitations are reduced with an increase in temperature and decrease in particle size.

10.7.5 Grape Seed Oil

Grape seeds, a byproduct of the wine industry, contain 7.5% oil. But hexane-extracted oil contains high content (33.8%) of FFA besides 2.89% unsaponifiable matter. SC CO_2 extraction was performed at 100 to 350 bar on hot water-treated grape seeds (Molero-Gomez et al., 1994). Hot water is added in order to get all the remaining sugars from the grape seeds in the wash water for alcohol production by fermentation. As a result, the total oil content is always less in the treated grape seeds than in the untreated ones. The yield of oil by SC CO_2 extraction at 350 bar and 40°C was found to be 6.9%, and the level of FFA and unsaponifiable matter were significantly lower (3.41% FFA and 0.27% unsaponifiables) than in the hexane-extracted product. Hexane extraction required 20 h vis-à-vis 3 h by SC CO_2 extraction and the quality of the oil extracted by SC CO_2 is equivalent to the refined (degummed, deacidified, and deodorized) oil obtained from the hexane-extracted crude grape seed oil.

10.7.6 Animal Lipids

Marine oils like cod and shark liver oils are rich sources of PUFA. Oily fish, such as anchovy, mackerel, sardine, menhaden, etc., are primarily used for fish oil production and constitute the conventional sources of PUFA. Most of these oils are presently hydrogenated and consumed in the form of margarine or shortening. Some nonconventional sources of PUFA include algae, fungus, krill, squids, etc.

Approximately 11 to 14% of fungal lipids are present in the organism *Saprolegnia parasitica*. These lipids constitute 75.8% neutral and 24.2% polar lipids based on Acetone Insolubles (AI). The neutral lipids comprise 36.3% TG, 19.6% DG, 12.3% MG, and 7.6% FFA. It is interesting to note that the EPA content is different in different fractions of fungal lipids of *S. parasitica* (Cygnrowicz-Provost et al., 1991), as shown in Table 10.5. SC CO_2 extraction of important PUFA, such as EPA, DHA, and other ω-3 fatty acids, from filamentous fungi *S. parasitica* was investigated at pressures in the range of 310 to 380 bar and temperatures in the range of 40 to 80°C with 10 wt% methanol or ethanol as entrainer. The total lipids recovered using SC CO_2 increases with increasing temperature, for example, recovery efficiency was 75% at 60°C, whereas at 40°C it was 47.5%. With increasing pressure from 310 bar to 380 bar, it increased from 49.8 to 75%. At a given temperature and pressure (60°C and 380 bar), the addition of 10% methanol or ethanol increased the recovery to 88.8 and 82.9%, respectively. However, the addition of an entrainer resulted in extracts containing 9.2 to 10.2% EPA and in lowering of selectivity for EPA. It was suggested that a two-step procedure would be preferable to use for recovering EPA, i.e., by extraction of saturated and lower molecular weight fatty acids with pure SC CO_2 at the beginning followed by extraction of more concentrated EPA fraction with SC CO_2 mixed with 10% ethanol, because EPA is present in the polar lipids of *S. parasitica* and not in the neutral lipids. SC CO_2 extraction of PUFA from Mortrielle genus yields ω-3 linolenic acid. It does not contain EPA.

Marine algae contain lipids stored within cells and in cell walls. These lipids consist mainly of acyl glycerols and free fatty acids. *Skeletonema costatum*, a marine alga, is known to have high content of PUFA. Another variety of algae, *Ochromonas*

TABLE 10.5
Amounts and Composition of Different Fractions of Total Fungal Lipids (11 to 14 wt% Dry Fungus) in *S. parasitica*

Fraction	Wt% of Total Lipids	Wt% EPA in the Fraction
TG	36.3	7
FFA	7.6	16
DG	19.6	28
MG	12.3	6
AI	24.4	25

Cygnarowicz-Provost et al., 1991.

danica, contains even higher levels of lipids. These algae can be easily cultured to increase their lipid content by genetic engineering. *O. danica* is a freshwater alga. Total lipid content of Skeletonema varies between 5.6 to 11.5% dry weight, whereas that in Ochromonas, 28% dry weight. SC CO_2 extraction of these two microalgae species was explored at pressures in the range of 170 to 310 bar at 40°C. It was observed (Polak et al., 1989) that the extraction pressure did not have much effect in improving the yield from Skeletonema algae, rather the pretreatment and culture age affected the yield and quality of the extracted oil. The SC CO_2 extracts of Ochromonas had higher total lipid content than that of Skeletonema, but the former had a lower concentration of EPA. SC CO_2 extracts had about 25% EPA in both algae extracts. Chlorophylls were not co-extracted, eliminating the need for bleaching, as in the case of conventional solvent extraction (Polak et al., 1989). Another variety of high yielding microalgae, *Chlorella vulgaris*, contains both carotenoids and lipids in substantial quantities. While this can produce canthaxanthin and astaxanthin, this can also be used for recovering high value oil. SC CO_2 extraction of finely ground powder of freeze-dried microalgae, *C. vulgaris*, at 350 bar and 55°C yielded as high as 55% lipids (Mendes et al., 1994), whereas slightly crushed microalgae resulted in 24.5% total lipids. Hence pretreatment is very important for improving the recovery efficiency of SC CO_2 extraction of lipids from both algae and fungi.

Other sources of EPA include a type of freeze-dried algae, such as *Palmaria palmata*, oil of which contains 45% EPA.

The lipids of the seaweed *P. palmata* constitute tryglycerides, free fatty acids, sterols, hydrocarbons, and phospholipids. These compounds have different solubilities in SC CO_2. The recovery of lipids is normally invariant with pressure and temperature variation in a relatively smaller range (Mishra et al., 1994). The fatty acid compositions of the seaweed extracts were determined and their contents are given in Table 10.6. It was observed that a pressure of 208 bar was sufficient to extract lipids from *P. palmata* seaweed and that EPA recovery decreased at higher pressures due to lowering of selectivity. The extracted EPA was distributed between neutral and polar lipids. Addition of ethanol not only led to an increase in the overall solubility, but coextraction of polar lipids also increased the amount of EPA in the

TABLE 10.6
EPA Content of Lipid Extracts from *P. palmata* by SC CO_2

Pressure (bar)	Temperature (°C)	Solubility (mg/l)	EPA (% Lipid)
207.8	35	3.47	0.59
	45	3.27	1.04
	55	3.37	0.78
207.8 (with ethanol)	45	9.35	
414.7	35	3.26	0.27
	45	2.88	0.23
	55	3.63	0.16
621.5	35	3.37	0.19
	45	3.28	0.06
	55	3.33	0.18

Mishra et al., 1994.

TABLE 10.7
Effects of Ethanol on Lipid Profile Extracted from *P. palmata* by SC CO_2 at 208 Bar and 45°C

Fatty Acid	Without Ethanol (wt%)	With 10% Ethanol (wt%)
14:0	3.3	4.2
16:0	15.7	18.3
16:1 (ω-5, ω-7)	4.5	4.9
18:0	1.8	3.1
18:1 ((ω-9, ω-7)	5.0	9.2
18:2 (ω-6)	1.3	1.3
18:4 (ω-3)	1.4	trace
20:4 (ω-6)	2.8	3.0
20:5 (ω-3)	64.1	56.1

Mishra et al., 1994.

extract. In spite of higher initial concentration of EPA (45%) in the algae, there was poor recovery of EPA by SC CO_2. When ethanol was added (10% w/w) to the seaweed for presoaking, the lipid recovery increased with a slight increase in the EPA recovery, as can be seen from the fatty acid profiles of lipids extracted with and without ethanol (Table 10.7). Due to increased coextraction of saturated fatty acids, the EPA content in the lipid decreased moderately (Mishra et al., 1994).

Edible snails are a rich source of lipids and extraction of fatty oils from edible snails and separation of concentrated Chondroitin fraction are of great interest for use in health foods. The lipids comprising C_{18} unsaturated triglycerides, mainly oleic and linoleic, and high-value DHA and EPA could be recovered by using SC CO_2 at

TABLE 10.8
Definition and Classification of Polar Lipids

Terminology	Definition
Polar lipids	Lipids insoluble in acetone or acetone insolube (AI)
Phospholipids	Lipids containing phosphorous (phospholipids with a fatty acid chain length less than C_{10} are soluble in acetone)
Lecithin (commercial)	Mixture of polar and neutral lipids with a polar lipid content of at least 60% (as per European community description)
Lecithin	Scientifically, it is referred to as a particular phospholipid, namely, phosphatidylcholine
Glycolipids	Glyceroglycolipids are compounds linked to the carbohydrate unit and the 3-hydroxy group by means of an α or β glycosidic bond; they are simply sugar containing lipids

Classification

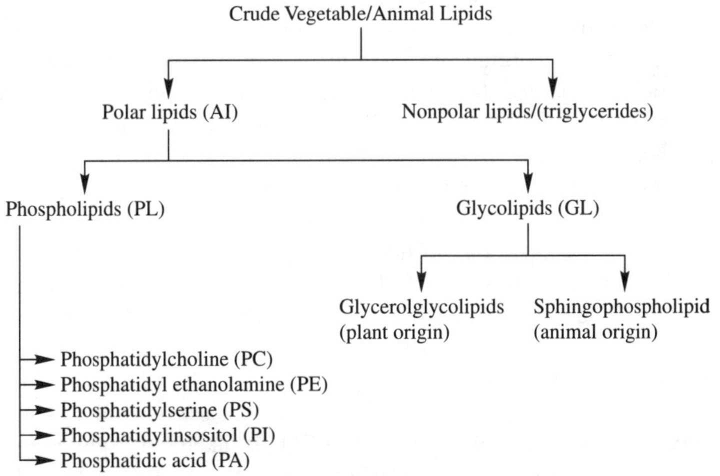

35°C and 124 bar with the highest recovery of 25% yield, but the highest concentrations of PUFA, EPA, and DHA could be obtained at a lower pressure of 103 bar at 35°C (Chun, 1998).

10.8 DEOILING OF LECITHIN BY SC CO_2

Lecithins are polar organic compounds consisting of glucosides on a phosphatide backbone and are mainly used as emulsifiers in the food industry. Some definitions and classifications are given in Table 10.8.

Commercial or crude lecithin is a complex mixture of phosphatides, triglycerides, phytoglycolipids, tocopherols, and fatty acids and is produced as a byproduct of the crude oil degumming process used for refining. This process is carried out by mechanically agitating the crude oil with water whereby the polar lecithin fraction is transferred to aqueous phase. Crude lecithin is obtained by water washing and drying of the hydrated gum. Pure lecithin is recovered from this natural mixture of neutral and polar lipids from both vegetables and animal sources. Polar lipids such as glycolipids and phospholipids, in contrast to neutral lipids, are almost insoluble in acetone. Neutral lipids are mainly triglycerides whereas polar lipids consist of glycolipids (sugar-containing lipids) and phospholipids (phosphorus-containing lipids). Crude vegetable oils may contain as high as 20% polar lipids or acetone-insoluble (AI) matter. Lecithin may also be manufactured by ultrafiltration of vegetable oils.

Presently pure lecithin is commercially produced from crude or commercial lecithin obtained from crude soybean, canola, cottonseed, corn, sunflower, and olive husk oils. Scientifically, pure lecithin refers to phosphatidylcholine (PC). Lecithin with high PC content has better emulsifying power than the phospholipid mixture, and hence it is necessary to enrich, refine, or fractionate polar lipids or phospholipids from deoiled flakes/seeds or from crude lecithin. Crude lecithin contains about 35% neutral lipids, which must be reduced to less than 2% in refined or pure lecithin for use in the food and pharmaceutical industries as emulsifier, surfactant, or antioxidant.

The conventional process for separation of neutral and polar lipids involves two steps, namely charging the mixture of crude lecithin and excess acetone into a separation tank, and then removing the insoluble polar lipid material which is later dried into the form of granules. The deoiled lecithin obtained is a light yellow solid. In another process, hexane solution of crude lecithin is passed through a silica column, when neutral lipids get adsorbed, whereas phospolipids pass through as they form aggregate in nonpolar solvent like hexane (Schneider, 1988). Alternatively SC CO_2 at a temperature of about 40°C and at a pressure of about 300 bar can be used to extract neutral lipids from the crude lecithin mixture, leaving the polar lipids as the raffinate. One disadvantage of the process is that during removal of the oil, the viscosity increases drastically and further extraction is hindered, making the continuous process difficult to operate. To overcome this difficulty a propane + carbon dioxide mixture was used to reduce the viscosity of deoiled lecithin (Alkio, 1998). Neutral-oil-loaded gas, rich in CO_2, is taken out from the top of the counter-current column of extraction, whereas liquid solution of deoiled lecithin rich in propane, is removed from the bottom. Although this process is superior to the acetone-treatment method, unfortunately the noninflammatory and environmental advantages of the SC CO_2 method are not prevalent when propane is mixed with it.

In order to overcome the problem of increasing viscosity with extraction of oil from lecithin, a spraying device was developed (Eggars and Wagner, 1994) and a process for the extraction from high viscous media in a turbulent two-phase flow was employed. When preheated crude lecithin is pumped into the spraying device and dispersed by the cocurrent high turbulent CO_2, the high velocity of CO_2 is utilized to create an extended surface area of droplets by reducing the drop sizes. The resulting two-phase zone flows down for a short path when extraction occurs. The heavier deoiled lecithin particles drop as powdery product into the collection vessel. The oil-loaded CO_2 is depressurized in a separator, where the oil separates out at a lower pressure.

Egg yolk contains about 35% lipids (on dry basis) comprising 65% TG, 30% PL, 4% cholesterol, and traces of carotenoids. The SC CO_2 extraction of freeze-dried egg yolk using 3 to 5% methanol or ethanol as an entrainer at 360 bar and

40°C resulted in fractionation of oil from phospholipids. After most of the oil was removed, the extract contained 7 to 17% phospholipids (Bulley and Labay, 1992). The role of ethanol addition as a cosolvent as well as for feed pretreatment is highlighted in the following discussions for the recovery and purification of lecithin from various natural sources.

10.8.1 Soya Phospholipids

A major part of lecithin is now manufactured as a byproduct of the soybean oil refining process. The crude soybean oil obtained by hexane extraction contains 2 to 3% polar lipids. As mentioned earlier, the commercial soya lecithin contains 35% nonpolar lipids and 64.3% polar lipids (AI) comprising 86% phospholipids (PL) and 14% glycolipids (GL), and 0.7% moisture. Heigel and Huescherts (1983) patented a process for deoiling crude soya lecithin with SC CO_2 at 400 bar and 60°C, which required 4 h to obtain light yellow-colored lecithin as the raffinate.

In the traditional process of hexane extraction, PL gets coextracted with the oil. However, if SC CO_2 is used for the extraction of oil from soybeans, then the deoiled meal retains most of the PL. When PL are selectively removed from the deoiled meal using SC CO_2 + ethanol mixture, phosphatidylcholine (PC) or pure lecithin and phosphatidyl ethanolamine (PE) are more readily solubilized in SC CO_2 + ethanol mixture, rather than phosphatidylinosital (PI). The extent of recovery of PC and PE is dependent on the ethanol content in SC CO_2 (Montanari et al., 1994). The SC CO_2 process for recovering PL directly from deoiled flakes is compatible with the end use of the meal protein for human consumption. The process has the flexibility of producing PL mixture with varying ratio of PC to PE, while leaving PI with the raffinate meal. The process can be described as:

10.8.2 Oat Lecithin

Polar lecithin of oats is of interest due to their superior emulsification and anti-oxidative properties. The oat grain contains 2 to 12% lipids. The oat oil is extracted with an alcohol, such as 2-propanol, and the composition of crude oat oil is as given in Table 10.9 (Alkio et al., 1991).

Oat oil markedly differs from soya and canola oils. Crude oat oil contains about 20% polar lipids and 6 to 7% phospholipids, whereas soya and canola oils contain 2 to 3% polar lipids and sunflower oil contains 1 to 2% polar lipids. Hence oat lecithin may be recovered from the crude oat oil directly. Unlike soya lecithin, oat

TABLE 10.9
Composition of Oat Oil

Nonpolar part (unbound form)	**(%)**
Triglycerides	50.6
Free fatty acids	11.1
Diglycerides	3.1
Sterols	0.5
Polar part (bound form)	
Glycolipids	11.8
Phospholipids	7.1

Alkio et al., 1991.

lecithin has a very pleasant aroma. The nonpolar fraction of oat oil was removed by SC CO_2 extraction, at 250 bar and 40°C (Alkio et al., 1991) and the dry raffinate (23.5%) contained the polar lecithin. The phospholipid content of this SC CO_2 raffinate was 30 to 35%. At a higher pressure of 300 bar and at a higher temperature 55°C, the yield of the raffinate was lower (21.6%). The antioxidative power of the oat lecithin was found superior to that of soya lecithin containing more than 50% phospholipids. Oat lecithins obtained directly from the crude oat oil by the single-step SC CO_2 extraction are crystalline pale brown needles or flakes.

10.8.3 Canola Lecithin

Soya lecithin and canola lecithin are very similar in their physicochemical characteristics. They can be extracted either from flakes after removal of oil, or alternatively from the crude lecithin obtained from the degumming process. Soybean and canola oil extracted with SC CO_2 from flakes contain less than 100 ppm of phospholipids (PL), (whereas hexane extracted oil contains 1% PL) because most of the phospholipids are left behind in the SC CO_2 processed raffinate. Accordingly, SC CO_2 deoiled flakes are considered a good feed material for extraction of phospholipids (PL).

SC CO_2 mixed with ethanol can be used to efficiently extract and fractionate polar lipids from deoiled canola flakes. Soaking the deoiled flakes with ethanol prior to extraction with SC CO_2 mixed with ethanol, increases the efficiency of PL recovery because the triglyceride (TG) concentration in the flakes is already reduced. The PL content of the extract increases by increasing the amounts of ethanol added to SC CO_2 and ethanol used for soaking. Phosphatidylcholine (PC) concentration in PL extracted from deoiled flakes can be as high as 78% of PL. This (PC) can be further enhanced, if PL is recovered from deoiled lecithin (AI), to as high as 90%, with 50% of the extract being PL (Dunford and Temelli, 1994), when SC CO_2-mixed ethanol is used as the extractant and the feed (AI) is presoaked with ethanol.

For deoiling, the full-fat canola flakes are extracted with SC CO_2 at 552 to 621 bar and 45 to 70°C to recover PL-free triglycerides, since SC CO_2 has higher selectivity for nonpolar triglycerides with respect to polar PL, when large amounts of triglycerides are present in the feed material. Even by using SC CO_2 mixed with ethanol as the extractant, the PL concentration in the extracted oil remains still less than 200 ppm. The oil content in the flakes is lowered as much as possible to increase the efficiency of the PL recovery and PL concentration in the extract in the subsequent step.

For PL recovery from the raffinate flakes, ethanol is added to the deoiled flakes for presoaking as well as to the SC CO_2 stream. At 552 bar and 70°C, increasing ethanol content in SC CO_2 from 5.0 to 10.1% (Dunford and Temelli, 1994), the PL content in the extract from the partially depleted canola flakes (containing 14% residual oil) increased from 0.5 to 5.4 wt%. On the other hand, when the deoiled canola lecithin (AI) was extracted with SC CO_2 mixed with ethanol at 552 bar and 45°C, it was possible to obtain a higher PL concentration of 42% in the extract.

It was reported that AI formed a hard cake after partial extraction if it was not mixed with ethanol. But when the deoiled lecithin (AI) sample was presoaked with ethanol and then extracted with SC CO_2 with 13% ethanol at 552 bar and 70°C, an increase in ethanol to AI ratio for soaking from 1 to 3, resulted PL concentration in the extract to increase nearly four times from 9 to 32% and PC content in PL from 76 to 86% (Dunford and Temelli, 1994). However, the extraction from the deoiled canola flakes using SC CO_2 mixed with 6.5% ethanol at 552 bar and 70°C, yielded 5.4% PL in the extract out of which only 4.2% was PC, whereas the extraction from deoiled canola crude lecithin (AI) presoaked with ethanol (1:3), with SC CO_2 mixed with 6.5% ethanol yielded extracts with 50% PL (90% of which was PC). This is attributed to the higher solubility of PC in ethanol-mixed SC CO_2 solvent compared to the other PL constituents. The effects of ethanol content in the solvent and in the feed on the content of PL in the extract and the concentration of PC in PL are summarized in Table 10.10. Thus, SC CO_2 + ethanol mixture has been proved to be an effective solvent for recovering phospholipids enriched with PC, from both pretreated flakes and presoaked deoiled lecithin (AI).

10.8.4 PC from Deoiled Cottonseed

SC CO_2 mixed with ethanol is found to selectively fractionate PC from polar lipids present in deoiled cottonseeds having 0.5% PC and 10% TG. The percent recovery and the concentration of PC in the extract are enhanced by increasing the ethanol concentration in SC CO_2 in the range from 1.5 to 6.5% at pressures in the range from 130 to 230 bar and at temperatures 50 and 55°C. It was reported by Sivik et al. (1994) that 80% PC from the deoiled cottonseed could be recovered at 230 bar and 50°C with 6.5% ethanol in SC CO_2. Extractability of PC increased with ethanol concentration, pressure, and presence of TG in SC CO_2. The loading of PC in the solvent phase was found to be 50 and 150 mg/kg at pressures of 350 and 250 bar, respectively, with 5% ethanol in SC CO_2 at 55°C. That is, the solubility of PC at 250 bar was found to be more than that at 350 bar; this may be attributed to the synergestic effect of TG as the loading of TG is 65 mg/kg at the higher pressure of 350 bar, compared to only 15 mg/kg at 250 bar. This implies that PC can be selectively extracted from deoiled cottonseeds with proper selection of pressure, temperature, ethanol concentration in the solvent, and pretreatment of the starting material.

TABLE 10.10
Effect of Ethanol Contents in the Solvent and in the Feed on PL Recovery and PC Concentrations in the Extract from Canola Flakes and AI

Feed	Solvent	P (bar)	T (°C)	PL Content % Extract	PC Content % PL
Flakes (43% oil) PC: 3.7% PL	$SCCO_2$	552–621	45–70	n.d.	n.d.
Flakes (43% oil) PC: 3.7% PL	$SCCO_2$ + 8% Ethanol	552	70	<0.02	—
Flakes (23% oil)	$SCCO_2$ + 10.1% Ethanol	552	55	0.7	—
Flakes (14% oil)	$SCCO_2$ + 5% Ethanol	552	70	0.5	—
Flakes (14% oil)	$SCCO_2$ + 10.1% Ethanol	552	70	5.4	78
Flakes (12% oil)	$SCCO_2$ + 10.3% Ethanol	552	55	1.3	—
AI (72% PL) PC: 42% PL	$SCCO_2$ + 13% Ethanol	552	45	42	90
AI (72% PL) PC: 42% PL	$SCCO_2$ + 13% Ethanol	552	70	8	50
AI + ethanol (1:1)	$SCCO_2$ + 13% Ethanol	552	70	9	76
AI + ethanol (1:1)	$SCCO_2$ + 6.5% Ethanol	552	70	8	70
AI + ethanol (1:2)	$SCCO_2$ + 13% Ethanol	552	70	28	88
AI + ethanol (1:2)	$SCCO_2$ + 6.5% Ethanol	552	70	30	86
AI + ethanol (1:3)	$SCCO_2$ + 13% Ethanol	552	70	32	86
AI + ethanol (1:3)	$SCCO_2$ + 6.5% Ethanol	552	70	50	90

Adapted from Dunford and Temelli, 1994.

10.8.5 PC FROM EGG YOLK

SC CO_2 mixed with ethanol has been found to be an effective solvent for recovering PC from low-fat egg yolk containing 22% PC and 10% TG and from whole-fat egg yolk containing 14% PC and 40% TG in the pressure range of 230 to 350 bar at 45 and 50°C with 5 to 6.5% ethanol in SC CO_2. The solubility of PC is reported to be dependent on pressure and ethanol concentration rather than the initial PC content in the feed material. The selectivity of PC with respect to PE and cholesterol was found to be very high (Sivik et al., 1994). It is reported that the solubility of PC is strongly dependent on the TG solubility because TG also acts as an entrainer when present in a smaller concentration in the feed. Increasing pressure and ethanol concentration improved the selectivity and yield of PC much more than those of nonpolar components. For example, in the case of whole-fat (dried) egg yolk, the PC concentration in the extract increased from 5.0 to 46%, whereas the concentration of a nonpolar lipid like cholesterol increased from 9 to 11%, and also the PE content in the extract increased from 1 to 4% by increasing ethanol content in SC CO_2 from 2 to 5%, respectively, at 250 bar and 45°C. The loading of PC in the solvent phase was found to be very close to 1 g/kg CO_2 for the whole-fat egg yolk, which was one order in magnitude lower for the low-fat egg in SC CO_2 with 5% ethanol at 40°C and was not found to be very different at the two pressures for either type of egg yolk (Sivik et al., 1994). This implies that the presence of more TG in the feed enhances the loading of both TG and PC synergistically. Thus PC can be selectively recovered from egg yolk, while simultaneously removing cholesterol using SC CO_2 mixed with 5% ethanol.

10.9 DILIPIDATION AND DECHOLESTERIFICATION OF FOOD

In recent years there has been increasing demand for "low fat" and "low cholesterol" food products by consumers who have become more conscious and concerned about maintaining good health and deriving maximum nutrition from food. Both cholesterol and saturated fat present in food adversely affect blood cholesterol and increase the risk of atherosclerosis and cardiovascular diseases. Accordingly several processes have been developed using SC CO_2 for delipidation and decholesterification of food products, resulting in low-fat wafers, defatted corn germs, deoiled lecithin, low-fat and low-cholesterol egg yolk powder, low-cholesterol high melting butter, defatted milk powder, etc. The basic principle behind these separation processes entails higher and variable selectivity of SC CO_2 and SC CO_2 mixed with a polar cosolvent for nonpolar triglycerides and slightly polar cholesterol, respectively, compared to proteins, carbohydrates, mineral salts, or polar lipids. Generally, the delipidation process is aimed at the partial removal of fat, since some residual fat is important from the point of view of flavor, texture, and taste of the residue. Selection of operating conditions, such as temperature, pressure, and cosolvent is made in order to preserve the thermosensitive bioactive compounds including avoiding protein denaturation. SC CO_2 delipidation process was first applied in deoiling potato chips by Hannigan (1981). Reduction of oil from 40 to 20% could be achieved while increasing the protein content from 1.5 to 2% without affecting the texture and flavor characteristics of the original product (Castera, 1994) at an optimum pressure of 410 bar and a temperature of 50 to 60°C. For a specified oil removal target of 66% of initial oil, extraction time could be less than 1 h at an optimum flow rate of CO_2 (Vijayan et al., 1994).

Corn germ is highly useful as a raw material for dietary fiber and is used for the fortification of health foods. Yamaguchi (1989) reported a process for deodorizing corn germ with SC CO_2 mixed with 3 to 15% ethanol at 250 to 400 bar, at a temperature of 35 to 40°C after deoiling the corn germ with SC CO_2 mixed with hexane, or with only SC CO_2. The processing conditions are given in Table 10.11. There was a reduction in oil and an increase in protein and fibrous material, as indicated by the water absorption results.

It is generally believed that decreasing food cholesterol decreases blood cholesterol. However, there are good cholesterols (e.g., high-density lipoprotein) as well, and cholesterol consumption of 300 mg/day is allowed by doctors. Cholesterol contents in various food items are shown in Table 10.12.

The conventional processes to remove cholesterol include molecular distillation, precipitation, solvent extraction, adsorption or complex formation because of the relatively high polarity, presence of free hydroxyl group and because of higher molecular weight (twice that of triglyceride) of cholesterol. However, SC CO_2

TABLE 10.11
Delipidation of Corn Germ Using SC CO_2

Initial Oil Content in Germs	Pressure (bar)	Temp. (°C)	Cosolvent	Oil in Product	Water Absorption (ml/g)
50	250	40	None	1.51	4.0
25	350	38	Hexane (3%)	0.67	5.5
10	250	40	Ethanol (3%)	0.29	11.0

Yamaguchi, 1989.

TABLE 10.12
Cholesterol Content in Food

Food Items	Cholesterol Content (%)
Lamb's brain	1.5–2.0
Egg yolk	1.5–2.0
Fish egg	0.4–0.7
Liver	0.4–0.7
Poultry egg	0.4–0.7
Butter	0.25–0.31
Tallow, lard, fish, cheese, milk	0.1

extraction offers the advantage of removing both cholesterol and triglycerides in a single step, with simultaneous fractionation of triglycerides on the basis of molecular weight and degree of unsaturation (Castera, 1994).

For a study on the commercial feasibility of utilizing SC CO_2 for decholesterolification of butter, beef tallow, and egg yolk, the distribution coefficients and selectivities were measured at different pressures and temperatures as given in Table 10.13 (Krukonis, 1988). It can be seen that the distribution coefficients of cholesterol for these three materials are of similar magnitude, but the selectivity of SC CO_2 for cholesterol is the highest in the case of egg yolk and the lowest in the case of butter. The analysis of fatty acid compositions of these materials revealed that egg lipids contain mostly C_{16} and higher-carbon-number fatty acids, whereas butter contains many fatty acids of C_{10} and lower, while the tallow is intermediate. Triglycerides of C_4 to C_8 fatty acids (as in butter) are more soluble than those with C_{16} or higher fatty acids (as in egg yolk). As a consequence, lower molecular-weight triglycerides can be easily extracted from butter by SC CO_2, resulting in a lower selectivity for cholesterol in butter than in tallow or egg yolk. However, the selectivity values of 3 to 5 are sufficient for complete (95% and more) removal of cholesterol from butter. Fractionation of triglycerides and selective reduction of cholesterol using SC CO_2 extraction will be discussed for the three most important high-nutrition food products, namely, butter, egg yolk, and fish/meat muscles in the following subsections.

TABLE 10.13
Selectivity and Distribution Coefficients for Decholesterolification

Temperature, Pressure	Distribution Coefficient	Selectivity
	Cholesterol from Butter	
60°C,150 bar	0.0014	3.89
80°C,170 bar	0.0019	4.57
80°C,155 bar	0.0012	5.73
	Cholesterol from Tallow	
60°C,150 bar	0.0013	9.64
80°C,155 bar	0.0011	11.44
	Cholesterol from Egg Yolk	
60°C, 150 bar	0.0014	12.21

Krukonis, 1988.

10.9.1 Butter

Butter or anhydrous milk fat (AMF) contains about 98% triglycerides with carbon numbers in the range from 24 to 54 besides minor constituents, like DG, MG, FFA, PL, and cholesterol. During the process of the fractionation of butter fat, the short-chain triglycerides (C_{24} to C_{34}) are preferentially extracted by SC CO_2 at pressures in the range of 100 to 150 bar and at temperatures in the range of 40 to 50°C; the medium-chain triglycerides (C_{36} to C_{42}) at 200 to 250 bar; while triglycerides with a carbon number more than C_{46} at pressures around 350 bar at 70°C. The cholesterol content in the butter can be reduced to about 0.1% from 0.31% by using SC CO_2 at 150 to 175 bar and 50°C, and 65% of cholesterol is coextracted with triglycerides (Majewski et al., 1994). Cholesterol is individually less soluble than triglyceride in SC CO_2, but since it possesses high affinity for short- and medium-chain triglycerides, its selectivity is enhanced by the presence of triglycerides and the extract gets enriched in cholesterol as compared to the original butter. The moisture content needs to be reduced below 15%, preferably in the range of 2 to 8% because it is difficult to extract cholesterol from food with a high moisture content. However, as shown in Figure 10.11, the solubility of triglycerides increases linearly with pressure. But in the presence of water in oil, the solubility of triglycerides increases with pressure up to only 200 bar (not shown in figure), and then decreases with increasing pressure. However, the extractability of cholesterol at 40°C is highest at 150 bar, and the extractability of cholesterol from dehydrated butter is less than that from normal butter at 40°C, as can be seen from Figure 10.11. The optimum condition for SC CO_2 extraction for obtaining the highest selectivity of cholesterol over triglycerides was thus reported to be 150 bar at 40°C for normal butter (Shishikura et al., 1986).

Figure 10.12 shows the selectivity behavior of cholesterol removal at 50 and 60°C as a function of pressure. It can be noted that the selectivity increases with pressure up to an optimum pressure beyond which it marginally decreases with pressure. The

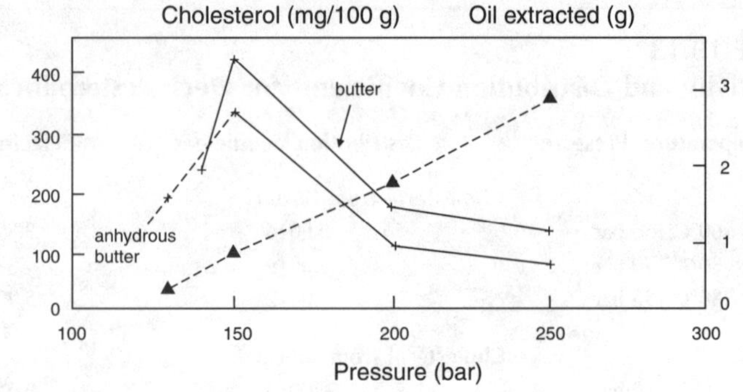

FIGURE 10.11 Effect of SC CO_2 pressure at 40°C on the extraction efficiency of the cholesterol and the oil from butter; (+) cholesterol; (----) butter oil (Castera, 1994).

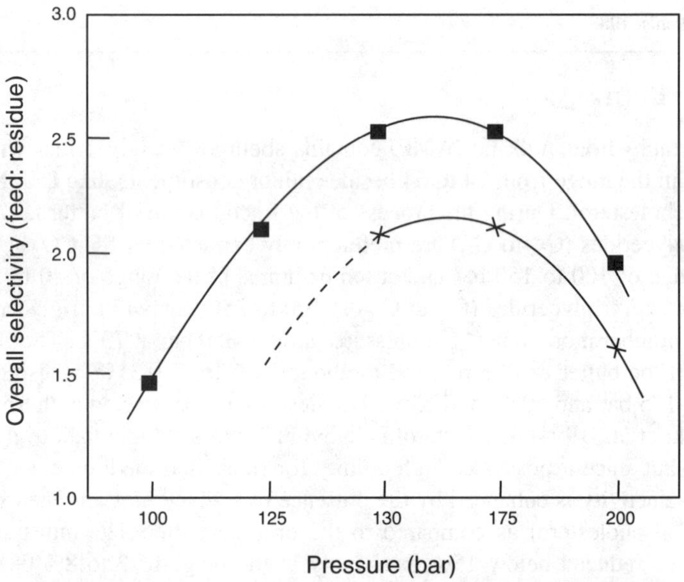

FIGURE 10.12 Selectivity of SC CO_2 for cholesterol removal at (■) 50°C; (+) 60°C in supercritical fractionation of butter oil (Majewski et al., 1994).

highest selectivity (2.6) for cholesterol reduction, by 66%, was achieved at pressures between 150 and 175 bar at 50°C, and a better value of selectivity could be expected at temperatures below 50°C, say, at 40°C (Majewski et al., 1994). This value of selectivity is, however, less than that reported by Krukonis (1988), as can be seen in Table 10.13. Along with cholesterol, some short-chain triglycerides with low melting points are coextracted. As a result, the melting characteristics of butter are improved, giving the butter good consistency and spreadability at ambient temperature.

Bradly (1989) employed a two-stage extraction process at a higher temperature. In the first stage, SC CO_2 at a lower pressure (less than 160 bar) allows extraction

TABLE 10.14
SC CO_2 Extraction of Butter Fat and Lard Using an Adsorption Column

Sample	Initial Cholesterol Content (%)	Adsorbent and Fat: Adsorbent Ratio (w/w)	Pressure Bar	Temp. (°C)	% TG Recovered	Cholesterol Adsorbed (%)
Butter	0.242	$Ca(OH)_2$:0.5/45	220	45	80.0	100
Butter	0.242	MgO:2/45	220	35	59.7	100
Lard	0.991	$Ca(OH)_2$:0.5/40	220	35	97.7	100
Lard	0.991	$CaCO_3$:0.4/40	220	45	100	99.7

MacLachlan and Catchpole, 1990.

of flavors and low molecular weight and unsaturated triglycerides from butter. The second stage extraction at a pressure higher than 170 bar, removes cholesterol, but high melting triglycerides are not coextracted. The light extract and the heavy residue with reduced cholesterol content are mixed together to obtain the final product (butter) with the desirable melting characteristics.

Cholesterol can be selectively fractionated from butter oil using a continuous countercurrent adsorbent (silica gel) column through which butter oil solubilized in SC CO_2 is passed. A basic adsorbent is preferable to an acid adsorbent, because triglycerides, flavor components, and pigments should not be adsorbed. Using a silica gel column, 94% of the initial cholesterol of anhydrous butter oil can be adsorbed from the butter oil extracted in SC CO_2 stream at 300 bar and 40°C, when a silica gel to butter ratio of 3:1 (w/w) is used (Sishikura et al., 1986). By using calcium hydroxide adsorbent, 100% of the cholesterol is retained in the adsorbent, while 80% of the triglycerides are recovered, as shown in Table 10.14. The extracted butter has a lower melting point than the initial butter due to fractionation of triglycerides.

10.9.2 Egg Yolk Powder

Egg yolk contains about 35% lipids, of which 65% is triglyceride, 30% phospholipids, 4% cholesterol, and rest carotenoids. SC CO_2 extraction of egg yolk powder at about 300 bar and 45°C removes approximately two thirds of the cholesterol and one third of the total neutral lipids (Castera, 1994). The moisture content, however, needs to be lowered below 15%, preferably to 2 to 8% prior to SC CO_2 extraction of cholesterol from egg yolk powder. This is also true for milk powder or dry meat, for which SC CO_2 extraction is normally carried out at 30 to 45°C and pressures in the range of 130 to 250 bar. At these conditions, cholesterol can be selectively removed without heat denaturation of proteins. Even vegetable oil can be used as the second cosolvent for improving the selectivity of separation for cholesterol.

Rossi et al., (1990) employed SC CO_2 mixed with ethanol at 150 bar and 50°C to extract cholesterol from egg yolk powder. Extraction of total lipids increased from 3% (when pure CO_2 is used) to about 30% when 7% ethanol was added to SC CO_2, while the protein level increased from 40 to 53%, and the cholesterol content decreased from 2.2 to 1.4% in the raffinate yolk powder.

TABLE 10.15
SC CO_2 Extraction of Fish Muscle at 275 bar and 40°C

Solvent	Time of Extraction (h)	Reduction of Lipids (%)	Reduction of Cholesterol (%)
Pure CO_2	9	78 (TG)	97.0
SC CO_2 + 10% ethanol	6	97 (TG+PL)	99.7

Hardardottir and Kinsella, 1988.

TABLE 10.16
SC CO_2 Extraction of Meat Samples at 220 bar and 40°C

Sample	Initial Moisture Content (%)	Total Lipids Decrease (%)	Cholesterol Decrease (%)
Chicken	76.2	38.0	3.5
Cooked beef	58.1	47.6	21.5
Dried lean meat	54.0	67.9	38.5
Dried lean meat	45.3	93.7	63.9

MacLachlan and Catchpole, 1990.

10.9.3 Fish and Meat Muscles

Some meats and fish have high levels of unsaturated fatty acids for which they have relatively short shelf life and give out rancid and unpleasant flavor. This requires delipidation. Decholesterolification is also desirable since there are also substantial amounts of cholesterol. SC CO_2 extraction with and without the addition of ethanol can be utilized for simultaneous delipidation, decholesterification, and protein enrichment of fish and meat samples, as shown in Tables 10.15 and 10.16. The raffinate fish extracted by neat SC CO_2 and SC CO_2 mixed with ethanol contain 97 and 99% protein, respectively, as a result of the simultaneous removal of fat, cholesterol, and water. Protein is not soluble in SC CO_2 and so is not removed during the process.

Defatting of meat samples involves three sequential steps: (1) reduction in particle size by slicing or flaking the frozen meat, (2) moisture removal, and (3) SC CO_2 extraction (MacLachlan and Catchpole, 1990) in the pressure range of 200 to 300 bar and temperature in the range of 30 to 50°C. Drying is preferably carried out by freezing in an inert atmosphere using SC CO_2 to reduce the water content to 30 to 50% w/w. In general, SC CO_2 extraction efficiency largely depends on the quantity of lipids, moisture, lipoprotein, and proteophospholipid complexes present in the meat. By proper selection of the process conditions of SC CO_2 extraction and fractionation, it is possible to fine tune the selectivity of separation of cholesterol and lipids simultaneously in order to get the best possible final product.

REFERENCES

Alkio, M., Aaltonen, O., Kervinen, R., Forssell, P., and Poutanen, K., Manufacture of lecithin from oat oil by supercritical extraction, Proc. 2nd Int. Symp. Supercritical Fluids, Boston, MA, 276, 1991.

Bamberger, T., Erickson, J. C., Cooney, C. L., and Kumar, S. K., *J. Chem. Eng. Data*, 33, 327, 1988.

Bharath, R., Adschiri, T., Inomata, H., Arai, K., and Saito, S., Separation of fatty acids with supercritical CO_2, Proc. 2nd Intl. Symp. Supercritical Fluids, McHugh, M., Ed., Boston, MA, 288–291, 1991.

Bondioli, P., Mariana, C., Lanzani, A., Fedeli, B., Mossa, A., and Muller, A., Olive refining with SC CO_2, J. A. O. C. S., 69, No. 5, 477, May, 1992.

Brunetti, L., Daghetta, A., Fedeli, B., Kikic, I., and Zanderight, L., Deacidification of olive oils by supercritical carbon dioxide, *J. Am. Chem. Oil. Soc.*, 66, 209, 1989.

Berger, C., Jusforgues, P., and Perrut, M., Purification of unsaturated fatty acids esters by preparative supercritical fluid chromatography, Proc. Intl. Symp. SCFs, Nice, France, 1, 397, 1988.

Borch-Jensen, C., Staby, J., and Mollerup, J., Partition coefficients of triglycerides and fatty acids methyl esters from fish oils in SCFs, Proc. 3rd Intl. Symp. SCFs, France, 2, 299, 1988.

Bradley, R. L., Removal of cholesterol from milk fat using supercritical carbon dioxide, *J. Diary Sci.*, 72, 2834, 1989.

Brunner, G. and Peter, S., *Chem. Eng. Tech.*, 53, 529, 1981.

Bulley, N. R., Labay, L., and Arntfield, J., *J. Supercritical Fluids*, 5, 13, 1992.

Catchpole, O. J., Kamp, J., and Grey, J. B., Extraction of squalane from shark liver oil using SC CO_2, *Ind. Eng. Chem. Res.*, 36, 4318–4324, 1997.

Chu, Y., Effect of soybean pretreatments on crude quality, *J. A. O. C. S.*, 72, 177, 1995.

Cygnarowicz-Provost, M. L., O'Brien, D. J., Maxwell, R. J., and Hampson, J. W., Supercritical fluid extraction of fungal lipids using mixed solvents, Proc. 2nd Intl. Symp. SCFs, Boston, MA, 17, May, 1991.

Chrastil, J., Solubility of solids and liquids in supercritical gases, *J. Phys. Chem.*, 86, 3016–3021, 1982.

Christiansen, D. D. and Friedrich, J. P., Production of Food Grade Corn Germ Product by Supercritical Fluid Extraction, U. S. Patent, 4,495,207, 1985.

Castera, A., Production of low fat and low-cholesterol food stuffs or biological products by supercritical CO_2 extraction: processes and applications, in *Supercritical Fluid Processing of Food and Biomaterials*, Rizvi, S. S. H., Ed., Blackie Academic & Professional, 186, 1994.

Calvo, L., Ramirez, A. I., and Cocero, M. J., Effect of the addition of butanol to SC CO_2 on the solubility and selectivity of the extraction of sunflower oil, Proc. 5th Meet. Supercritical Fluids, Nice, France, 699, 1998.

Carmelo, P. J., Simoes, P., and Ponte, M. N. D., SCFE of olive oil in countercurrent extraction columns: experimental results and modeling, Proc. 3rd Intl. Symp. Supercritical Fluids, 2, 107, 1994.

Choo, Y. M., Ma, A. N., Yahaya, H., Yamauchi, Y., and Saito, M., Separation of crude palm oil components by semipreparative supercritical fluid chromatography, *J. A. O. C. S.*, 73, No. 4, 523, 1996.

Cocero, M. J. and Calvo, L., Supercritical fluid extraction of sunflower seed oil with CO_2 — ethanol mixtures, *J. A. O. C. S.*, 73, No. 11, 1573, 1996.

de Lucas, A., Rincon, J., and Gracia, I., Supercritical extraction of husk oil, Proc. 5th Meet. Supercritical Fluids, Nice, France, 2, 711, 1998.

Dunford, N. T. and Temelli, F., Extraction of canola phospholipids with supercritical carbon dioxide and ethanol, Proc. 3rd Intl. Symp. Supercritical Fluids, Strasbourg, France, 3, 471, 1994.

Eisenbach, W., Supercritical fluid extraction: a film demonstration, *Ber.-Bungsenges, Phys. Chem.*, 88, 882, 1984.

Esquivel, M. M., Fontan, I. M., and Gil, M. G. B., The quality of edible oils extracted by compressed CO_2, Proc. 3rd Intl. Symp. Supercritical Fluids, Strasbourg, France, 429, 1994.

Friedrich, J. P., List, G. R., and Heakin, A. J., Petroleum free extraction of oil from soybean with SC CO_2, *J. A. O. C. S.*, 59, No. 7, 288, July, 1982.

Favati, F., Fiorentini, R., and Vitis, V., Supercritical fluid extraction of sunflower oil: extraction dynamics and process optimization, Proc. 3rd Symp. SCFs, France, 2, 305, 1994.

Fontan, I. M., Esquivel, M. M., and Bernardo-Gil, M. G., Supercritical CO_2 extraction of corn germ oil, Proc. 3rd Intl. Symp. SCFs, France, 2, 423, 1994.

Goncalves, M., Vascoonecelos, A. M. P., Gomes, D. A. E. J. S., Chaves, D. N. H. J., and Ponte, M. N. D., *J. A. O. C. S.*, 68, No. 7, 474, 1991.

Eggars, R. and Wagner, H., Jet extraction of high viscous material, Proc. 3rd Intl. Symp. Supercritical Fluids, Strasbourg, France, 2, 125, 1994.

Heigel, W. and Hueschens, R., Process for the Production of Pure Lecithin Directly Usable for Physiological Purposes, U. S. Patent, 4,367,178, 1983.

Hannigan, K. J., Extraction process creates low fat potato chips, *Food Eng.*, 7, 77, 1981.

Hardardottir, I. and Kinsella, J. E., Extraction of lipid and cholesterol from fish muscle with supercritical fluids, *J. Food Sci.*, 53, 1656, 1988.

Ikushima, Y., Saito, N., and Goto, T., Selective extraction of oleic, linoleic and linolenic acid methyl esters from their mixtures with SC CO_2-entrainer, *Ind. Eng. Chem.*, 28, 1364, 1989.

Krukonis, V. J., Supercritical fluid processing: current research and operations, Proc. Intl. Symp. SCFs, 2, 541, 1988.

Krukonis, V. J., Supercritical fluid processing of fish oils: extraction of polychlorinated biphenyls, *J. Am. Chem. Oil Soc.*, 66, 818, 1989.

Kumar et al., *J. Chem. Eng. Data*, 33, 327–333, 1988.

List, G. R. and Friedrick, J. P., *J. Am. Oil Chem. Soc.*, 62, 82, 1985.

MacLachlan, C. N. S. and Catchpole, O. J., Separation of Sterols from Lipids, World Patent, 90/02788, 1990.

Majewski, M., Mengal, P., Perrut, M., and Ecalard, J. P., Supercritical fluid fractionation of butter oil, in *Supercritical Fluid Processing of Food and Biomaterials*, Rizvi, S. S. H., Ed., Blackie Academic & Professional, Glasgow, 129, 1994.

Mendes, R. L., Fernandes, H. L., and Cyugnarowicz-Provost, M., Supercritical CO_2 extraction of lipids from micro algae, Proc. 3rd Intl. Symp. SCFs, France, 2, 477, 1994.

Molero-Gomez, A. M., Huber, W., Lopez, C. P., and Martinez de La Ossa, E., Extraction of grape seed oil with liquid and SC CO_2, Proc. 3rd Intl. Symp. SCFs, 2, 412, 1994.

Montanari, L., King, J. W., List, G. R., Rennick, K. A., Selective extraction and fractionation of natural phospholipid mixtures by supercritical CO_2 and cosolvent mixtures, Proc. 3rd Intl. Symp. Supercritical Fluids, Strasbourg, France, 2, 497, 1994.

Mishra, V. K., Temelli, F., and Ooraikul, B., Supercritical CO_2 extraction of oil from a seaweed *Palmaria palmata, Supercritical Fluid Processing of Food and Biomaterials*, Rizvi, S. S. H., Ed., Blackie Academic & Professional, Glasgow, 214–222, 1994.

Mukhopadhyay, M. and Nath, M. K., Removal of FFA from rice bran and cotton seed oils using supercritical carbon dioxide, *Ind. Chem. Eng.*, Sect. A, 37, 53, 1995.

Nisson, W. B., Ganglitz, E. J., Hudson, J. K., Stout, U. F., and Spinelli, J., Fractionation of menhaden oil ethyl esters using supercritical fluid CO_2, *J. Am. Oil Chem. Soc.*, 65, 109, 1988.

Nisson, W. B., Ganglitz, E. J., and Hudson, J. K., Supercritical fluid fractionation of fish oil esters using incremental pressure programming and a temperature gradient, *J. Am. Oil Chem. Soc.*, 66, 1596, 1989.

Nisson, W. B., Ganglitz, E. J., and Hudson, J. K., Solubilities of methyl oleate, oleic acid, oleyl glycerols, and oleyl glycerol mixtures in SC CO_2, *J. Am. Oil Chem. Soc.*, 68, 3016–3021, 1991.

Ooi, C. K., Bhaskar, A., Yenar, M. S., Tuan, D. Q., Hsu, J., and Rizvi, S. S. H., Continuous supercritical carbon dioxide processing of palm oil, *J. Am. Chem. Oil Soc.*, 73, No. 2, 233, 1996.

Penedo, P. L. M. and Coelho, G. L. V., Optimization of deacidification of vegetable oils using SC CO_2, Proc. 4th Intl. Symp. SCFs, Japan, Vol. B, 503, 1997.

Polak, J. T., Balaban, M., Peplow, A., and Philips, A. J., Supercritical CO_2 extraction of lipids from algae, in *Supercritical Fluid Science and Technology*, Johnston, K. P. and Penninger, M. L., Eds., ACS Symp. Series, chap. 28, 449, 1989.

Reverchon, E. and Osseo, L. S., Comparison of processes for the supercritical CO_2 extraction of oil from soybean seeds, *J. Am. Oil Chem. Soc.*, 71, No. 9, 1007, 1994.

Riha, V. and Brunner, G., Separation of fatty acid methyl esters by chain length and degree of saturation, a chemical engineering design analysis, Proc. 3rd Intl. Symp. Supercritical Fluids, 2, 119, 1994.

Rizvi, S. S. H., Chao, R. R., and Liew, Y. J., Concentration of omega-3 fatty acids from fish oil using supercritical carbon dioxide, in *Supercritical Fluid Extraction and Chromatography*, Charpentier, B. A. and Sevenanats, M. R., Eds., ACS Symp. Series, No. 366, ACS, Washington, D.C., 1988.

Ronyai, E., Simandi, B., Tomoskozi, S., Deak, A., Vigh, L., and Weinbrenner, Zs., Supercritical fluid extraction of corn germ with carbon dioxide-ethyl alcohol mixture, Proc. 4th Intl. Symp. Supercritical Fluids, Sendai, Japan, 227, 1997.

Rossi, M., Spedicato, E., and Schiraldi, A., Improvement of supercritical CO_2 extraction of egg lipids by means of ethanolic entrainer, *Ital. J. Food Sci.*, 4, 249, 1990.

Schneider, M., *Fractionation and Purification of Lecithin in Lecithins: Sources, Manufacture and Uses*, Szuhaj, B. F., Ed., American Oil Chemists' Society, Champaign, Illinois, chap. 7, 108, 1988.

Shishikura, A., Fujimoto, K., Kaneda, T., Arai, K., and Saito, S., Modification of butter oil by extraction with supercritical carbon dioxide, *Agric. Biol. Chem.*, 50, 1209, 1986.

Sahle-Demessie, E., King, J. W., and Temlli, F., Packed column fractionation of glycerides using supercritical carbon dioxide, Proc. 4th Intl. Symp. Supercritical Fluids, Sandai, Japan, 621, 1997.

Simoes, P., Da Ponte, M. N., and Brunner, G., Deacidification of olive oil by supercritical fluid extraction: phase equilibria and separation experiments in a counter-current packed column, 3rd Intl. Symp. Supercritical Fluids, 2, Strasbourg, France, 481, 1994.

Sivik, B., Gunnlaugsdottir, H., Hammam, H., and Lukaszynski, D., Supercritical extraction of polar lipids by carbon dioxide and a low concentration of ethanol, Proc. 3rd Intl. Symp. SCFs, France, 2, 311, 1994.

Vijayan, S., Byskal, D. P., and Buckley, L. P., Separation of oil from fried chips by supercritical extraction process, in *Supercritical Fluid Processing of Food and Biomaterials*, Rizvi, S. S. H., Ed., Blackie Academic & Professional, Glasgow, 75, 1994.

Yamaguchi, K., Murqakami, N. M., Nakano, H., Konosu, T., Yamamoto, H., Kosaka, M., and Hatta, K., Supercritical carbon dioxide extraction of oils from Antarctic krill, *J. Agric. Food Chem.*, 34, 904, 1986.

Yamaguchi, M., Method for the Preparation of Defatted Corn Germs, European Patent O,367,128, 1989.

Zanderight, L., Deacidification of olive oils by supercritical carbon oxide, *J. Am. Chem. Oil. Soc.*, 66, 209, 1989.

Zhu, H., Yang, J., and Shen, Z., Supercritical fluid extraction and fractionation of fish oil esters using incremental pressure programming and a temperature gradient process, Proc. 3rd Intl. Symp. Supercritical Fluids, France, 2, 95, 1994.

11 Natural Pesticides

11.1 IMPORTANCE OF RECOVERY

Over the last decade, there has been an increasing consumer demand for food devoid of synthetic pesticides and a growing farmer concern about the immunity being developed by pests to some synthetic pesticides. Consequently, there has been a serious spurt of activity all over the world to gradually switch from synthetic to natural pesticides. Among common pesticides, DDT and BHC are considered carcinogenic and are already banned. Another synthetic pesticide, monochrotophos, has been placed on the watch list. In a recent report of the World Health Organization (WHO), the annual number of acute poisonings caused by synthetic pesticides has been estimated at 3 million, with 20,000 deaths every year. The number of insects resistant to these pesticides has increased significantly over the years. Some pests may soon be "beyond effective chemical control" as some experts believe. Several species of mites, nematodes, desert and migrating locusts, rice and maize borers, pulse beetle, rice weevil, citrus red mint, and white flies have developed resistance to synthetic pesticides, but have yet to develop resistance to a natural pesticide. Natural extracts from citronella, sassafras, aloe vera, chrysanthemums, and neem have been established for several years and are being marketed as alternative ecofriendly solutions to insect and pest controls. Pyrethrins isolated from chrysanthemums, piperonyl butoxide from sassafras, and azadirachtin from neem are some of the active ingredients used in the formulations of natural pesticides.

A formulation based on 10% citronella in aloe vera and water medium is used to control mosquitoes, black flies, gnats, and fleas. Pyrethrum, an extract from chrysanthemum flowers, is a highly toxic but biodegradable natural insecticide used against fleas and ticks. Nicotine, an alkaloid, occurs naturally in many plant leaves and is used in making formulations that are applied in the form of sprays and dusts. Commercial nicotine-based insecticides have long been marketed as Black Leaf 40. Rotenone is a natural insecticidal compound present in the roots of many plants, e.g., barbasco and derris. It is used in formulations with less than 5% concentration and are applied as dusts, powders, and sprays in gardens, lakes, ponds, and on food crops. Rotenone is insoluble in water. Rotenone degrades rapidly and is mixed with natural synergist piperonyl butoxide in order to retain its activity until it is applied.

Sabadilla is a naturally occurring insecticidal alkaloid of veratrin type. It is obtained from powdered ripe seeds of a South American lily. It is used to kill parasites living on domestic animals and humans. It is applied in the form of a dust powder or mixed with kerosene for spraying. The commercial formulations are made to contain less than 0.5% sabadilla and are rarely used in the U.S., though it is still in use in some other countries.

From prehistoric times, neem has been used primarily against household and storage pests and to some extent against pests related to field crops in the Indian subcontinent. It was a common practice in rural India to mix dried neem leaves

(2 to 5%) with grains for storage, which is practiced even today in several parts of the subcontinent. Neem cake was applied as early as 1930 to rice and sugarcane fields against stem borers and white ants. For the last decade, the pest control potential of neem, which does not kill pests, but affects their behavior and physiology, has been utilized in neem-based pest management for enhancing agricultural productivity in Asia and Africa (Saxena, 1996). The potential of neem has been established against pests in stored products, e.g., grain, legumes, maize, sorghum, wheat, rice, and paddy and potato tubers. Annual losses of stored food grains worldwide amount to 10%, i.e., 13 million tonnes due to insects which can be controlled by treatment with neem.

Locusts, gypsy moths, aphids, and medflies are among the 200-odd pests that are repelled by neem seed extracts (Rajan, 1993). Neem also has a systemic action in some plants which could prove extremely useful against stem- and root-feeding pests that are difficult to control. Neem-based pesticides have a significant potential in view of the abundant availability of the raw material, its ecofriendly nature, and its nontoxicity to vertebrates and safety to the human population.

There are several other insecticides based on extracts of natural origin which are less toxic to mammals, but natural enemies to insects, bacteria, fungi, virus, etc. Biopesticides are thus becoming effective alternatives to synthetic pesticides as can be substantiated from the steady increase in sales and market share. In the U.S., the sale of natural pesticides is increasing at the rate of 10% per annum (Saxena, 1996). For commercial viability, natural pesticides are required to be effective over a broad spectrum of pests and should be price competitive. Considerable efforts are however needed to popularize them through field demonstration and educational programs for farmers and general public.

11.2 BIOACTIVITY OF NEEM

The neem tree *Azadirachta indica*, a native to India, is such a valuable tree with immense potential to nourish and protect the environment and agricultural crops that it is considered the wonder-tree of India. It is ordinarily found in the tropics and accordingly is grown outside India in small pockets of Africa and Southeast Asia, particularly in Bangladesh and Myanmar. There are about 18 million neem trees in India. A fully grown tree can yield about 25 kg of seeds and 6 to 9 kg of neem oil annually. The evergreen neem tree grows very fast, even in poor soil, up to a height of 9 m in 6 years and a height of 25 and 2.5 m in girth in its lifetime.

Every part of the plant finds use in applications such as cosmetics, personal hygiene, and medicines for itching, skin disease, leprosy, blood purification, worms, diabetes, piles, dysentry, jaundice, AIDS, all kinds of fever and eye diseases, and even as contraceptives. The neem tree, in total, could emerge as a universal pharmacy and an omnipotent panacea of the 21st century.

More than 150 compounds have so far been isolated from neem, of which 101 bioactive compounds are present in seeds, 37 in leaves, and the rest in flowers, bark, and root. There is no tree other than neem which has gained so much importance world over in the recent past, as reflected by large-scale plantation drive in the U.S., Germany, India, and several African and other countries, in addition to

TABLE 11.1
Pharmacological Activity of Neem Extracts

Activity	Active Ingredient	Part from Which Isolated
Inhibition of insect feeding and hormone production	Azadirachtin	Seed, leaves
Antiinflammatory	Nimbidin	Seed
Antipyretic	—	Leaf and bark
Antimalarial	Gedunin, Quercetin	Leaf
Antimicrobial	Gedunin	Leaf
Antiulcerous	Nimbidin	Seed
Antidiabetic	—	Leaf
Antifertility	—	Seed, leaf

Rajan, 1993.

the establishment of a full-fledged research institute in Germany and the Neem Foundation in India. Although neem is a wonder-tree, its advantages in biopesticides and health care products have not yet reached the large spectrum of human population, as well as animal pets, due to the lack of scientific awareness. Environmentally, neem has a reliable reputation as an air purifier. There are no known toxic effects of neem extracts, but there are several pharmacological effects on human beings as outlined in Table 11.1.

Neem kernels have about 45% oil in addition to 3 to 8% active ingredients, which are isolated from the deoiled neem kernels. Neem seeds from different geographic regions are known to vary considerably with respect to active ingredient concentrations. These active ingredients are completely biodegradable and anticancerous to mammals and other warm-blooded organisms. They have indicated diverse bioactivity as insecticidal, insect-repellent, antifeedant, growth inhibiting and/or regulatory and sterilant (impairs hatching of eggs). Two to three neem trees may have sufficient bioactivity to protect one hectare of crop. Neem extracts can prevent the fungus Aspergillus from producing aflatoxins, a highly carcinogenic substance. No plant is known to have control over such a diverse spectrum of pests as neem and a number of pests that can be controlled by neem are given in Table 11.2.

Neem is bitter in taste due to the presence of an array of complex compounds called triterpenes, or more specifically limonoids. Nimin, a ready-to-use neem product with 15% neem bitter is used to coat urea (1 part Nimin with 100 parts urea, wt/wt). The product can delay nitrification up to 40 to 45 days and also increases the crop yield over plain urea. Neem-coated urea results in a savings of up to 20% of urea and an increase in crop yields by up to 25% (Vyas et al., 1996). Neem oil can be used as a preservative also, replacing methyl bromide, which is being phased out.

11.3 NEEM-BASED PESTICIDES

Neem-based pesticides contain mainly two most useful active ingredients, namely, azadirachtin and marrangin, which are triterpenes present in the kernels of

TABLE 11.2
Agricultural Pests Controlled by Neem

Cowpea weavil	Green rice leaf hopper
Spotted cucumber beetle	Brown plant hopper
Rice hispa	Rice leaf folder
Mexican bean beetle	Cotton boll worm
Colorado potato beetle	Spiny boll worm
Lesser grain borer	Gypsy moth
Rice weevil	Tobacco horn worm
Red flour beetle	Pink boll worm
Khapra beetle	Potato tuber moth
Sorghum shoot fly	Diamond-back moth
Vegetable leaf miner	Potato tuber moth
Rice gall midge	Tobacco caterpillar
Red cotton bug	Migrating locust
White fly	Desert locust

Rajan, 1993.

Azadirachta indica and *Azadirachta excelsa*, respectively. Azadirachin (molecular formula: $C_{35}H_{44}O_{16}$) is a highly oxidized triterpenoid and is the most widely publicized bioactive molecule in neem. It is a well-known insect growth inhibitor of botanical origin and inhibits feeding and molting in a wide variety of insects. Neem contains several other bioactive ingredients, such as salanin, nimbin, nimbidin, meliantriol, etc. belonging to the tetranortriterpenoids (TNTT), and exhibits biological activities both as pesticides and pharmaceuticals. When present in pure form, azadirachtin degrades with moisture and light. Azadirachtin is stable when present in neem oil medium with its natural neighbors. Hence it is preferable to use neem oil enriched with azadirachtin as the stable feed stock for making pesticide formulations. Neem formulations have proved as potent as many commercially available synthetic pesticides (Kumar, 1997). The larvicidal activity, namely, blocking the larvae from molting, is presumed to be the most important quality of neem pesticides and is used to kill off many pest species. The active ingredients of neem exhibit high antifeedant and repellent action against a number of insects. Neem-based pesticide formulations are safe, natural, biodegradable, manageable at the farmers level, and environment friendly, unlike synthetic pesticides. However, neem pesticides cannot completely replace, but only supplement, synthetic pesticides used in the preservation of stored foods. Nevertheless, the amounts of synthetic pesticides needed could be reduced, thereby decreasing the synthetic pesticide load on food grains. Field trials of neem insecticide for controlling pests of crucifiers, corn, and potato indicate that neem insecticide can provide pest control as effective as or better than pyrethrum, the current botanical pesticide of choice for organic growers.

11.3.1 Azadirachtin-Based Formulations

Considering its significant utility for controlling a variety of pests, a neem pesticide formulation was cleared by the U.S.D.A., named Margosan-O, and standardized to contain 3000 ppm (0.3%) azadirachtin. This is diluted 150 fold to 20 ppm to be used as a spray solution which is adequate to control a large number of pests (Govindachari, 1992). The Bureau of Indian Standards prescribed two standards in 1995, namely, IS 14299:1995 (Specification for Neem Extract Concentrate Containing Azadirachtin) and IS 14300:1995 (Specification for Neem-Based Emulsion Concentrate (EC) Containing Azadirachtin). Since then, many more standards have come into effect. Most of the neem oil formulations presently used in India are produced by the addition of some dispersing agent to oil and are not standardized in terms of azadirachtin content. Neem oil produced by the cold expeller technique or extraction with petroleum ether contains only traces or negligible azadirachtin, yet they are quite effective due to the presence of several other active ingredients, such as salanin and nimbin. Formulations based on this neem oil are presently being marketed and distributed to farmers and are found to be useful. For example, the formulation containing 2% neem oil was found to be effective for different crops susceptible to virus diseases transmitted by white flees (Roychoudhury and Jain, 1996). However, they can never match the efficacy of neem formulations with higher concentration of azadirachtin, such as 300 ppm azadirachtin (oil based) and 1500 ppm azadirachtin (solvent based) neem formulations which are already in extensive usage.

11.3.2 Commercial Production

Currently neem-based pesticides are in production in more than 40 commercial plants under the Central Insecticides Control Board's (CIB) registration. An estimated 1500 kl of pesticides valued at $4 million U.S. were sold during 1996 to 1997. Neem pesticides are mainly used on cotton, rice, redgram, groundnut, soybean, vegetables, fruits, and tea (Venkatashwarlu, 1997). Research investigations are going on to find out its bioactivity on other pests and insects and it is expected that it will widen its applicability to other crops. Accordingly neem research has acquired the focus of worldwide attention. There is a huge market not only for neem-based pesticide formulations, but also for pharmaceutical and cosmetic products based on the active constituents of neem. However, time required for a new product to be evaluated and registered for field use is extensive and the process is elaborate making it difficult to be marketed. From 2000 on, neem-based pesticides are likely to have a significant market in the U.S., Europe, Far East, and Australia. Currently a milligram of practical grade (95%) pure azadirachtin is sold (by Aldrich-Sigma Chemicals) as an analytical reagent for $90 U.S. Hence there is a huge value addition in concentrating azadirachtin in neem extract. In 1997, the largest market for neem pesticides was in the U.S., estimated at $10 million U.S. European registration takes a little more time because they insist on tolerence tests unlike in the U.S, where neem's efficacy has been established and accepted by the USDA. The global market for neem-based pesticide is estimated to be worth $160 million by the year 2000 (Shenoy, 1997), as more synthetic pesticidal chemicals are being banned in a phased manner. Recently a plant has been set up in India which has a turnover of 6 million U.S. dollars. The potential azadirachtin production from all 18 million neem trees in India is estimated at 450 to 720 MTPA (with azadirachtin content of kernel at 0.5 to 0.8%) (Ketkar, 1995). The domestic market for neem pesticides in India in 1997 was equivalent to \$5 million U.S. and is expected to grow now to \$25 million U.S. by the year 2000, as the awareness for ecofriendly systems spreads. In India field trials have been conducted on tobacco, cotton, and tea, the three critical crops on which the highest quantities of pesticides are presently used.

11.4 RECOVERY OF AZADIRACHTIN FROM NEEM KERNEL

Quality of neem seeds is an important criterion for the selection of raw material for azadirachtin recovery. The seeds should be absolutely free from aflatoxin and lead contamination in order to produce neem extracts of export quality. The azadirachtin content varies in the range of 0.3 to 0.8% of kernel. It is reported that about 1.2 to 4.0 g (0.12 to 0.40%) of azadirachtin can be recovered from 1 kg of seeds, depending on the soil, climate, storage, and maturity of seeds. Simple, nonhazardous, and inexpensive methods of extraction of neem oil and its active ingredients are required to facilitate practical use of neem extracts for pest control.

11.4.1 Conventional Processes

Conventional recovery methods for azadirachtin are quite complex, which involve a multistage and multisolvent liquid–liquid extraction process. It is recovered from neem kernel after the removal of oil from it. Neem oil is produced by passing neem kernel through a series of mechanically driven expellers. The oil so obtained is dark in color and contains only traces of azadirachtin (up to a maximum concentration of 0.03%) and is not ordinarily suitable for making pesticidal formulations, because it is too dilute and requires a large volume of neem oil to be used for acquiring the desired activity.

Alternatively, the neem kernel is first deoiled by using a nonpolar solvent, such as petroleum ether or hexane. Azadirachtin and other active ingredients are polar and are not dissolved in hexane or petroleum ether. They are recovered from the deoiled neem cake or raffinate using either aqueous methanol (95%) or azeotropic mixture of ethanol, or a mixture of methyl tertiary butyl ether and methanol. The neem oil obtained during deoiling of the kernel by solvent extraction contains hardly any azadirachtin. It is a dark colored liquid with an offensive odor. It is unsuitable for usage as a cooking oil or a pharmaceutical oil for application in cosmetics and medicines. This neem oil is mostly used in the soap industries.

The extract from the deoiled neem kernel containing the active ingredients is partitioned with hexane and aqueous methanol (95%) and then with ethanoic acetate and water in succession. The ethanoic acetate layer is filtered and the solvent is evaporated by distillation to obtain the active extract which contains the active ingredients. Azadirachtin is isolated from the active extract (AE) by vacuum and flash chromatographic techniques (Schroeder and Nakanishi, 1987). Though high-purity azadirachtin could be obtained by this elaborate procedure, the yield of azadirachin is very low (0.24% of the kernel). The concentration of azadirachtin is low (2.6%) in the AE due to coextraction of other polar components. The active extract is not stable as azadirachtin and other active ingredients degrade in the presence of moisture and light.

In a second process for concentrating azadirachtin, it is isolated from the active extract solution by first eluting it in a silica column with hexane-ethyl acetate mixture in the ratio of 1:3 and by evaporating and then again dissolving it in highly pure (>99.9%) methanol for the preparative high performance liquid chromatography (HPLC) (Kumar et al., 1995). The yield of azadirachtin varied between 0.39 to 0.55%, depending on the source of the neem kernel.

Alternatively, azadirachtin is isolated from the active extract by preparative HPLC technique with methanol-water in the ratio 3:2 (Govindachari et al., 1990). The yield of azadirachtin by this process is also quite low (0.3% of the kernel), with a low concentration (3.0%) of azadirachtin in the crude active extract.

The solvent extraction techniques for recovering azadirachtin are time consuming and require large amounts of solvent and consequently require considerable thermal energy. Besides, azadirachtin obtained by these processes is not stable and has a poor shelf life. It is usually stabilized by mixing it with neem oil particularly for insecticidal, larvicidal, and pesticidal applications and during storage and transportation.

11.4.2 SC CO_2 Extraction

Most of the disadvantages of the conventional processes can be eliminated by using SC CO_2 as the extractant which dissolves and elutes the solutes from neem kernel in order of increasing polarity or decreasing volatility or both. This separation process exploits the complex thermodynamic behavior of SC CO_2 and specific interactions between the constituent molecules and SC CO_2 to recover the complex molecules of azadirachtin and other polar active ingredients from the solid substrate of ground neem kernel. Cernia et al. (1994) reported that the yield of azadirachtin from neem "seeds" was low, only 0.02% of the feed using SC CO_2 extraction under static condition. However the yield could be improved to merely 0.15% by addition of a modifier (methanol) to the ground neem kernel prior to rapid decompression followed by extraction with SC CO_2. Only 50% azadirachtin could be recovered by SC CO_2 at 55°C and 350 bar with addition of an entrainer (methanol) to SC CO_2. The process involved either addition of an entrainer to neem seeds or rapid decompression of initially pressurized neem seeds, followed by sweeping of the extraction cell volumes by CO_2 in successive stages.

Subsequently, Cernia et al. (1997) demonstrated that initial sudden decompression prior to the two-stage sequential static SC CO_2 extraction could significantly improve the yield of active ingredients, more so of nimbin and salanin compared to azadirachtin. On the other hand, the addition of an entrainer like methanol directly on the neem kernel could enhance the selectivity of azadirachtin with respect to salanin and salanin + nimbin further to 6.36 and 2.68, respectively, but the yields were quite low. These values were quite high compared to the corresponding selectivities of 2.70 and 1.27 reported in the case of two-stage sequential sweeping extraction with prior depressurization. It was observed that salanin and nimbin could

TABLE 11.3
Comparison of Yield and Concentration of Azadirachtin by Various Processes

Process	Max. Conc. Aza in % AE	Max. Conc. Aza in % oil	AE % Yield	Aza % Yield	Ref.
Conventional partitioning	2.63	traces	9.2	0.24	(Schroeder and Nakanishi, 1987)
Dual solvent	3.0	traces	10.0	0.30	(Govindachari et al., 1990)
SC CO_2	—	—	—	0.15	(Cernia et al., 1994)
SC CO_2 extracted mid fraction	21.4	3.0	4.5	0.42	(Mukhopadhyay and Ram, 1997)
Dual solvent[a] extraction	2.56	traces	18.9	0.47	

[a] Using the same starting neem kernel.

be selectively extracted to some extent by SC CO_2 in the first stage of extraction at 237 bar and 55°C (0.8 g/ml density) while azadirachtin was not at all extracted, which could only be selectively extracted in the second stage of static extraction at a higher pressure of 375 bar and 55°C (0.9 g/ml density). On the other hand, when rapid depressurization preceded the extraction, both single-stage and two-stage sequential sweeping extractions rendered selectivity of azadirachtin with respect to salanin + nimbin almost invariant at 1.27. Thus it is possible to selectivity fractionate azadirachtin from the other active ingredients by proper feed pretreatment and selection of the process scheme.

Recently a process has been developed and patented (Mukhopadhyay and Ram, 1997) for dynamic sequential SC CO_2 extraction and fractionation of neem oil enriched with azadirachtin and other active components. The process requires much lower pressures for extraction due to a scheme of pretreatment of the substrate followed by fractionation.

The concentration of azadirachtin in some fractions can be as high as 3% (Ram, 1997) and the concentration of azadirachtin in the corresponding active extract (which is obtained by methanol extraction of the SC CO_2 extract) can be as high as 21% as compared to 3% by the dual solvent-partitioning process. The results of the dynamic SC CO_2 extraction process are compared with the conventional dual solvent partitioning process in Table 11.3. However, without fractionation, the total yield of the extracted oil can be as high as 48% with 1% azadirachtin content and 89% recovery of azadirachtin. It is observed that the concentration of azadirachtin in the SC CO_2 extracted neem oil varied between 0.3 to 3% by weight in different fractions. The concentration of azadirachtin in the active extract (which is obtained by methanol extraction of the SC CO_2 extract) varies between 5 to 21% with an average value of 11%, as compared to 2.5% by conventional methanol extraction of deoiled neem. In another process, similar results could be obtained in the 10-l SC CO_2 extraction pilot plant even without the addition of an entrainer, and it was possible to obtain a high-quality, azadirachtin-enriched deodorized neem oil. The SC CO_2 extracted neem oil has a sparklingly clear (transparent), light yellowish color due to a very mild temperature of operation and has negligible offensive odor. Both fixed oil and essential oil extracted from neem by SC CO_2 can find wide-ranging applications in the cosmetic and pharmaceutical industries, besides natural pesticides. The oil required for making the pesticidal formulation is already present in the SC CO_2 extract along with the active ingredients.

The SC CO_2 process has one other advantage over the conventional process in rendering the raffinate neem cake a clean and dry byproduct which can be used as a very effective fertilizer and as a solid pesticide. Thus, the usage of the multistep, tedious process of extraction, partitioning, and chromatography, using expensive multiple solvents, can be avoided by means of a single step, using the safe and clean process of SC CO_2 extraction. The most attractive feature of the process is that azadirachtin, along with a host of other bioactive ingredients, are recovered along with neem oil in concentrated form. It is also stable, being in its natural environment even after its extraction.

11.5 PYRETHRUM-BASED PESTICIDES

It is interesting to find how Nature protects attractive flowers and plants from the harsh encounters of insects. Beautiful chrysanthemum flowers are known to contain some insecticidal constituents. At least six naturally occurring pesticidal closely related organic compounds can be extracted from *Chrysanthemum cinerariaefolium* flowers. The six insecticidal constituents, are pyrethrin I and II, cinarin I and II, and jasmolin I and II, which are known as pyrethrins. They are keto-alcohol esters comprising an acid containing a 3-carbon-ring joined to an alcohol containing a 5-carbon ring. Pyrethrin I and II differ by only the presence of a terminal methyl ester group in the latter, while cinerin I and II are related compounds having the same structure as pyrethrin except that the terminal methylene in pyrethrins is replaced by a methyl group. Jasmolin I and II differ from the pyrethrins only in that one double bond in the side chain of the alcohol of pyrethrins is saturated. The II series (e.g., pyrethrin II, cinerin II, and jasmolin II) are diesters having both a terminal methyl ester group and the ester linkage between the two rings (Weiboldt and Smith, 1988). The acid present in compounds designated by the I series is called chrysanthemic acid while that in II series is called pyrethric acid. These two acids can be joined to alcohol possessing an unsaturated side chain containing five carbon atoms (pyrethrolone and jasmolone) or four carbon atoms (cinerole). These characteristics of the compounds present and their relative proportions are summarized in Table 11.4.

These keto alcoholic esters of chrysanthemic and pyrethric acids act on the sodium channels that exist on the nerve axons of insects. By opening these channels, the nerve is initially excited, soon paralyzing the central nervous systems, ultimately killing them at the end. These compounds are however, susceptible to enzymatic degradation by heat and light and accordingly they are mixed (1:4) with natural synergists, such as piperonyl butoxide derived from sassafras or n-octyle bicyclo-heptane dicarboximide, in order to retain their activity until they are dispersed on crops. In order to slow down their quick action on the insects and make them more effective, they are often mixed with organophosphatic or carbamate insecticides. Pyrethrins, if injested by humans/mammals, are effectively hydrolyzed to inert products by liver enzymes, as a result they are among the least toxic to mammals.

TABLE 11.4
Proportions of the Six Active Ingredients in a Typical Pyrethrum Product

Alcohol	Acid: Chrysanthemic	Acid: Pyrethric
Pyrethrolone	Pyrethrin I (35%)	Pyrethrin II (32%)
Cinerolone	Cinerin I (10%)	Cinerin II (14%)
Jasmolone	Jasmolin I (5%)	Jasmolin II (4%)

Hassal, 1982.

The oleoresin extract of dried chrysanthemum flowers is known as pyrethrum, half of which constitute the insecticidal compounds. These insecticidal compounds of natural origin are so effective to a large numbers of species of insects that chemically identical molecules having similar insecticidal properties have been synthesized, which are called pyrethroids.

Pyrethrum has an edge over pyrethroids in specialized applications where selective toxicity and low environmental hazard are the most important considerations; for example, for insect control in stored foods, for insect control in animals, and for spraying the insecticide in the food-processing industry. As pyrethrum has been produced and used in a relatively small scale over the last one and a half centuries, there has been relatively less insect resistance developed against pyrethrum. The advantages of pyrethrum include its effectiveness at low dosages, its minimal harm to man and animals, its rapid degradation, and the lack of its accumulation in food chains or groundwater. However, commercially, the marketing of pyrethrum has to face competition with pyrethroids being produced by the well-organized and efficiently coordinated and financed synthetic pesticide industry.

Commercially available pyrethrum contains 25% pyrethrum. It is known to have instant knock-down effect on cockroaches, white fly, aphids, mosquitoes, caterpillars, and most insects on flowers, vegetables, and fruit trees and is used for controlling other domestic pests both as insecticides and as insect repellant. It is the world's safest insecticide known today and is used for the protection and preservation of food grains and other food items from pests. Pyrethrum is decomposed by sunlight and its residual killing effect usually lasts only 2 to 3 days.

11.5.1 Recovery of Pyrethrins

During the 18th century, the crushed dried flowers of the daisy (*Chrysanthemum coccineum*) were used by the Persians (now called Iranians). The powder, then termed Persian insect powder and now known as pyrethrum, contains 68% of the very effective pyrethrins. Since 1840, however, pyrethrum has been produced from the more potent *C. cinerariafolium*, a species native to the Adriatic coastal mountain regions and widely grown in Kenya, Ecuador, and Japan. Even though pyrethins are of natural origin, there are regulations on these biopesticides as well, necessitating their separation, isolation, identification, and quantification as the key active ingredients of the pyrethrum extract.

Pyrethrins are soluble in petroleum ethers, liquid CO_2, and SC CO_2. Pyrethrins, particularly the II series, are sensitive to temperature, because they rapidly decompose, making them difficult to recover by solvent extraction even at a slightly higher temperature. Petroleum ether extraction of pyrethrum from dried inflorescences of *C. cinerariafolium* containing about 0.9% pyrethrum gives as high as 3% yield of extract, as a large number of nonpolar components are co-extracted with petroleum ether, which has less selectivity for polar pyrethrins. Subcritical or supercritical CO_2 extraction is therefore a logical choice for selective separation of the pyrethrins from dried chrysanthemum flowers. SC CO_2 extraction from the dried flowers at a pressure of 100 bar and temperature of 40°C yields 1% pyrethrum (Bunzenberger et al., 1984). However, with subcritical (liquid) CO_2 at 65 bar and 22°C, 88% of the pyrethrins in the flowers could be recovered with 20% pyrethrin in the extract in 2 h. On the other hand, with increasing temperature and pressure to 80 bar and 40°C, the pyrethrin concentration in the extract increased up to 30%, but the pyrethrin recovery decreased to merely 50% (Marentis, 1988). The yield and concentration of pyrethrins can both be improved by the addition of a polar entrainer, like methanol or ethanol to SC CO_2, even at a density of up to 0.7 g/ml (Wieboldt and Smith, 1988).

11.6 NICOTINE-BASED PESTICIDES

Nicotine and its close relatives nornicotine and anabasine are other botanical insecticides. Nicotine becomes dark upon exposure to light, its boiling point is 247°C and it is miscible with water. It is a weak base and forms salts with acids. Nicotine is present with tobacco leaves to the extent of 2 to 3% in the form of salt of citric acid or malic acid. It is an alkaloid extracted as a byproduct from denicotinization of tobacco leaves. A concentrated (95%) solution of this alkaloid is used as a fumigant. Even 40% aqueous solution of nicotine sulfate is effective as a fumigant. Commercial formulations based on nicotine are available in the form of powder and liquid aerosol sprays, though not used much in the U.S. Nicotine may however be absorbed by the human body system and biologically transformed to products which are excreted in a couple of hours. They are toxic and should be used with caution, though most of the incidents of nicotine poisoning reported are due to consumption of tobacco products.

11.6.1 Recovery of Nicotine by SC CO_2

Nicotine and nornicotine occur in the *Nicotiana* genus of the Solanaceae family. Anabasine occurs in *Anabasis aphylla*. They are usually extracted by steam distillation. But then both flavor and nicotine are simultaneously extracted. For reducing the nicotine content in tobacco leaves without removing its flavor components, SC CO_2 extraction is considered to be one of the most appropriate methods. SC CO_2 is more selective for the removal of nicotine than other conventional liquid organic solvents, such as alcohol, acetone, and hexane, since they extract more associated components along with nicotine and flavor from tobacco for having not much selectivity. In order to save energy and to selectively remove only nicotine from tobacco, the SC CO_2 extraction process is operated without any pressure reduction in a closed circulation isobaric system comprising an extractor and an absorber. The denicotization is performed on tobacco presoaked with water. The extract can be recovered and CO_2 can be regenerated by absorption of nicotine with an absorbent, such as water and alcohol, or an adsorbent, such as activated carbon, at the same operating pressure. The SC CO_2 extraction of nicotine was performed at a pressure of 250 bar and 70°C with 0.32 wt% water in SC CO_2 and absorption of nicotine in water at 53°C and 250 bar (Uematsu et al., 1994). The nicotine content could be reduced to 0.3% in 2 h from the initial nicotine content of 2%. However, the nicotine recovery increases with an increase in the amount of water in the absorber, because nicotine gets distributed in three phases, namely, SC CO_2, water, and solid tobacco leaves. For example, for an increased ratio of water to tobacco of 2.3, the nicotine content in the treated tobacco could be reduced to 0.2% (Uematsu et al., 1994). Almost 98% of nicotine can be recovered from tobacco by increasing the ratio of water to nicotine to 5:6, without affecting the flavor of tobacco. The nicotine dissolved in water (0.9 to 1.3%) can be easily concentrated or used as such for making formulations.

REFERENCES

Bunzenberger, G., Lack, E., and Marr, R., CO_2 extraction: comparison of supercritical and sub-critical conditions, *German Chem. Eng.*, 7, 25–31, 1984.

Cernia, E., Pallocci, C., D'Andrea, A., Ferri, D., Maccioni, O., and Vitali, F., Fractionation by SFE of active substances from *Azadirachta indica* seeds, Proc. 3rd Int. Symp. Supercritical Fluids, 317, 1994.

Cernia, E., D'Andrea, A., Nocera, R., Palocci, C., and Pedrazzi, E., SFE/SFC of secondary metabolities from plant tissue culture of *azadirachta indica*, Proc. 4th Intl. Symp. SCFs, Vol. A, 1, 1997.

Govindachary, T. R., Chemical and biological investigations on Azadirachtin Indica (the Neem Tree), *Current Science*, Vol. 63, 117, August 1992.

Govindachary, T. R., Sandhya, G., and Ganesh Raj, S. P., Simple method for the isolation of Azadirachtin by preparative HPLC, *J. Chromatogr.*, 513, 389–391, 1990.

Hassall, K. A., *The Chemistry of Pesticides*, chap. 7, 148, The McMillan Press, London, 1982.

Ketkar, C. M., Neem: opportunities in the 21st century, *Chem. Weekly*, Jan. 17, 1995.

Kumar, Kaushal, Neem products and standardisation — an overview, *Neem Update*, 2, No. 3, 5–6, 1997.

Kumar, M. G. and Kumar R. J., Regupathy, A., and Rajasekaran, B., Liquid chromatographic determination and monitoring of azadirchtin in neem ecotypes, *Neem Update*, 1, No. 1, July, 5, 199 .

Marentis, R. T., Steps to developing a commercial supercritical CO_2 processing plant, in *Supercritical Extraction and Chromatography*, Charpentier, B. A. and Sevenants, M. R., Eds., ACS Symp. Series No. 366, Washington, D. C., chap. 7, 127, 1988.

Mukhopadhyay, M. and Ram, T. V. K., A Process for Sequential Supercritical CO_2 Extraction and Fractionation of Neem Oil Enriched with Azadirachtin for Neem Kernels, Indian Patent No. 182587,428/Bom/97, 1997.

Rajan, T. P. S., Neem — nature's boon to humanity, *Chem. Weekly*, 38, No. 37, 51, May 18, 1993.

Ram, T. V. K., Extraction and Fractionation of Active Ingredients from Neem Seeds by Using SC CO_2, M. Tech. dissertation, Indian Institute of Technology, Bombay, India, 1997.

Roy Choudhury, R. and Jain, R. K., Evaluation of neem-based formulations, Proc. Intl. Conf. Plants and Environmental Pollution, held in Lucknow, India, Nov. 17, 1996.

Shenoy, M., *Business India*, Sept. 8–21, 1997.

Saxena, R., Agricultural technologies of 21st century: neem for sustainable agriculture and environmental conservation, Proc. Annu. Sustainable Dev. Conf., Islamabad, August 4–9, 1996.

Schroeder, D. R. and Nakanishi, N., *J. Nat. Prod.*, 50, No. 2, 241–244, March–April, 1987.

Uematsu, H., Itoh, J., Takeuchi, M., and Yonei, Y., Selective separation of nicotine from tobacco with supercritical CO_2, Proc. 3rd Intl. Symp., France, 2, 335, 1994.

Venkataswarlu, B., It's neem cake walk, *Neem Update*, 2, No. 3, 1997.

Vyas, B. N., Godrej, N. B., and Highway, K. B., Potential of nimin-coated urea for enhancing nitrogen efficiency of crops, Proc. Intl. Neem Conf., Australia (Abstr.), 1996.

Wieboldt, R. C. and Smith, J. A., Supercritical fluid chromatography with fourier transform infrared detection, in *Supercritical Fluid Extraction and Chromatography, Techniques and Applications*, Charpentier, B. A. and Sevenants, M. R., Ed., ACS Symp. Series No. 366, chap. 13, 241, 1988.

Appendix A

Thermophysical Properties of Carbon Dioxide

TABLE A.1
Densities (gm·cm^{-3}) of Pure CO_2 Calculated by P-R EOS

T, °C/P, bar	50	55	60	65	70	75	80	85	90	95	100
–10	1.014	1.018	1.021	1.025	1.029	1.032	1.036	1.039	1.043	1.046	1.049
–5	0.978	0.983	0.987	0.992	0.996	1.000	1.004	1.008	1.012	1.016	1.020
0	0.940	0.944	0.950	0.955	0.960	0.965	0.970	0.975	0.979	0.984	0.988
5	0.894	0.901	0.907	0.914	0.920	0.926	0.932	0.938	0.944	0.949	0.954
10	0.117	0.856	0.860	0.867	0.876	0.884	0.891	0.898	0.905	0.911	0.918
15	0.156	0.208	0.800	0.813	0.824	0.834	0.844	0.853	0.860	0.869	0.877
20	0.144	0.171	0.235	0.744	0.761	0.776	0.789	0.801	0.813	0.823	0.832
25	0.134	0.160	0.195	0.269	0.673	0.700	0.722	0.740	0.755	0.769	0.781
28	0.130	0.153	0.183	0.225	0.586	0.639	0.671	0.695	0.715	0.732	0.747
30	0.127	0.145	0.171	0.205	0.272	0.578	0.629	0.661	0.685	0.705	0.723
35	0.121	0.140	0.163	0.191	0.226	0.281	0.417	0.534	0.589	0.625	0.651
40	0.116	0.130	0.149	0.171	0.197	0.231	0.276	0.356	0.497	0.516	0.630
45	0.111	0.127	0.145	0.165	0.188	0.215	0.247	0.287	0.337	0.398	0.457
50	0.107	0.119	0.135	0.153	0.172	0.194	0.219	0.249	0.285	0.328	0.386

T, °C/P, bar	110	120	130	140	150	160	170	180	190	200
0	0.997	1.005	1.012	1.020	1.027	1.033	1.039	1.045	1.051	1.057
5	0.964	0.973	0.982	0.990	0.998	1.006	1.013	1.020	1.026	1.032
10	0.929	0.940	0.950	0.959	0.968	0.977	0.985	0.992	1.00	1.007
15	0.891	0.904	0.916	0.927	0.937	0.946	0.955	0.964	0.972	0.980
20	0.849	0.865	0.879	0.891	0.903	0.914	0.925	0.934	0.943	0.952
25	0.803	0.822	0.839	0.854	0.868	0.880	0.892	0.903	0.913	0.923
30	0.751	0.775	0.796	0.814	0.830	0.845	0.858	0.870	0.882	0.893
35	0.692	0.723	0.749	0.770	0.790	0.807	0.822	0.836	0.849	0.862
40	0.685	0.665	0.698	0.724	0.747	0.767	0.785	0.801	0.816	0.829
50	0.504	0.584	0.583	0.622	0.655	0.682	0.705	0.726	0.745	0.762
55	0.397	0.467	0.524	0.569	0.606	0.637	0.664	0.687	0.708	0.727
60	0.359	0.436	0.469	0.518	0.559	0.593	0.623	0.648	0.671	0.692
65	0.317	0.371	0.423	0.472	0.514	0.550	0.582	0.610	0.635	0.657
70	0.294	0.347	0.386	0.431	0.473	0.511	0.544	0.573	0.600	0.623
75	0.274	0.314	0.357	0.398	0.438	0.475	0.508	0.539	0.566	0.591
80	0.248	0.265	0.332	0.371	0.408	0.443	0.476	0.506	0.534	0.560
85	0.244	0.279	0.318	0.358	0.399	0.438	0.474	0.507	0.537	0.563
90	0.232	0.265	0.299	0.336	0.373	0.410	0.445	0.477	0.507	0.534
100	0.214	0.242	0.271	0.301	0.333	0.365	0.396	0.426	0.455	0.482

TABLE A.1 (continued)
Densities (gm·cm^{-3}) of Pure CO_2 Calculated by P-R EOS

T, °C/P, bar	220	240	260	280	300	320	340	360	380	400
5	1.044	1.055	1.066	1.075	1.085	1.093	1.101	1.109	1.117	1.124
10	1.020	1.032	1.043	1.054	1.063	1.073	1.082	1.090	1.098	1.106
15	0.992	1.008	1.020	1.032	1.042	1.052	1.062	1.071	1.080	1.088
20	0.968	0.983	0.996	1.009	1.020	1.031	1.041	1.051	1.060	1.069
25	0.941	0.957	0.972	0.985	0.998	1.010	1.021	1.031	1.041	1.050
30	0.913	0.930	0.946	0.961	0.975	0.987	0.999	1.011	1.021	1.031
35	0.884	0.903	0.921	0.937	0.952	0.965	0.978	0.990	1.001	1.012
40	0.854	0.875	0.894	0.912	0.928	0.942	0.956	0.969	0.981	0.992
50	0.792	0.818	0.840	0.861	0.879	0.896	0.912	0.926	0.940	0.952
60	0.728	0.758	0.785	0.809	0.830	0.849	0.867	0.883	0.898	0.912
70	0.665	0.700	0.730	0.757	0.780	0.802	0.821	0.839	0.856	0.872
80	0.605	0.643	0.676	0.706	0.732	0.755	0.777	0.797	0.815	0.832
90	0.581	0.620	0.652	0.675	0.703	0.724	0.743	0.760	0.775	0.789
100	0.530	0.571	0.606	0.636	0.662	0.685	0.706	0.724	0.741	0.756

TABLE A.2
Liquid, ρ_L and Vapor Densities, ρ_V (g·cm^{-3}) at Boiling Point, T_B(°C) and Saturation Pressure, P_S (bar) of CO_2

T_B	P_S	ρ_L	ρ_V	P_S	T_B	ρ_V	ρ_L
–10	26.50	0.983	0.033	30	–5.57	0.082	0.960
–5	30.47	0.957	0.084	35	0.15	0.098	0.928
0	34.86	0.928	0.097	40	5.30	0.115	0.896
2	36.73	0.916	0.104	42	7.23	0.123	0.883
4	38.68	0.904	0.111	44	9.08	0.130	0.870
6	40.72	0.891	0.118	46	10.88	0.139	0.857
8	42.825	0.877	0.126	48	12.62	0.147	0.843
10	45.01	0.863	0.135	50	14.03	0.156	0.830
12	47.28	0.848	0.143	52	15.94	0.165	0.816
14	49.63	0.832	0.154	54	17.52	0.175	0.801
16	52.07	0.815	0.166	56	19.06	0.185	0.786
18	54.61	0.797	0.178	58	20.56	0.197	0.769
20	57.24	0.776	0.193	60	22.02	0.209	0.753
22	59.97	0.753	0.209	62	23.43	0.223	0.735
24	62.81	0.727	0.229	64	24.81	0.238	0.715
26	65.77	0.696	0.254	66	26.15	0.256	0.693
28	68.85	0.655	0.287	68	27.46	0.277	0.667
30	72.06	0.592	0.346	70	28.73	0.304	0.636
31.1	73.8	0.468	0.468	73.8	31.1	0.468	0.468

Appendix B

Definition of Fatty Acids and Compositions in Various Oils

Fatty acids are carboxylic acids derived from vegetable oils or animal fats. These lipids are composed of a chain of alkyl groups containing usually an even number (4 to 22) of carbon atoms and are characterized by a terminal carboxyl group COOH. The generic formula is $CH_3(CH_2)_xCOOH$, the carbon atom count includes the carboxyl group. Fatty acids may be saturated or unsaturated, the latter containing one or more double bonds between the carbon atoms in the alkyl chain. The carbon atoms of the alkyl chain in saturated fatty acids are connected by single bonds. The most important of these are butyric (C_4), lauric (C_{12}), palmitic (C_{16}), and stearic (C_{18}). The unsaturated fatty acids derived from vegetable oils contain 18 or more carbon atoms with the characteristic end group –COOH. The most common unsaturated acids are oleic, linoleic, and linolenic (all C_{18}). Oleic acid is a monounsaturated fatty acid (MUFA) and is designatged as $C_{18:1}$. Similarly, linoleic acid is a fatty acid with two double bonds in both conjugated and unconjugated forms and is designated as $C_{18:2}$. Similarly, linolenic acid is a polyunsaturated fatty acid (PUFA) with three double bonds and is designated as $C_{18:3}$ of α-, β-, and γ- forms depending on the locations of the double bonds.

Like oleic, linoleic, and linolenic acids, all PUFA are also designated by the location of the last methyl group of the fatty acid chain, which is called "omega" (ω) followed by the number of carbon atoms counted before the first double bond.

For example, $CH_3\text{-}(CH_2)_7\text{-}CH = CH\text{-}(CH_2)_7\text{-}COOH$, (oleic acid) is known as (18:1 ω-9) fatty acids; $H_3C\text{-}(CH_2)_3\text{-}CH_2\text{-}CH = CH\text{-}CH = CH\text{-}CH\text{-}CH_2\text{-}(CH_2)_7\text{-}COOH$, (linoleic acid) is known as (18:2, ω-6) fatty acid; and $H_3C\text{-}CH_2\text{-}CH = CH\text{-}[CH_2\text{-}CH = CH]_2\text{-}(CH_2)_7\text{-}COOH$, (linolenic acid) is known as (18:3, ω-3) fatty acid. The fatty acid compositions in various common oils are given in Table B.1.

TABLE B.1
Fatty Acid Compositions of Various Oils (www.nutrition.com/nuts.html)

Oil	Satd. fat %	MUFA (%)	PUFA (%)	$C_{18:3}$ (%)	$C_{18:2}$ (%)
Almond	8.2	69.9	17.4	0.0	17.4
Canola	7.1	58.9	29.6	9.3	20.3
Coconut	86.5	5.8	1.8	0.0	1.8
Corn	12.7	24.2	58.7	0.7	58.0
Olive	13.5	73.7	8.4	0.6	7.9
Palm	49.3	37.0	9.3	0.2	9.1
Peanut	16.9	46.2	32.0	0.0	32.0
Safflower	9.6	12.6	73.4	0.2	73.0
Sesame	14.2	39.7	41.7	0.3	41.3
Soybean	14.4	23.3	57.9	6.8	51.0
Walnut	9.1	22.8	63.3	10.4	52.9

Appendix C

Some Statistics on Major Vegetable Oils and Oilseeds

The world's oilseed production in 1996 to 1997 is estimated at 259.1 million tons and the world's soybean production is estimated at 134 million tons. Major vegetable oils and marine oils produced and consumed in 1996 to 1997 are presented in Table C.1 and the amounts of oil present in various oilseeds are given in Table C.2.

TABLE C.1
World Supply and Consumption (MMT) of Vegetable and Oils in 1996–1997 (www.fas.usda.govt/oilseeds/circular/1997)

Oil	Production	Consumption
Soybean	20.47	17.22
Palm	16.60	12.41
Sunflower	8.80	7.52
Canola	10.49	8.56
Cottonseed	3.80	3.65
Peanut	4.05	3.69
Coconut	3.33	2.82
Olive	2.00	1.89
Fish	1.25	1.13

TABLE C.2
The Amounts of Oil in Oilseeds and Distribution of Triglycerides (TG) and Phosphpolipids (PL) in Crude Seed Oil (Bailey's Oil and Fats, 1981)

Oil Seed	Oil (%)	TG (% of oil)	PL (% of oil)
Corn	38–44	94.4–99.1	<2.5
Castor	45		
Coconut	34	99.8	
Corn germ	45		
Cottonseed	19	98	0.7–0.9
Peanut	38		
Poppy seed	40		
Oat	12	50.6	7.1
Canola	35	91.8–99.0	<3.5
Rice bran	14–20	88–89	4–5
Safflower	11.5–47.5	—	—
Sesame	47		0.03–0.13
Soybean	18	93.0–99.2	<4.0
Sunflower	29		0.02–1.5
Tea seed	48		

Source: *Bailey's Industrial Oil and Fat Products*, Vol. 2, 4th ed., Swern, D., Ed., John Wiley & Sons, New York, 1981.

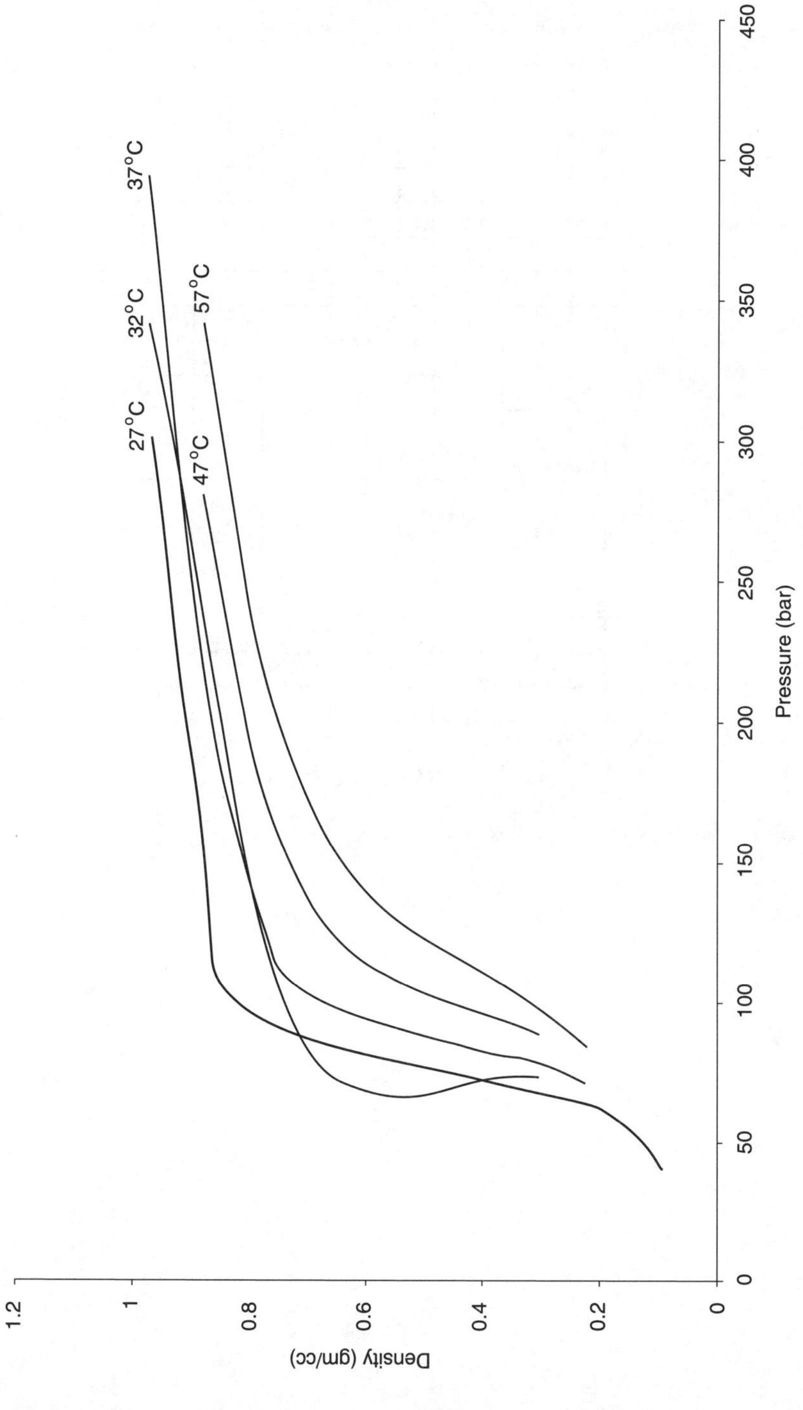

FIGURE A.1 Density vs. pressure isotherm for carbon dioxide.

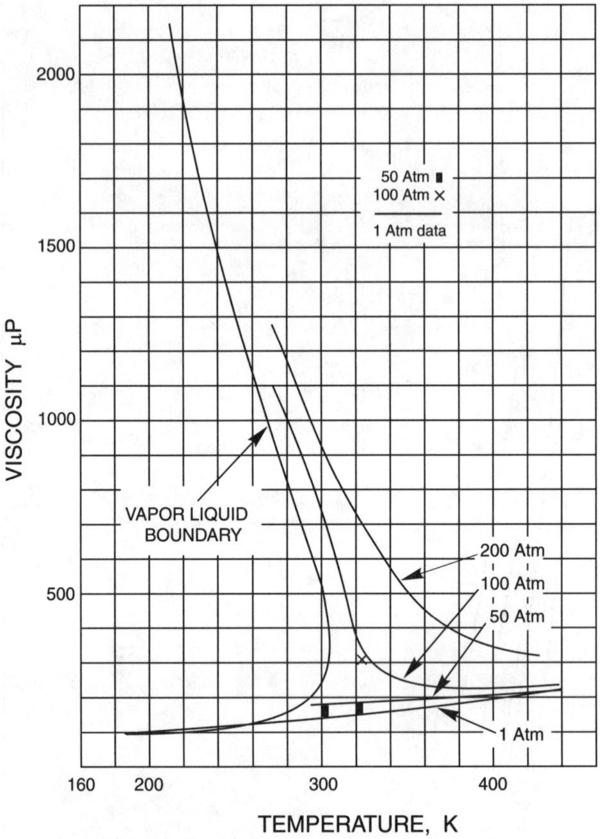

FIGURE A.2 Viscosity of carbon dioxide (Haselden, 1971).

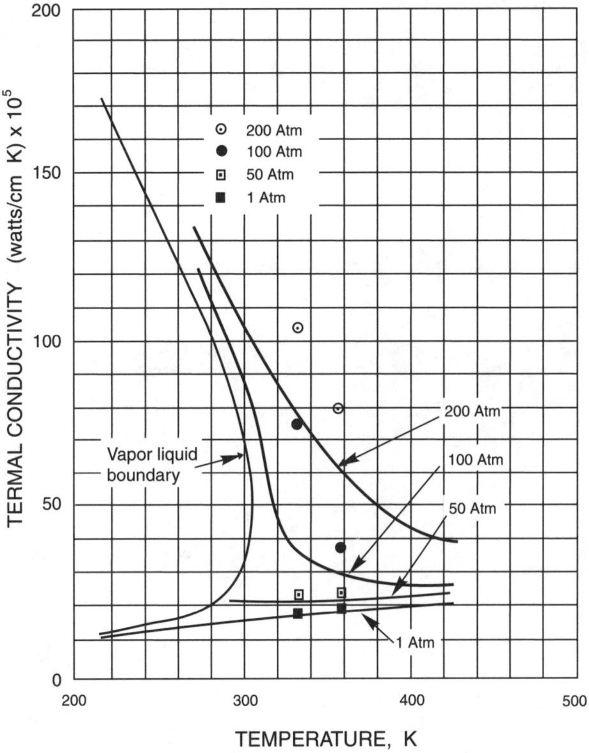

FIGURE A.3 Thermal Conductivity of carbon dioxide (Haselden, 1971).

Index

A

B

D

E

F

R

S

T